R. Rajesh, B. Mathivanan (Eds.)
Communication and Power Engineering

Communication and Power Engineering

Edited by
R. Rajesh, B. Mathivanan

DE GRUYTER
OLDENBOURG

Editors

Dr. R. Rajesh
Central University of Kerala
India
kollamrajeshr@gmail.com

Dr. B. Mathivanan
Sri Ramakrishna Engg. College
India
mathivanan.bala@srec.ac.in

ISBN 978-3-11-046860-1
e-ISBN (PDF) 978-3-11-046960-8
Set-ISBN 978-3-11-046961-5

Library of Congress Cataloging-in-Publication Data
A CIP catalog record for this book has been applied for at the Library of Congress.

Bibliographic information published by the Deutsche Nationalbibliothek
The Deutsche Nationalbibliothek lists this publication in the Deutsche Nationalbibliografie;
detailed bibliographic data are available on the Internet at http://dnb.dnb.de.

© 2016 Walter de Gruyter GmbH, Berlin/Boston
Printing and binding: CPI books GmbH, Leck
cover image: Thinkstock/tStockbyte
♾ Printed on acid-free paper
Printed in Germany

www.degruyter.com

Committees

Honorary Chair
Dr. Shuvra Das (University of Detroit Mercy, USA)
Dr. Jiguo Yu (Qufu Normal University, China)

Technical Chair
Dr. Sumeet Dua (Louisiana Tech University, USA)
Dr. Amit Banerjee (The Pennsylvania State University, USA)
Dr. Narayan C Debnath (Winona State University, USA)
Dr. Xiaodi Li (Shandong Normal University, China)

Technical Co-Chair
Dr. Natarajan Meghanathan (Jackson State University, USA)
Dr. Hicham Elzabadani (American University in Dubai)
Dr. Shahrokh Valaee (University of Toronto, Canada)

Chief Editors
Dr. Rajesh R (Central University of Kerala, India)
Dr. B Mathivanan (Sri Ramakrishna Engg. College, India)

General Chair
Dr. Janahanlal Stephen (Matha College of Technology, India)
Dr. Yogesh Chaba (Guru Jambeswara University, India)

General Co-Chair
Prof. K. U Abraham (Holykings College of Engineering, India)

Publicity Chair
Dr. Amit Manocha (Maharaja Agrasen Institute of Technology, India)

Finanace Chair
Dr. Gylson Thomas (Jyothi Engineering College, India)
Dr. Ilias Maglogiannis (University of Central Greece)

Publicity Co-Chair
Prof. Ford Lumban Gaol (University of Indonesia)
Dr. Amlan Chakrabarti (University of Culcutta, India)
Prof. Prafulla Kumar Behera, PhD(Utkal University, India)

Publication Chair
Dr. Vijayakumar (NSS Engg. College, India)
Dr. T.S.B.Sudarshan (BITS Pilani, India)
Dr. KP Soman (Amritha University, India)
Prof. N.Jaisankar (VIT University, India)
Dr. Rajiv Pandey (Amity University, India)

Program Committee Chair
Dr. Harry E. Ruda (University of Toronto, Canada)
Dr Deepak Laxmi Narasimha (University of Malaya, Malaysia)
Dr.N.Nagarajan (Anna University, Coimbatore, India)
Prof. Akash Rajak (Krishna Institute of Engg. & Tech., UP, India)
Prof. M Ayoub Khan (CDAC, NOIDA, India)

Programming Committee
Prof. Shelly Sachdeva (Jaypee Institute of Information & Technology University, India)
Prof. PRADHEEP KUMAR K (SEEE, India)
Mrs. Rupa Ashutosh Fadnavis (Yeshwantrao Chavan College of Engineering, India)
Dr. Shu-Ching Chen (Florida International University, USA)
Dr. Stefan Wagner (Fakultät für Informatik Technische Universität München, Boltzmannstr)
Prof. Juha Puustjärvi (Helsinki University of Technology)
Dr. Selwyn Piramuthu (University of Florida)
Dr. Werner Retschitzegger (University of Linz, Austria)
Dr. Habibollah Haro (Universiti Teknologi Malaysia)
Dr. Derek Molloy (Dublin City University, Ireland)
Dr. Anirban Mukhopadhyay (University of Kalyani, India)
Dr. Malabika Basu (Dublin Institute of Technology, Ireland)
Dr. Tahseen Al-Doori (American University in Dubai)
Dr. V. K. Bhat (SMVD University, India)
Dr. Ranjit Abraham (Armia Systems, India)
Dr. Naomie Salim (Universiti Teknologi Malaysia)
Dr. Abdullah Ibrahim (Universiti Malaysia Pahang)
Dr. Charles McCorkell (Dublin City University, Ireland)
Dr. Neeraj Nehra (SMVD University, India)

Dr. Muhammad Nubli (Universiti Malaysia Pahang)
Dr. Zhenyu Y Angz (Florida International University, USA)
Dr. Keivan Navi (Shahid Beheshti University,

Preface

It is my proud privilege to welcome you all to the joint International Conferences organized by **IDES**. This conference is jointly organized by the **IDES** and the **Association of Computer Electrical Electronics and Communication Engineers (ACEECom).** The primary objective of this conference is to promote research and developmental activities in Computer Science, Electrical, Electronics, Network, Computational Engineering, and Communication. Another objective is to promote scientific information interchange between researchers, developers, engineers, students, and practitioners working in India and abroad.

I am very excited to see the research papers from various parts of the world. This proceeding brings out the various Research Papers from diverse areas of Computer Science, Electrical, Electronics, Network, Computational Engineering, and Communication. This conference is intended to provide a common platform for Researchers, Academicians and Professionals to present their ideas and innovative practices and to explore future trends and applications in the field of Science and Engineering. This conference also provides a forum for dissemination of Experts' domain knowledge. The papers included in the proceedings are peer-reviewed scientific and practitioners' papers, reflecting the variety of **Advances in Communication, Network,** Electrical, Electronics, **and Computing.**

As a Chief Editor of this joint Conference proceeding, I would like to thank all of the presenters who made this conference so interesting and enjoyable. Special thanks should also be extended to the session chairs and the reviewers who gave of their time to evaluate the record number of submissions. To all of the members of various Committees, I owe a great debt as this conference would not have not have been possible without their constant efforts. We hope that all of you reading enjoy these selections as much as we enjoyed the conference.

Dr. B Mathivanan
Sri Ramakrishna Engineering College, India

Table of Contents

Foreword

The Institute of Doctors Engineers and Scientists (IDES) (with an objective to promote the Research and Development activities in the Science, Medical, Engineering and Management field) and the Association of Computer Electrical Electronics and Communication (ACEECom) (with an objective t**o disseminate knowledge and to promote the research and development activities in the engineering and technology field**) has both joined hands to hand together for the benefit of the society. For more than a decade, both IDES and ACEECom are well established in organizing conferences and publishing journals.

This joint International Conference organized by IDES and ACEECom in 2016, aiming to bring together the Professors, Researchers, and Students in all areas of Computer Science, Information Technology, Computational Engineering, Communication, Signal Processing, Power Electronics, Image Processing, etc. in one platform, where they can interact and share the ideas.

A total of 35 eminent scholars/speakers have registered their papers in areas of Computer Science and Electrical & Electronics discipline. These papers are published in a proceedings by **De Gruyter Digital Library** and are definitely going to be the eye-opening to the world for further research in this area.

The organizations (IDES and ACEECom) will again come together in front of you in future for further exposure of the unending research.

Dr. Rajesh R
Central University of Kerala, India

Pawan Kumar Singh[1], Iman Chatterjee[2], Ram Sarkar[3] and
Mita Nasipuri[4]

Handwritten Script Identification from Text Lines

Abstract: In a multilingual country like India where 12 different official scripts
are in use, automatic identification of handwritten script facilitates many important applications such as automatic transcription of multilingual documents, searching for documents on the web/digital archives containing a particular script and for the selection of script specific Optical Character
Recognition (OCR) system in a multilingual environment. In this paper, we propose a robust method towards identifying scripts from the handwritten documents at text line-level. The recognition is based upon features extracted using
Chain Code Histogram (CCH) and Discrete Fourier Transform (DFT). The proposed method is experimented on 800 handwritten text lines written in seven
Indic scripts *namely, Gujarati, Kannada, Malayalam, Oriya, Tamil, Telugu, Urdu*
along with *Roman* script and yielded an average identification rate of 95.14%
using Support Vector Machine (SVM) classifier.

Keywords: Script Identification, Handwritten text lines, *Indic* scripts, Chain
Code Histogram, Discrete Fourier Transform, Multiple Classifiers

1 Introduction

One of the major Document Image Analysis research thrusts is the implementation of OCR algorithms that are able to make the alphanumeric characters present in a digitized document into a machine readable form. Examples of the
applications of such research include automated word recognition, bank check

---

1 Department of Computer Science and Engineering, Jadavpur University, Kolkata, India
pawansingh.ju@gmail.com
2 Department of Computer Science and Engineering, Netaji Subhash Engineering College,
Kolkata, India
imanchatterjee9@gmail.com
3 Department of Computer Science and Engineering, Jadavpur University, Kolkata, India
raamsarkar@gmail.com
4 Department of Computer Science and Engineering, Jadavpur University, Kolkata, India
mitanasipuri@gmail.com

processing, and address sorting in postal applications etc. Consequently, the vast majority of the OCR algorithms used in these applications are selected based upon a priori knowledge of the script and/or language of the document under analysis. This assumption requires human intervention to select the appropriate OCR algorithm, limiting the possibility of completely automating the analysis process, especially when the environment is purely multilingual. In this scenario, it is very necessary to have the script recognition module before applying such document into appropriate OCR system.

In general, script identification can be achieved at any of the three levels: (a) Page-level, (b) Text-line level and (c) Word-level. In comparison to page or word-level, script recognition at the text line-level in a multi-script document may be much more challenging but it has its own advantages. To reliably identify the script type, one needs a certain amount of textual data. But identifying text words of different scripts with only a few numbers of characters may not always be feasible because at word-level, the number of characters present in a single word may not be always informative. In addition, performing script identification at word-level also requires the exact segmentation of text words which is again an exigent task. On the contrary, identifying scripts at page-level can be sometimes too convoluted and protracted. So, it would be better to perform the script identification at text line-level than its two counterparts.

A detailed state-of-the-art on *Indic* script identification described by P. K. Singh *et al.* [1] shows that most of the reported studies [2-8], accomplishing script identification at text line-level, work for printed text documents. G. D. Joshi *et al.* [2] proposed a hierarchical script classifier which uses a two-level, tree based scheme for identifying 10 printed *Indic* scripts *namely, Bangla, Devanagari, Gujarati, Gurumukhi, Kannada, Malayalam, Oriya, Tamil* and *Urdu* including *Roman* script. A total of 3 feature set such as, statistical, local, horizontal profile are extracted from the normalized energy of log-Gabor filters designed at 8 equi-spaced orientations (0^0, 22.5^0, 45^0, 77.5^0, 90^0, 112.5^0, 135.5^0 and 180^0) and at an empirically determined optimal scale. An overall classification accuracy of 97.11% is obtained. M. C. Padma *et al.* [3] proposed to develop a model based on top and bottom profile based features to identify and separate text lines of *Telugu, Devnagari* and *English* scripts from a printed trilingual document. A set of eight features (i.e. bottom max-row, top-horizontal-line, tick-component, bottom component (extracted from the bottom-portion of the input text line), top-pipe-size, bottom-pipe-size, top-pipe-density, bottom-pipe-density) are experimentally computed and the overall accuracy of the system is found to be 99.67%. M. C. Padma *et al.* [4] also proposed a model to identify the script type of a trilingual document printed in *Kannada, Hindi* and *Eng-*

lish scripts. The distinct characteristic features of said scripts are thoroughly studied from the nature of the top and bottom profiles. A set of 4 features *name-ly*, profile_value (computed as the density of the pixels present at top_max_row and bottom_max_row), bottom_max_row_no (the value of the attribute bot-tom_max_row), coeff_profile, top_component_density (the density of the con-nected components at the top_max_row) are computed. Finally, *k*-NN (*k*-Nearest Neighbor) classifier is used to classify the test samples with an average recogni-tion rate of 99.5%. R. Gopakumar *et al.* [5] described a zone-based structural feature extraction algorithm for the recognition of South-*Indic* scripts (*Kannada, Telugu, Tamil* and *Malayalam*) along with *English* and *Hindi*. A set of 9 features such as number of horizontal lines, vertical lines, right diagonals, left diago-nals, normalized lengths of horizontal lines, vertical lines, right diagonals, left diagonals and normalized area of the line image are computed for each text line image. Finally, the classification accuracies of 100% and 98.3% are achieved using *k*-NN and SVM (Support Vector Machine) respectively. M. Jindal *et al.* [6] proposed a script identification approach for *Indic* scripts at text line-level based upon features extracted using Discrete Cosine Transform (DCT) and Prin-cipal Component Analysis (PCA) algorithm. The proposed method is tested on printed document images in 11 major Indian languages (*viz., Bangla, Hindi, Gujarati, Kannada, Malayalam, Oriya, Punjabi, Tamil, Telugu, English* and *Urdu*) and 95% recognition accuracy is obtained. R. Rani *et al.* [7] presented the effec-tiveness of Gabor filter banks using *k*-NN, SVM and PNN (Probabilistic Neural Network) classifiers to identify the scripts at text-line level from trilingual doc-uments printed in *Gurumukhi, Hindi* and *English*. The experiment shows that a set of 140 features based on Gabor filter with SVM classifier achieve the maxi-mum recognition rate of 99.85%. I. Kaur *et al.* [8] presented a script identifica-tion work for the identification of English and Punjabi scripts at text-line level through headline and characters density features. The approach is thoroughly tested for different font size images and an average accuracy of 90.75% is achieved. On the contrary, researches made on handwritten documents are only a few in number. M. Hangarge *et al.* [9] investigated texture pattern as a tool for determining the script of handwritten document image, based on the observa-tion that text has a distinct visual texture. A set of 13 spatial spread features of the three *Indic* scripts *namely, English, Devanagari* and *Urdu* are extracted using morphological filters and the overall accuracies of the proposed algorithm are found to be 88.67% and 99.2% for tri-script and bi-script classifications respec-tively using *k*-NN classifier. P. K. Singh *et al.* [10] proposed a texture based ap-proach for text line-level script identification of six handwritten scripts *namely, Bangla, Devanagari, Malayalam, Tamil, Telugu* and *Roman*. A set of 80 features based on Gray Level Co-occurrence Matrix (GLCM) is used and an overall recognition rate of 95.67% is achieved using Multi Layer Perceptron (MLP) classifier. To the best of our knowledge, script identification at text line-level considering large number of *Indic* handwritten scripts does not exist in the literature. In this paper, we propose a text line-level script identification technique written in *seven* popular official *Indic* scripts *namely, Gujarati, Kannada, Malayalam, Oriya, Tamil, Telugu, Urdu* along with *Roman* script.

2 Data Collection and Preprocessing

At present, no standard database of handwritten *Indic* scripts are available in public domain. Hence, we created our own database of handwritten documents in the laboratory. The document pages for the database are collected by different persons on request under our supervision. The writers are asked to write inside A-4 size pages, without imposing any constraint regarding the content of the textual materials. The document pages are digitized at 300 dpi resolution and stored as gray tone images. The scanned images may contain noisy pixels which are removed by applying Gaussian filter [11]. It should be noted that the handwritten text line (actually, portion of the line arbitrarily chosen) may contain two or more words with noticeable *intra-* and *inter-word* spacings. Numerals that may appear in the text are not considered for the present work. It is ensured that at least 50% of the cropped text line contains text. A sample snapshot of text line images written in eight different scripts is shown in Fig. 1. Otsu's global thresholding approach [12] is used to convert them into two-tone images. However, the dots and punctuation marks appearing in the text lines are not eliminated, since these may also contribute to the features of respective scripts. Finally, a total of 800 handwritten text line images are considered, with exactly100 text lines per script.

3 Feature Extraction

The feature extraction is based on the combination of Chain Code Histogram (CCH) and Discrete Fourier Transform (DFT) which are described in detail in the next subsection.

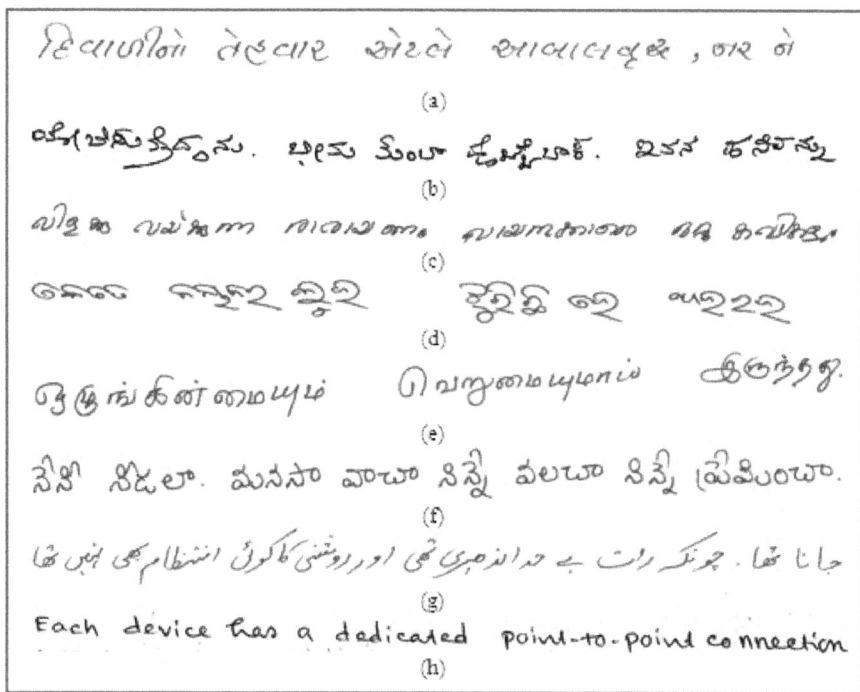

Figure 1. Sample text line images taken from our database written in: (a) *Gujarati*, (b) *Kannada*, (c) *Malayalam*, (d) *Oriya*, (e) *Tamil*, (f) *Telugu*, (g) *Urdu*, and (h) *Roman* scripts respectively

3.1 Chain Code Histogram

Chain codes [11] are used to represent a boundary by a connected sequence of straight-line segments of specified length and direction. It describes the movement along a digital curve or a sequence based on the connectivity. Two types of chain codes are possible which are based on the numbers of neighbors of a pixel, *namely*, four or eight, giving rise to 4- or 8-neighbourhood. The corresponding codes are the 4-directional code and 8-directional code, respectively. The direction of each segment is coded by using a numbering scheme as shown in Fig. 2. In the present work, the boundaries of handwritten text lines written in different scripts can be traced and allotted the respective numbers based on the directions. Thus, the boundary of each of the text line is reduced to a sequence of numbers. A boundary code formed as a sequence of such directional numbers is referred to as a *Freeman chain code*.

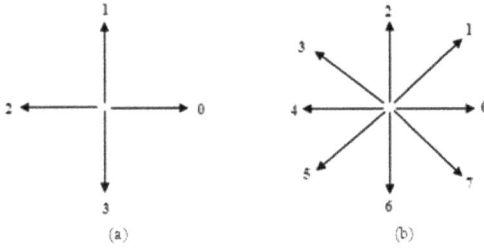

Figure 2. Illustration of numbering the directions for: (a) 4-dimensional, and (b) 8-dimensional chain codes

The histogram of *Freeman chain codes* are taken as feature values F1-F8 and the histogram of first difference of the chain codes are also taken as feature values F9-F15. Let us denote the set of pixels by R. The perimeter of a region R is the number of pixels present in the boundary of R. In a binary image, the perimeter is the number of foreground pixels that touches the background in the image. For an 8-directional code, the length of perimeter of each text line (F16) is calculated as: |P| = Even count + $\sqrt{2}$ *(Odd count). A circularity measure (F17) proposed by Haralick [13] can be written as:

$$C_2 = \frac{\mu_R}{\sigma_R} \tag{1}$$

where, μ_R and σ_R are the mean and standard deviation of the distance from the centroid of the shape to the shape boundary and can be computed as follows:

$$\mu_R = \frac{1}{K} \sum_{k=0}^{K-1} \|(x_k, y_k) - (\bar{x}, \bar{y})\| \tag{2}$$

$$\sigma_R = \left(\frac{1}{K} \sum_{k=0}^{K-1} [\|(x_k, y_k) - (\bar{x}, \bar{y})\| - \mu_R]^2 \right)^{1/2} \tag{3}$$

where, the set of pixels (x_k, y_k), $k = 0, \ldots, K - 1$ lie on the perimeter P of the region. The circularity measure C_2 increases monotonically as the digital shape becomes more circular and is similar for digital and continuous shapes. Along the circularity, the slopes are labeled in accordance with their chain codes which are shown in Table 1.

Table 1. Labeling of slope angles according to their chain codes

Chain code	0	1	2	3	4	5	6	7
θ	0	45^0	90^0	135^0	180^0	-135^0	-90^0	-45^0

The count of the slopes having θ values 0^0, $|45^0|$, $|90^0|$, $|135^0|$, 180^0 for each of the handwritten text line images are taken as feature values (F18-F22).

3.2 Discrete Fourier Transform

The Fourier Transform [11] is an important image processing tool which is used to decompose an image into its sine and cosine components. The output of the transformation represents the image in the Fourier or frequency domain, while the input image is the spatial domain equivalent. In the Fourier domain, each point in the spatial domain image represents a particular frequency.

The Discrete Fourier Transform (DFT) is the sampled Fourier Transform and therefore does not contain all frequencies forming an image, but only a set of samples which is large enough to fully describe the spatial domain image. The number of frequencies corresponds to the number of pixels in the spatial domain image, i.e., the images in the spatial and Fourier domains are of the same size. The DFT of a digital image of size MxN can be written as:

$$G(u,v) = \frac{1}{MN} \sum_{m=0}^{M-1} \sum_{n=0}^{N-1} g(m,n) e^{-j2\pi\left(\frac{mu}{M}+\frac{nv}{N}\right)} \tag{4}$$

where, $g(m,n)$ is the image in the spatial domain and the exponential term is the basis function corresponding to each point $G(u,v)$ in the Fourier space. The value of each point $G(u,v)$ is obtained by multiplying the spatial image with the corresponding base function and summing the result. The Fourier Transform produces a complex number valued output which can be displayed with two images, either with the real and imaginary parts or with the *magnitude* and *phase*, where *magnitude* determines the contribution of each component and *phase* determines which components are present. The plots for magnitude and phase components for a sample *Tamil* handwritten text-line image are shown in Fig. 3. In the current work, only the *magnitude* part of DFT is employed as it contains most of the information of the geometric structure of the spatial domain image. This in turn becomes easy to examine or process certain frequencies of the image. The *magnitude* coefficient is normalized as follows:

$$G'(u, v) = \frac{|G(u, v)|}{\sqrt{\sum_{u,v}|G(u, v)|^2}} \tag{5}$$

The algorithm for feature extraction using DFT is as follows:

Step 1: Divide the input text line image into nxn non-overlapping blocks which are known as *grids*. The optimal value of n has been chosen as 4.

Step 2: Compute the DFT (by applying Eqn. (4)) in each of the *grids*.

Step 3: Estimate only the *magnitude* part of the DFT and normalize it using Eqn. (5).

Step 4: Calculate the mean and standard deviation of the *magnitude* part from each of the *grids* which give a feature vector of 32 elements (F23-F54).

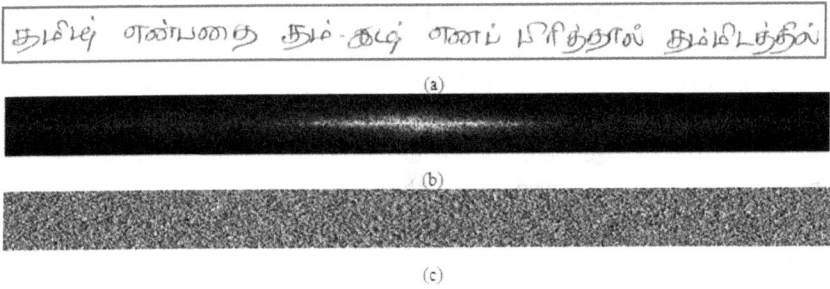

Figure 3. Illustration of: (a) handwritten *Tamil* text-line image, (b) its magnitude component, and (c) its phase component after applying DFT

4 Experimental Results and Discussion

The performance of the present script identification scheme is evaluated on a dataset of 800 preprocessed text line images as described in Section 2. For each dataset of 100 text line images of a particular script, 65 images are used for training and the remaining 45 images are used for testing purpose. The proposed approach is evaluated by using seven well-known classifiers *namely*, Naïve Bayes, Bayes Net, MLP, SVM, Random Forest, Bagging and MultiClass Classifier. The recognition performances and their corresponding scores achieved at 95% confidence level are shown in Table 2.

Table 2. Recognition performances of the proposed script identification technique using seven well-known classifiers (best case is shaded in grey and styled in bold)

	Classifiers						
	Naïve Bayes	Bayes Net	MLP	SVM	Random Forest	Bagging	MultiClass Classifier
Success Rate (%)	89.33	90.09	95.14	**97.03**	94.6	91.25	92.74
95% confidence score (%)	91.62	93.27	96.85	**99.7**	97.39	93.54	95.52

As observed from Table 2 that SVM classifier produces the highest identification accuracy of 97.03%. In the present work, detailed error analysis of SVM classifier with respect to different well-known parameters *namely*, Kappa statistics, mean absolute error, root mean square error, True Positive rate (TPR), False Positive rate (FPR), precision, recall, F-measure, Matthews Correlation Coefficient (MCC) and Area Under ROC (AUC) are also computed. The values of Kappa statistics, mean absolute error, root mean square error of SVM classifier for the present technique are found to be 0.9661, 0.0074 and 0.0862 respectively. Table 3 provides a statistical performance analysis of the remaining parameters for each of the aforementioned scripts.

Table 3. Statistical performance measures along with their respective means (shaded in grey and styled in bold) achieved by the proposed technique for eight handwritten scripts

Scripts	TP rate	FP rate	Precision	Recall	F-measure	MCC	AUC
Gujarati	1.000	0.000	1.000	1.000	1.000	1.000	1.000
Kannada	0.970	0.025	0.845	0.970	0.903	0.891	0.972
Malayalam	0.950	0.000	1.000	0.950	0.975	0.972	0.975
Oriya	1.000	0.000	1.000	1.000	1.000	1.000	1.000
Tamil	0.990	0.000	1.000	0.990	0.995	0.994	0.995
Telugu	0.980	0.000	1.000	0.980	0.990	0.989	0.990
Urdu	0.941	0.004	0.969	0.941	0.955	0.949	0.968
Roman	0.931	0.004	0.969	0.931	0.949	0.943	0.963
Weighted Average	**0.970**	**0.004**	**0.973**	**0.970**	**0.971**	**0.967**	**0.983**

Though Table 2 shows encouraging results but still some of the handwritten text lines are misclassified during the experimentation. The main reasons for the same are: (a) presence of speckled noise, (b) skewed words present in some text lines, and (c) occurrence of irregular spaces within text words, punctuation symbols, etc. The structural resemblance in the character set of some of the *Indic* scripts like *Kannada* and *Telugu* as well as *Malayalam* and *Tamil* causes similarity in the contiguous pixel distribution which in turns misclassifies them among each other. Fig. 4 shows some samples of misclassified text line images.

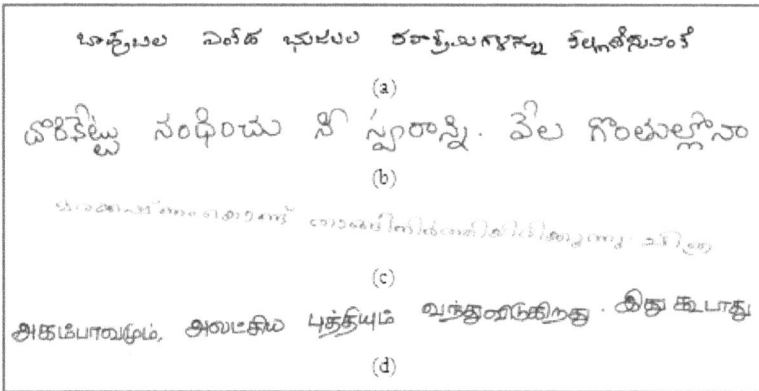

Figure 4. Samples of text line images written in (a) *Kannada*, (b) *Telugu*, (c) *Malayalam*, and (d) *Tamil* scripts misclassified as *Telugu*, *Kannada*, *Tamil* and *Malayalam* scripts respectively

Conclusion

In this paper, we have proposed a robust method for handwritten script identification at text line-level for eight official scripts of India. The aim of this paper is to facilitate the research of multilingual handwritten OCR. A set of 54 feature values are extracted using the combination of CCH and DFT. Experimental results have shown that an accuracy rate of 97.03% is achieved using SVM classifier with limited dataset of eight different scripts which is quite acceptable taking the complexities and shape variations of the scripts under consideration. In our future endeavor, we plan to modify this technique to perform the script identification from handwritten document images containing more number of Indian languages. Another focus is to increase the size of the database to incorporate larger variations of writing styles which in turn would establish our technique as writer independent.

Acknowledgment

The authors are thankful to the Center for Microprocessor Application for Training Education and Research (*CMATER*) and Project on Storage Retrieval and Understanding of Video for Multimedia (SRUVM) of Computer Science and Engineering Department, Jadavpur University, for providing infrastructure facilities during progress of the work. The current work, reported here, has been partially funded by University with Potential for Excellence (UPE), Phase-II, UGC, Government of India.

References

1 P.K. Singh, R. Sarkar, M. Nasipuri: *"Offline Script Identification from Multilingual Indic-script Documents: A state-of-the-art"*, In: Computer Science Review (Elsevier), vol. 15-16, pp. 1-28, 2015.

2 G. D. Joshi, S. Garg, J. Sivaswamy, *"Script Identification from Indian Documents"*, In: Lecture Notes in Computer Science: International Workshop Document Analysis Systems, Nelson, LNCS-3872, pp. 255-267, Feb. 2006.

3 M. C. Padma, P. A. Vijaya, *"Identification of Telugu, Devnagari and English scripts using discriminating features"*, In: International Journal of Computer Science and Information Technology (IJCSIT), vol. 1, no.2, Nov.2009.

4 M. C. Padma, P. A. Vijaya, *"Script Identification from Trilingual Documents using Profile based Features"*, In: International Journal of Computer Science and Applications (IJCSA), vol. 7, no. 4, pp. 16-33, 2010.

5 R. Gopakumar, N. V. SubbaReddy, K. Makkithaya, U. Dinesh Acharya, *"Script Identification from Multilingual Indian documents using Structural Features"*, In: Journal of Computing, vol. 2, issue 7, pp. 106-111, July 2010.

6 M. Jindal, N. Hemrajani, *"Script Identification for printed document images at text-line level using DCT and PCA"*, In: IOSR Journal of Computer Engineering, vol. 12, issue 5, pp. 97-102, 2013.

7 R. Rani, R. Dhir, G. S. Lehal, *"Gabor Features Based Script Identification of Lines within a Bilingual/Trilingual Document"*, In: International Journal of Advanced Science and Technology, vol. 66, pp. 1-12, 2014.

8 I. Kaur, S. Mahajan, *"Bilingual Script Identification of Printed Text Image"*, In: International Journal of Engineering and Technology, vol. 2, issue 3, pp. 768-773, June 2015.

9 M. Hangarge, B. V. Dhandra, *"Offline Handwritten Script Identification in Document Images"*, In: International Journal of Computer Applications (IJCA), vol.4, no.6, pp. 1-5, July 2010.

10 P. K. Singh, R. Sarkar, M. Nasipuri, *"Line-level Script Identification for six handwritten scripts using texture based features"*, In: Proc. of 2nd Information Systems Design and Intelligent Applications, AISC, vol. 340, pp. 285-293, 2015.
11 R. C. Gonzalez, R. E. Woods, *"Digital Image Processing"*, vol. I. Prentice-Hall, India (1992).
12 N. Ostu, *"A thresholding selection method from gray-level histogram"*, In: IEEE Transactions on Systems Man Cybernetics, SMC-8, pp. 62-66, 1978.
13 R. M. Haralick, *"A Measure of Circularity of Digital Figures"*, In: IEEE Transactions on Systems, Man and Cybernetics, vol. SMC-4, pp. 394-396, 1974.

Neelotpal Chakraborty[1], Samir Malakar[2], Ram Sarkar[3] and
Mita Nasipuri[4]

A Rule based Approach for Noun Phrase Extraction from English Text Document

Abstract: This paper is an attempt to focus on an approach that is quite simple to implement and efficient enough to extract Noun Phrases (NPs) from text document written in English. The selected text documents are articles of reputed English newspapers of India, namely, The Times of India, The Telegraph and The Statesman. A specific column (sports) has been taken into consideration. The proposed approach concentrates on the following objectives: First, to explore and exploit the grammatical features of the language. Second, to prepare an updated stop list classified into conjunctions, prepositions, articles, common verbs and adjectives. Third, to give special characters due importance.

Keywords: Noun Phrase, Rule-based Approach, Natural Language Processing, Data Mining

1 Introduction

In the past few decades, world has witnessed a huge text data explosion in the form of printed and/or handwritten form. This data growth would increase exponentially as time will pass. Also it is well known paradigm that the searching time for some document is directly proportional to the size of the database where it belongs to. Therefore, such abrupt increase of data is eventually increasing the searching time. But the technology enabled society demands a fast and efficient way to reduce the searching time. The searching time can be optimized only when

1 Department of Computer Science and Engineering, Jadavpur University, Kolkata, India
neelotpal_chakraborty@yahoo.com
2 Department of Master of Computer Applications, MCKV Institute of Engineering, Howrah, India
malakarkarsamir@gmail.com
3 Department of Computer Science and Engineering, Jadavpur University, Kolkata, India
raamsarkar@gmail.com
4 Department of Computer Science and Engineering, Jadavpur University, Kolkata, India
mitanasipuri@gmail.com

the data are maintained using proper some structure. One of the ways to accomplish this is the document clustering which is an application of Natural Language Processing (NLP). The job of NLP is to understand and analyze the *Natural Language* (the language spoken by humans). The increasing nature of documents motivates a section of the research fraternity throughout the world to direct their research into NLP. The process of document clustering is carried out through sequential processes comprises of *Noun Phrase (NP) extraction, Key Phrase (KP)* selection and *document ranking*.

1.1 Noun Phrase, Key Phrase and Document Clustering

Any text document irrespective of the content comprises of certain terminology (words or phrases) using which, out of several documents, that particular document can be identified (or classified) as describing a particular subject or topic. The process of assignment of any text document into a predefined class or subject is known as document clustering. The terminologies used for tagging the text document into a predefined class are usually termed as *Keyword* or KP which is comprised of single / multiple word(s). Traditionally, a *Named Entity (NE)*, a special type of NP, is an obvious choice of KP.

The research approaches on document clustering, till date, have mainly focused on developing statistical model to identify *NPs* [1] i.e., identifying quantitative features [1]. However, there are certain aspects of any natural language that requires understanding of the subjective/qualitative features of that language. Each particular natural language has its own specific grammatical structure. In general, the vocabulary of the same can be classified into two types, entitled as *closed class type* and *open class type* [1]. In first category, new words are added frequently, whereas words are rarely added in the other type.

The *NPs* fall under the open class, and the prepositions, articles, conjunctions, certain common verbs, adjectives, adverbs are of closed class types. However, certain verbs, adverbs, adjectives are derived from the NPs, example: *Pasturization (noun)* → *Pasturize (verb)*. Also the appearance of preposition before a particular word determines the type of words. For example, consider the following sentences where the word "take" has been used as noun in the first sentence and as verb in the second sentence.

Sentence 1: Just one take is enough for this scene.

Sentence 2: I have come to take my books.

However, it is worth noting that some verbs, adverbs or adjectives can be derived from the nouns and vice versa. The English language also comprises of upper-case/lowercase letters that add to relevancy of terminology for any particular text document. Therefore in the present work, apart from maintaining a different stop list at different level, conversion of these specific words into their respective noun forms is conducted and then they have been considered for *NP* extraction. Again, it is well known that human brain possesses the capability of understand the sub-jective or *aesthetic* features of a natural language. But they might not always be dealt by some statistical or probabilistic models. Therefore, these characteristics of natural language deserve special consideration. The proposed work is devel-oped to extract NPs from text document considering the *aesthetic* features of the English language.

2 Related Work

A number of works [2-19] found in literature aims to extract NPs from text docu-ment. These works can broadly be classified in three categories such as *Rule Based, Statistical Model Based and Hybrid Approach*. The first category of works [2-7] is mainly employed if adequate data is unavailable. It uses the linguistic model of the language. The second category of works [1, 8-13] does not require linguistic information of language. They are language independent and need suf-ficient data for its successful execution. The third category of works [14-19] de-scribes some hybrid approaches where linguistic features of language along with the statistical information are used for extraction of NPs. The present work be-longs to first category of work.

Rule based approaches [2-7] have mainly used two different approaches: top-down [2-3] and bottom-up [4-7] approach. In the first category of works, the sen-tences are divided in continual way and the NPs get extracted whereas in the sec-ond category of works, words, the fundamental unit of sentence, are extracted first and then rules are applied to form the phrasal forms. The work in [2] has considered 7 word length adaptive windows to extract NPs. The approach in [3] is based on sequence labeling and training by kernel methods that captures the non linear relationships of the morphological features of Tamil language. In the work [4], the authors have used morpheme based augmented transition network to construct and detect the NPs form words. The work described in [5] has used CRF-based mechanism with morphological and contextual features. Another method mentioned in [6] has extracted the words from the sentence first and then uses Finite State Machine to combine the words to form NPs using Marathi language Morphology. N-gram based machine translation mechanism has been applied in [7] to extract NPs form English and French language.

In statistical / quantitative models, the words are first extracted and then they are combined to form phrase using some probabilistic model using knowledge from large scale of data. The work described in [1] has used Hidden Markov Model (HMM) to extract NPs. Probabilistic finite-state automaton has been used in [8]. A Support Vector Machine (SVM) based method to perform NP extraction has been used in [9]. The work as described in [10] has used a statistical natural language parser trained on a nonmedical domain words as a NP extractor. Feed-forward neural network with embedding, hidden and softmax layers [11], long short-term memory (LSTM) recurrent neural networks [12] and multiword Expressions using Semantic Clustering [13] have been introduced in to parse the sentence and tag the corresponding NPs therein.

The methods belonging of hybrid approach exploits some rule-based approach to create a tagged corpus first and then uses some statistical model to confirm as NPs or vice-versa. The works as described in [15-16] uses Part of speech tagger to create tagged corpus and then used Artificial Immune Systems (IAS) to confirm final list of NPs for English and Malayalam Language respectively whereas the work mentioned in [17] has employed handmade rule for corpus preparation and then memory based training rule for NP extraction from Japanese language. The method [18] first exploits rule-based approach to create a tagged corpus for training and then a multilayer perceptron (MLP) based neural network and Fuzzy C-Means clustering have been used. Ref. [19] first employed HMM to extract the NPs in initial level then has used rule to purify the final result.

3 Corpus Description

The corpus is prepared here to conduct experiment on NP extraction from English text document. The text documents comprises of news articles from the sports column of different well known English News papers. 50 such articles are collected from popular English newspapers in India namely, The Telegraph, The Times of India and The Statesman. The distribution of the document paper wise is shown in Fig. 1. The database contains total 20378 words. A stop list has been prepared to include words that are highly frequent in all text documents. The stop list includes 49 prepositions, 26 conjunctions, 3 articles, 6 clitics and 682 other stop words that includes common verbs, adjectives, adverbs, etc.

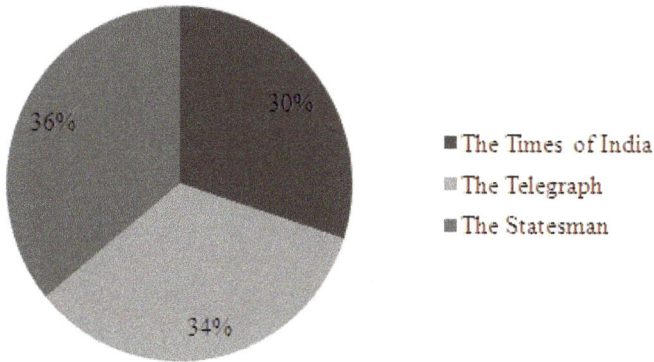

Fig 1. Distribution of the collected data from different newspapers

4 English Morphology

Morphology [1] for a particular language describes a way by which small meaningful units (morphemes) collectively generate words. For example, the morphemes ball and s together make up the word *balls*. Similarly, the word *players* is made up of three morphemes *play*, *er* and *s*. Morphemes can be broadly classified into two major classes. They are *stem* and *affix*. The main/fundamental meaning of the word is carried by its stem and affix provides the additional meanings to the word. Affixes can further be categorized as prefix (precedes a stem e.g., $un - do$), suffix (follows a stem e.g., $look - ing$), infix (within stem e.g., $fan - flaming - tastic$) and circumfix (stem is in the middle e.g., en $-$ light $-$ en). English morphological methods are classified into 4 major types: Inflectional, Derivational, Cliticization and Compounding which along with additional terminology (ies) are detailed in the following subsections. The morphological processes are concatenative in nature.

4.1 Inflectional Morphology

A stem combines with a grammatical morpheme to generate a word with the same stem class and syntactic function like agreement (see Agreement section). Plural form of noun is usually formed by adding s or *es* as suffix to its singular form. English comprises a relatively small number of possible inflectional affixes, i.e. only nouns, verbs, and certain adjectives can be inflected. Table 1 contains some examples of Noun inflection in the form of regular (having suffix –*s* or -*es*) and irregular (having different spelling to form new word) plurals.

Table 1. Regular and Irregular plurals

Morphological Class	Regular		Irregular	
Singular form	player	ball	man	child
Plural form	players	balls	men	children

On the other hand, the verbal inflection in English is more complex rather than Noun inflection. The English language contains generally three types verbs like *main verbs* (e.g., bowl, play), *modal verbs* (e.g., shall, can), and *primary verbs* (e.g., has, be). The majority of main verbs are regular since by knowing their stem only, one can form their other forms easily by concatenating suffixes like -*s*, -*ed*, -*ing*. However, the irregular verbs have idiosyncratic inflectional forms. Some evident of such morphological forms for regular / irregular verbs is depicted in Table 2.

Table 2. Morphological forms of Regular / Irregular verbs

Morphological Class	Regular Verb Inflection		Irregular Verb Inflection		
Stem	play	kick	catch	hit	go
Singular form	plays	kicks	catches	hits	goes
Present participle form	playing	kicking	cat-ching	hit-ting	going
Past form	played	kicked	caught	hit	went
Present / Past participle form	played	kicked	caught	hit	gone

4.2 Derivational Morphology

In Derivational Morphology a stem combines with a grammatical morpheme to generate a word of different class. Its class belongingness is difficult to determine in automatic way. It is quite complex than inflection. In English, it is often found that new nouns can be derived from verbs or adjectives. Such kind of derivations is termed as *nominalization* [1]. Some examples of such types of derivational nouns are depicted in Table 3.

Table 3. Example of Different Derivations

Stem	Stem Type	Suffix	Derived into	
			Noun	Adjective
organization	Noun	-al	-	Organizational
spine	Noun	-less	-	Spineless
modernize	Verb	-ation	modernization	-
appoint	Verb	-ee	appointee	-
bowl	Verb	-er	bowler	-
depend	Verb	-able	-	Dependable
sharp	Adjective	-ness	sharpness	

4.3 Cliticization

A stem is combined with a clitic, reduced form of a syntactic word like morpheme (e.g., have is reduced to 've'), is termed as Cliticization in English morphology. The new string or word thus formed often acts like a pronoun, article, conjunction, or verbs. Clitics may precede or follow a word. In the former case, it is called a *proclitic* and in the latter case it is called an *enclitic*. Some examples are depicted in Table 4.

Table 4. Examples of Clitics and their full forms

Actual Form	am	are	have	not	will
Clitic form	'm	're	've	n't	'll

In English, usage of clitics is often ambiguous. For example, *he'd* can be expanded to *he had* or *he would*. However, the apostrophe simplifies the proper segmentation of English clitic.

4.4 Compounding

Multiple stems are sometime combined to generate a new word. The word oversee is generated by combining the stems over and see.

4.5 Agreement

In English language, the noun and main verb are required to agree in numbers. Hence, plural markings are important. These markings are also required to signify the gender. English has the masculine and feminine genders to represent male and female respectively. Other genders include any object or thing that cannot generalize into male or female. When the class number is very large, they are usually referred to as noun classes instead of gender.

5 Proposed System

The work as described here is a rule based mechanism to extract NPs from text document. At first it accepts the whole text document and then extracts sentence(s) from it. The extracted sentences based on *full stop* as delimiter and are passed through two modules. The first module is *Phrase extractor* which splits each sentence into a number of simple sentence like phrases and then it continues to extract fundamental phrases. The splitting delimiters are punctuation and bracket symbols, conjunctions, prepositions and other stop words. The final list of phrases becomes the input to the second module. The second module is designed to finalize the NPs from the set of phrases. Therefore, the present work uses top-down approach to extract NPs. The modules are detailed in the following subsections.

5.1 Phrase Extractor

In this phase, a text document is broken down into a list of phrases. It first split the sentences into simple sentence like phrases. Then it continues to extract fundamental phrases form it. It is also noteworthy, ambiguity is found for some stop words. For example, *Jammu and Kashmir*, a name of an Indian state, where *and* cannot be considered as conjunction since it is used to join two nouns. Also, the issue of uppercase and lowercase letters is quite prevalent in English language. The first letter in a sentence is in uppercase form in most of the cases. Obviously, a sentence may begin with a stop word or number or symbol. It may not be a noun. Other such issues get addressed here during simple sentence like phrase extraction and also for NP selection (described in NP Finalization section). The detail mechanism of phrase extraction is described in Algorithm 1.

Algorithm 1:

Input: A text document
Output: List of NEs

Step 1:   Extract sentences from text document using full stop as delimiter
Step 2:   If sentence's first character = Upper case letter
                   {
                   If (first word is a stop word)
                   {
                   Change the first character to lower case
                   }
                   Else
                   {
                   No change
                   }
                   }
Step 3:   Split each sentence into sub sentences using punctuation and bracket symbols as delimiters.
Step 4:   Split each sub sentence into parts using conjunction as delimiter
        If conjunction is 'and':
        If string before 'and' has verb/verb phrase
        {
        If string after 'and' has verb/verb phrase
        {
        Split using 'and' as delimiter
        }
        Else
        {
        No change
        }
        }
        Else
        {
        No change

}

Step 5:	Split each part into sub-parts using preposition as delimiter
Step 6:	Split sub-parts using clitics to get sub sub-parts
Step 7:	Split sub sub-parts into phrases using other stop words such as common verbs, common adjectives and adverbs, pronouns as delimiters.

5.2 NP Finalization

The list obtained the first module (Phrase Extractor) may contain phrases that have some stop words attached to it either at the beginning or at the end. Furthermore, the phrase itself may be a non NP. There a purifying mechanism has been designed to confirm the final list of NPs from phrase list. The mechanism is described in Algorithm 2.

Algorithm 2:

Input: List of phrases
Output: List of NPs

For (all phrases in the list)
{

Step 1:	If phrase contains only stop word(s) or unwanted symbol(s), delete phrase.
Step 2:	If phrase starts with an article or upper case letter, no change.
Step 3:	If phrase contains stop word or unwanted symbol at the beginning or end, prune it from the phrase.
Step 4:	If word in a phrase has suffices and no capital letter at word's first position split the phrase.

}

6 Result and Discussion

For experimental purpose 50 text document from 3 popular News paper has been collected. The detail of data has already been described in Corpus Description section. The intermediate and final result of the designed process is described using a text line as example which is described using an example. Note that we are not considering the articles or pronouns.

The quantitative result of the described mechanism has been prepared in manual way. All the valid NPs from the documents have been selected first. The designed process is employed on the same text document. Finally, human generated list of NPs is compared with the NP list generated by the proposed system. The final result is quantizing using the statistical measures like *recall, precision* and *F-measure*. The average recall, precision and F-measure for these 50 text documents are 97%, 74% and 84% respectively. Sample result of the same is given in Table 5.

Example

Sentence: The former Australian batsman has been a part of the South African support staff during the World Cup in Australia and New Zealand, and T20 captain Faf Du Plessis feels that his presence will only help youngsters.

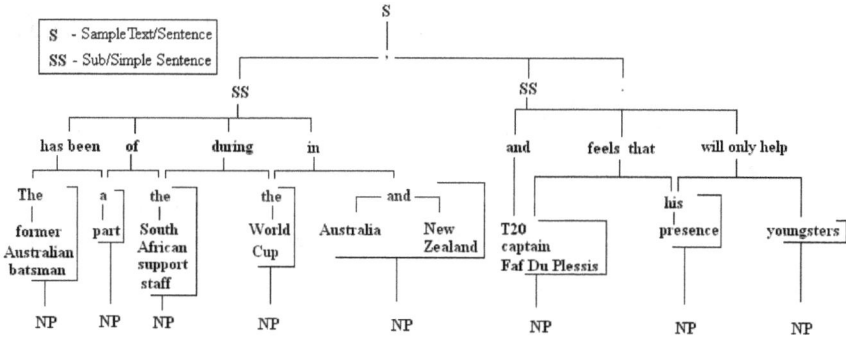

Fig 2. Successive breaking/splitting of sample text/sentence to get NPs

Desired result (NP List)	Result from the proposed system
1. former Australian batsman	1. former Australian batsman
2. part	2. South African support staff
3. South African support staff	3. World Cup
4. World Cup	4. Australia and New Zealand
5. Australia and New Zealand	5. T20 captain Faf Du Plessis
6. T20 captain Faf Du Plessis	6. presence
7. presence	7. youngsters
8. youngsters.	

Table 5. Depiction of detail result for 5 sample text documents

Document #	# of Words	TP	FP	FN	Precision	Recall	F-measure
1	692	165	57	0	0.743	1	0.853
2	572	111	54	2	0.673	0.982	0.799
3	308	94	20	1	0.825	0.989	0.897
4	175	35	14	1	0.714	0.972	0.823
8	280	84	16	0	0.840	1	0.913

Conclusion

Any natural language follows certain rules or grammar. Although the rules may vary from language to language, still most languages currently being communicated (by significant number of humans) have more or less the same grammatical syntax and structure. The present work proposes a mechanism to extract NPs from English text documents using these rules. The English morphological rules are considered here. The algorithm is extremely simple and although it may seem rather primitive in nature, the method provides some vital benefits since it includes some subjective or *aesthetic* features of a natural language. Therefore the proposed system has its tendency towards universality. Also, the mechanism extracts the NPs in admissible time. The average recall, precision and F-measure for these 50 English text documents are 97%, 74% and 84% respectively.

In English, the word count is 1.2 billion and still counting since English has many words derived from various languages namely, Latin, Sanskrit, French, etc. as a result, the number of nouns may also increase. This approach uses storage of a significant number of stop words but this stop list is not ultimate. So, number of stop words may constrain overall performance. Incorporating more composite rules to phrase extraction can enhance the model.

References

1   Daniel Jurafsky and James H. Martin, "Speech and Language Processing: An Introduction to Natural Language Processing, Computational Linguistics and Speech Recognition", *Pearson, 2nd Edition.*
2   Bennett, Nuala A., et al. "Extracting noun phrases for all of MEDLINE." *Proceedings of the AMIA Symposium.* American Medical Informatics Association, 1999.

3 Dhivya, R., Dhanalakshmi, V., Kumar, M. A., & Soman, K. P. (2012). Clause boundary identification for tamil language using dependency parsing. In *Signal Processing and Information Technology* (pp. 195-197). Springer Berlin Heidelberg.

4 Nair, L. R., & Peter, S. D. (2011, October). Shallow parser for Malayalam language using finite state cascades. In Biomedical Engineering and Informatics (BMEI), 2011 4th International Conference on (Vol. 3, pp. 1264-1267). IEEE.

5 El-Kahlout, I. D., & Akın, A. A. (2013). Turkish constituent chunking with morphological and contextual features. In *Computational Linguistics and Intelligent Text Processing* (pp. 270-281). Springer Berlin Heidelberg.

6 Bapat, M., Gune, H., & Bhattacharyya, P. (2010, August). A paradigm-based finite state morphological analyzer for Marathi. In *Proceedings of the 1st Workshop on South and Southeast Asian Natural Language Processing (WSSANLP)* (pp. 26-34).

7 Marino, J. B., Banchs, R. E., Crego, J. M., de Gispert, A., Lambert, P., Fonollosa, J. A., & Costa-Jussà, M. R. (2006). N-gram-based machine translation. *Computational Linguistics*, *32*(4), 527-549.

8 Serrano, J. I., & Araujo, L. (2005, September). Evolutionary algorithm for noun phrase detection in natural language processing. In *Evolutionary Computation, 2005. The 2005 IEEE Congress on* (Vol. 1, pp. 640-647). IEEE.

9 Dhanalakshmi, V., & Rajendran, S. (2010). Natural Language processing Tools for Tamil grammar Learning and Teaching. *International journal of Computer Applications (0975-8887)*, *8*(14).

10 Huang, Y., Lowe, H. J., Klein, D., & Cucina, R. J. (2005). Improved identification of noun phrases in clinical radiology reports using a high-performance statistical natural language parser augmented with the UMLS specialist lexicon. *Journal of the American Medical Informatics Association*, *12*(3), 275-285.

11 Coppola, C. A. D. W. G., & Petrov, S. Improved Transition-Based Parsing and Tagging with Neural Networks.

12 Ballesteros, M., Dyer, C., & Smith, N. A. (2015). Improved transition-based parsing by modeling characters instead of words with LSTMs. *arXiv preprint arXiv:1508.00657*.

13 Chakraborty, Tanmoy, Dipankar Das, and Sivaji Bandyopadhyay. "Identifying Bengali Multiword Expressions using Semantic Clustering" *Lingvisticæ Investigationes* 37.1 (2014): 106-128.

14 Pattabhi R K Rao T, Vijay Sundar Ram R, Vijayakrishna R and Sobha L, "A Text Chunker and Hybrid POS Tagger for Indian Languages", *Proceedings of IJCAI-2007, SPSAL-2007*.

15 Kumar, A., & Nair, S. B. (2007). An artificial immune system based approach for English grammar checking. In *Artificial Immune Systems* (pp. 348-357). Springer Berlin Heidelberg.

16 Bindu, M. S., & Idicula, S. M. (2011). A Hybrid Model For Phrase Chunking Employing Artificial Immunity System And Rule Based Methods. *International Journal of Artificial Intelligence & Applications*, *2*(4), 95.

17 Park, S. B., & Zhang, B. T. (2003, July). Text chunking by combining hand-crafted rules and memory-based learning. In *Proceedings of the 41st Annual Meeting on Association for Computational Linguistics-Volume 1* (pp. 497-504). Association for Computational Linguistics.

18 Kian, S., Akhavan, T., & Shamsfard, M. (2009, October). Developing a persian chunker using a hybrid approach. In *Computer Science and Information Technology, 2009. IMCSIT'09. International Multiconference on* (pp. 227-234). IEEE.

19 Ibrahim, A., & Assabie, Y. (2013). Hierarchical Amharic Base Phrase Chunking Using HMM
    With Error Pruning. In *Proceedings of the 6th Conference on Language and Technology, Poz-
    nan, Poland* (pp. 328-332).

Jaya Gera[1] and Harmeet Kaur[2]

Recommending Investors using Association Rule Mining for Crowd Funding Projects

Abstract: Many projects fail to meet their funding goal due to lack of sufficient funders. Crowd Funders are the key component of crowdfunding phenomenon. Their monetary support makes a project's success possible. Their decision is based on project's quality and their own interests. In this paper, we aim to promote projects by recommending promising projects to potential funders so as to help projects meet their goal. We have developed a recommendation model that learns funders' interests and recommends promising projects that match their profiles. A profile is generated using funders backing history. This experiment is conducted using Kickstarter dataset. Projects are analysed on several aspects: various project features, funding pattern during its funding cycle, success probability etc. Initially, recommendations are generated by mining funders' history using association rule. As few backers have backed multiple projects, data is sparse. Though, association rule mining is quiet efficient and generates important rules but is not able to promote promising projects. So, recommendations are refined by identifying promising projects on the basis of percentage funding received, pledge behaviour and success probability.

Keywords: crowdfunding; recommender systems; association rule mining; user interest; success probability; pledge behaviour.

1 Introduction

One of the most challenging tasks for setting up a new venture is to arrange sufficient funds. Although crowdfunding has emerged as viable alternative solution for raising funds for new ventures; not all of them are successful to raise sufficient funds. One of the most common reasons for failure is that venture initiators are novice and have difficulty in understanding and leveraging their social network

1 Department of Computer Science, Shyama Prasad Mukherji College, University of Delhi, Delhi, India
jayagera@spm.du.ac.in
2 Department of Computer Science, Hans Raj College, University of Delhi, Delhi, India
hkaur@hrc.du.ac.in

to reach to correct audience for their product [1][2]. Audience is diverse in nature and spread across the globe. Audience is not just consumer of product but turning to the role of wise investors/funders [3]. Crowd funders not only provide monetary support but also motivate and influence other funders' decision and help in promoting projects that is essential for projects' success. Capturing wisdom of funders and finding funders matching with the project profile cannot be done by the project initiator or creator alone. This leads to requirement of emergence of crowdfunding intermediators or crowdfunding platforms.

Crowdfunding Platforms act as intermediators between project initiators and potential funders [4]. It provides certain functionality and acts as an electronic matching market that overcomes information asymmetry and costs [5]. Their objective is to maximize the number of successful projects [6]. To achieve this, they need to design policies and strategies to motivate funders to fund [7] so that the site can get more projects funded. This can be achieved by analysing funders' trend and timing of contribution and via coordination among them [7].

With the increase in popularity of crowdfunding, crowdfunding platforms have also grown like mushrooms. Most platforms do not do more than providing a platform to present the projects and a mechanism for online payment to collect pledge [8]. But, some of them do provide value added services such as provide suggestions [8], help in expanding network [6], building trust and much more. The efforts put by crowdfunding platforms have positive influence and help creators in raising funds.

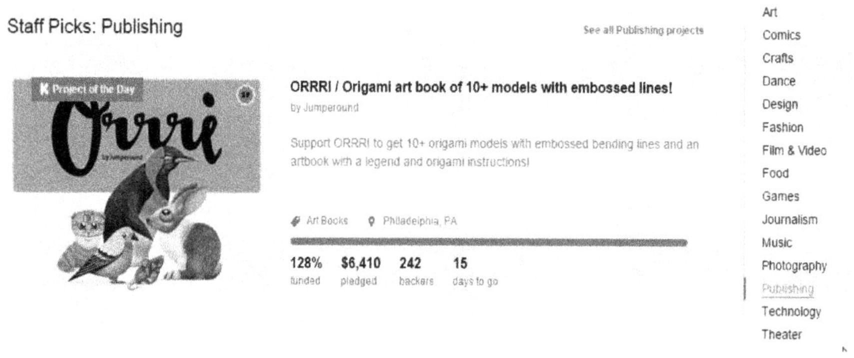

Staff Picks: Publishing   See all Publishing projects

ORRRI / Origami art book of 10+ models with embossed lines!
by Jumperound

Support ORRRI to get 10+ origami models with embossed bending lines and an artbook with a legend and origami instructions!

Art Books   Philadelphia, PA

128% $6,410   242   15
funded   pledged   backers   days to go

Art
Comics
Crafts
Dance
Design
Fashion
Film & Video
Food
Games
Journalism
Music
Photography
Publishing
Technology
Theater

Figure 1. source: https://www.kickstarter.com/

Some also assess and promote projects on their sites, for example, Kickstarter platform promotes projects in various ways: staffs pick projects, project of the day, popular projects, allows to discover projects popular among friend circle etc. Figure 1 shows one such snapshot of Kickstarter (retrieved on 5 Jan 2016).

In Literature, attention has not been paid towards matching projects and funders to promote them among suitable funders. Proposed work has developed a method that generates rules using backers' funding pattern, learns funders' profile and trend of funding and matches projects with profile and recommends them to funders. The aim is to assist project initiator in raising funds and to improve performance and add functionality to the platforms. Rest of paper discusses literature work, then dataset and its characteristics, followed by proposed work and conclusion.

2 Related Work

Though, crowdfunding is now a mature domain, understanding about dynamics of crowdfunding platform is lacking [4]. Most of the literature work focuses on analysing various factors and their impact on success of crowdfunding projects, role of social media and geography, impact of social network size on success, motivation behind investment decisions, effect of timing and coordination of investors etc. However, less emphasis is given on understanding role of various platforms, ways of making policies, understanding and adding to existing functionalities. Some researchers have paid attention to these dimensions and brought new insights about crowdfunding intermediators.

Ref. [5] developed an empirical taxonomy of crowd funding platform that characterizes various crowdfunding intermediation models on the basis of Hedonism, Altruism, and Profit. They also focused on how crowdfunding intermediaries manage financial intermediation and how do they transform relations between initiator and funder in two-sided online markets.

Ref. [9] proposed different ways of recommending investors based on twitter data. Recommendation is generated on the basis of pledge behaviour of frequent and occasional investors. Research suggested that frequent investors are attracted by ambitious projects whereas occasional investors act as donors and mainly invest in art related projects.

Ref. [10] categorized Kickstarter projects' features as social, temporal and geographical features and analysed these features' impact on project success. This analysis also build recommendation model using gradient boosting tree that uses these features to recommend set of backers to Kickstarter projects.

Ref. [7] analysed donors' contribution, their timing of contribution and coordination among donors and impact on funding of projects. Ref. [11] suggests donors funding decision play an important role in the ultimate success of a crowdfunding project. Potential donors see the level of support from other project backers as well as their timing before making their own funding decision. Ref. [12] observed temporal distribution of customer interest and concluded that there exist strong correlation between a crowd funding projects early promotional activities and its final outcome. They also discussed importance of concurrent promotion of projects from multiple sources.

Ref. [1] revealed that interacting and connecting with certain key individuals provide advantage of acquiring resources and spreading information about projects. This study also disclosed that a small portion of fund comes from strong ties (family, friends, colleagues, relatives etc.) but large portion of funds come from weak ties i.e. from people on network whom creator rarely met or interacted with. Ref. [5] also suggested that matching projects with its potential investors enables successful funding.

Various research studies suggest that crowdfunding market is growing fast in all respects i.e. volume of projects launched every day, number of funders turning up and number of platforms rising up. But an increase in volume does not mean increase in performance. So, there is need to develop a mechanism to match projects and funders. The contributions of this work are:

    i.   Add to platform functionality by automatically matching projects with potential funders

    ii.   Understanding funders and their interests by maintaining their profiles

    iii.   Assist initiators in evaluating success prospects of their projects

3 Data Set

This experiment is conducted on Kickstarter data. This dataset consists of data about projects, funding history of projects and project backers. Projects and their funding history are obtained from kickspy[1] website and backers' data for each of the project in this dataset is obtained by crawling backers' pages from kickstarter[2] website. This dataset consists of data of 4862 projects launched in the month of April 2014 and backing history of 97,608 backers who backed these projects. Project data includes project id, name of project, goal amount, pledged amount, status, category, subcategory, duration, rewards, facebook friends, facebook shares, start date, end date, duration etc. Pledge data consists of amount pledged on each day during funding cycles by these projects. This dataset also contains data of live, suspended and cancelled projects. These projects and their backing transactions are removed for analysis purpose. After removing these projects, dataset is left with 4,121 projects and 92,770 backers. Out of 4,121 projects, 1,899 (46%) are successful and 2,232 (54%) are unsuccessful.

To have better understanding of individual project characteristics, Mean Value Analysis is done. Table 1 lists mean value for some of project characteristics.

Table 1. Mean Value Analysis

	All	Successful	Unsuccessful
Projects	4,121	1,899 (46%)	2,232 (54%)
Goal Amount	54537.69	10882.21	91484.46
Pledged Amount	11393.78	22166.31	2276.70
Backers	139.70	272.80	27.06
Rewards	9.82	11.30	8.57
Updates	3.18	5.25	1.42
Comments	19.62	38.95	3.26
Duration	31.81	30.14	33.21
Facebook Friends	711.25	823.66	613.07

In nutshell, successful campaigns on an average have low goal to achieve, less duration, a significantly large number of funders and facebook friends to support, offers a good number of rewards and have a better interaction between creators and funders through updates and comments.

4 Proposed work

There are a large number of projects that are unable to complete because they fail to publicize and attract sufficient number of funders. For a project to be successful, it must reach its funding goal. To reach its goal, there should be sufficient number of investors, who are willing to invest and take risk. Research reveals that 20-40% of initial funding comes from family and friend [13]. But, a large number of funders are unknown to the creator and fund for various reasons. With the

---

1 http://www.kickspy.com. This web site is currently shut down.

2 https://www.kickstarter.com/ * Backers' page has now been removed by Kickstarter website.

growth of technology and security aspects, large number of creators as well funders are participating online. These funds are small in amount and spread across various projects [14]. As the funding amounts are not very large and come from large network of unknown people, there is a need to coordinate investors funding [14] to have more number of successful projects. To assist potential funders, we have developed a recommender model that learns through funders backing history using association rule mining and recommend and promote projects among potential funders.

4.1 Method

Our aim is to assist initiators, funders as well as platforms such that overall success rate of platform is increased and all the stakeholders are benefitted. Some important issue are: which projects need to be promoted? What criteria should be used to identify such projects? Projects that signal high quality and popular in social network get funded soon. Projects that signal low quality raise nothing or very less. Such projects may not get funded even by friends and family. Projects that possess good quality and perform well initially but lose their track later on are the best candidates for promotion. This model identifies such projects by analysing their quality and funding pattern. Fig. 2 shows model components. Recommendation model has five modules:
i)   Predictor
ii)   Trend Monitor
iii) Profile Modeller
iv)   Rule Generator
v)   Recommender

Predictor: Some projects perform well on monetary front and attract large amount than required. Some projects perform poorly and attract nothing or little monetary investment. Project success is also influenced by project quality [15]. Project quality is assessed by project preparation and presentation. Assessing true status of project preparation is not feasible, because creators disclose as much as they wish to. Crowdfunding suffers from information asymmetry [6] i.e. creator knows actual situation whereas funder can assess using information disclosed. So, in this module, project success is evaluated based on project presentation. Project is characterized by various features such as Category, has video, number of videos, number of images, goal amount, duration, facebook friends etc. These features are good indicator of project quality. Project success is predicted by feeding these attributes to logistic regression. This module predicts project success with 81.5% accuracy.

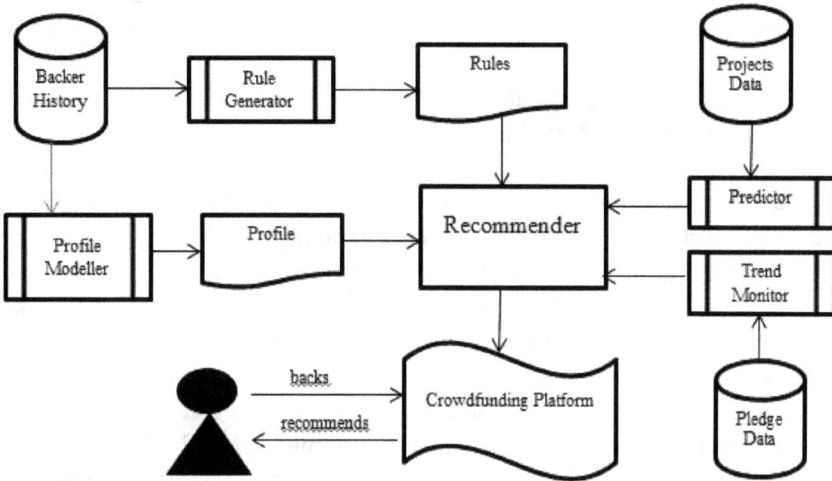

Figure 2. Recommender model

Trend Monitor: Predictor's prediction is based on static features available at the time of launch of projects such as goal amount, category etc. This does not assess performance of project after launch. Our aim is to promote projects that are of good quality but could not raise enough and lack by a little margin. We need to identify such projects whose project presentations are as good as successful ones but grow slow during their funding cycle. This can be done by monitoring their funding behaviour. We need to understand nature of successful and unsuccessful funding pattern. Successful projects generally grow faster than unsuccessful one. Pledge analysis [16] states, if a campaign has raised approximately 20% of its goal within the first 15% of funding cycle, its success probability is high. Unsuccessful initially starts well but fails to retain this growth after sometime. So, campaigns that could raise 20% of funds within 20% of funding time are good candidates to be promoted. Module Trend monitor performs analysis of funding behaviour of project and identifies such projects.

Profile Modeller: This module learns backer profile by analysing backer's funding history. Profile of a backer B_i is defined as

$$B_i = \{Backer\_id_i, Name_i, Location_i, CategoryPref_i\}$$

Each backer is assigned a unique identification i.e. backer_id. Name attribute contains Name of backer and Location attribute contains address and city of backer. CategoryPref is generated by scanning backing history and finding category and subcategory of each project backed. $CategoryPref_i$ is a set that is a defined as:

$CategoryPref_i = \{\{Cat_{j1}, subcat_{k1}, n_{k1}\}, \{Cat_{j2}, subcat_{k2}, n_{k2}\}, \ldots \{Cat_{jm}, subcat_{km}, n_{km}\}\}$ i.e. Backer has supported n_{ka} number of projects of subcategory *ka* of category *ja*.

Rule Generator: Recommender system not only identifies projects to be promoted but also understands trend of backers and learns which projects, backers are frequently backing. To understand behaviour pattern of backers, we used Association rule mining technique of data mining. Association rule mining aims to extract interesting correlations, frequent patterns, associations or casual structures among sets of items in the transaction databases or other data repositories [17]. Association rule mining is generally used in Market Basket Analysis. It mines transactions history and tells which items are frequently bought. As we are interested in knowing which projects are backed together by different backers, we have used association rule mining technique.

Rules are generated by applying Apriori algorithm of Association rule mining technique. Two parameters support and confidence are used to measure interestingness of rules. Rules that satisfy minimum support and minimum confidence value are refereed as strong association rule and are of interest [17][18]. For association rules of the form $X \Rightarrow Y$ where X and Y are sets of items, support and confidence formulas are defined as:

$$Support(X \Rightarrow Y) = \frac{Number of records containing both X \wedge Y}{Total Number of records}$$

$$Confidence(X \Rightarrow Y) = \frac{Number of records containing both X \wedge Y}{Total Number of records containing Y}$$

Association rule mining has two phases: i) finding frequent item sets ii) generating rules. First phase finds itemsets that satisfy minimum support count value. Second phase generates rule using itemsets that satisfy confidence threshold value.

This dataset consists of list of projects backed by backers. Let us understand Apriori Algorithm with the help of an example.

Assume, there are 15 backers (named b_1, b_2,...,b_{15}) and 5 projects (named p_1, p_2,...,p_5) and data of projects backed by them. Table 2 shows steps of Apriori algorithm.

Table 2. Apriori Algorithm Steps

Dataset	Frequent item-set1	Frequent item-set2	Frequent item-set3	Rules
$b_1 = \{p_2, p_5\}$	**p_1:5,**	**(p_1,p_3):2,**	**(p_1,p_3,p_4):2**	p_1=>p_3
$b_2 = \{p_1\}$	**p_2:4,**	**(p_1,p_4):2,**		p_3=>p_1
$b_3 = \{p_3, p_4\}$	**p_3:5**	(p_2,p_3):1,		p_1=>p_4
$b_4 = \{p_1, p_3, p_4\}$	**p_4:6,**	**(p_2,p_5):2,**		p_4=>p_1
$b_5 = \{p_2\}$	**p_5:4}**	**(p_3,p_4):3**		p_2=>p_5
$b_6 = \{p_2\}$		(P_3,p_5):1		p_5=>p_2
$b_7 = \{p_2, p_3, p_5\}$		(p_4,p_5):1		p_3=>p_4
$b_8 = \{p4\}$				p_4=>p_3
$b_9 = \{p1\}$				(p_1,p_3)=>p_4
$b_{10} = \{p_1\}$				(p_1,p_4)=>p_3
$b_{11} = \{p_5\}$				(p_3,p_4)=>p_1
$b_{12} = \{p_4, p_5\}$				
$b_{13} = \{p_1, p_3, p_4\}$				
$b_{14} = \{ p_4\}$				
$b_{15} = \{p_3\}$				

Assumption: support count is 2. Items in **Bold** satisfy support count value and are selected for next step.

Recommender: Recommender module generates recommendations for backers using association rules and backer profile and ranks them on basis of pledge behaviour and prediction value. Suppose a backer has backed projects $\{p_1, p_5\}$. Now we need to identify rules that can match with any of the subsets :$\{\{p_1\}, \{p_5\}, \{p_1, p_5\}\}$. Two rules (left hand side) $\{p_1 => p_4, p_5=>p_2\}$ match so recommender system will generate list of projects to be recommended as $\{p_2, p_4\}$. Now suppose these two projects are of music and video categories respectively. Then recommender system finds other projects that are of these two categories and have probability value 0.5 and have raised between 10%-20% of goal amount in initial period. Suppose such projects (new projects launched) are p_6, and p_9. Now recommendation list is expanded to $\{p_2, p_4, p_6, p_9\}$. Now ranking will be done on basis of percentage money raised. Lower the pledge percentage, higher the rank.

This model of recommender system resolves "cold start problem". New projects are selected by matching backer profile (not by rule). As projects here are time bound and have to raise required goal amount within given time frame. Projects that are closed cannot be recommended. So, recommendation process is dynamic and needs to update project database and rules on daily basis, as everyday a good number of projects are launched on Kickstarter platform. A new project (new item) is selected if it matches with backer's profile.

4.2 Experiment & Result

Experiment is conducted using SPSS and Python. This dataset consists of projects and backer data. There are 4,121 projects and 92,770 backers who backed these projects. Project data is rich in terms of project features. Each project is characterized by more than 35 features.

Predictor Analysis: Predictor identifies projects to be promoted by using features available at the time of launch. It evaluates projects on the basis of predicted success probability of projects. Success probability is generated using logistic regression in SPSS. Predictor uses following features for evaluating project probability of success.

Category, Subcategory, Goal Amount, Duration, Facebook Connection, Facebook friends, Facebook Shares, Has Video, Number of Videos, Number of Images, Number of Words in Description, Number of Words in Risks and Challenges, Number of frequently asked questions (FAQs).

Kickstarter classifies projects in 15 categories. These categories further classified in subcategories, for example, category film & video is sub classified as horror, short, documentary. The accuracy of prediction is 81.5%. As per logistic regression analysis, Prediction value 0.5 is sound indicator of good project quality. Table 3 shows confusion matrix.

Trend Monitor Analysis: This analysis is done using python in ref [19]. Projects are classified into four categories on the basis of percentage of goal amount raised: overfunded (raised >= 120%), funded (raised 100-120%), potential (>=60%), lesshopeful (<60%). Overfunded projects are projects that raised more than 120%.They grow very fast and attract large number of backers, are popular on social media and achieve goals soon after launched. Funded projects are projects that get 100% or little more than that. Their growth is stable. Potential projects are projects that could raise more than 60 % of goal amount but could not succeed. These are the projects that can be successful, if

Table 3. Confusion matrix

		Predicted		Percentage
	State	Failed	Successful	Correct
Observed	Failed	1,929	303	86.4
	Successful	460	1,429	75.6
Overall Percentage				81.5

*Cut off value is 0.5

identified timely and promoted. Lesshopeful projects are projects that raised nothing or little but less than 60%. Figure 3 shows funding pattern followed by various type of projects. This analysis is done using median of all projects of different categories.

Our current focus is on promoting potential projects. By closely monitoring funding pattern of funded and potential projects, we identified that both start well but after some time, potential projects growth slows down, as it fails to attract more backers. It is suggested that the projects that have achieved 20% funding within 20% of the funding cycle come under this category should be identified and should be promoted [16].

Rule Generator Analysis: This part of experiment is conducted using 5-fold cross validation and ratio of training and test data was 80% and 20% respectively. Dataset contained backing history of 92,770 backers. Each iteration of cross validation used backing history of 74,216 backers for training and backing history of 18,554 backers for testing.

To generate backing history dataset, each projects webpage was crawled from Kickstarter website and details of backers of each project were extracted in csv file. Then, backer's details of all projects were combined into one csv file. This file was then processed and list of projects backed by each backer was stored in the following format:

$$\{b_1: (p_2, p_5), b_2 :(p_1), b_3 : (p_3, p_4)\ldots\ldots\}$$

This data was used for generating rules. Support count was set as 3 and confidence was set to 1 for generating rules. Support count is very low, because itemsets are not frequent as less number of backers had 3 or more projects in common and are less significant. Out of 92,770 backers, 85,112 backers have backed only one project and 7,658 backers have backed more than one project. Higher support count results in less number of rules. Rule mining with support

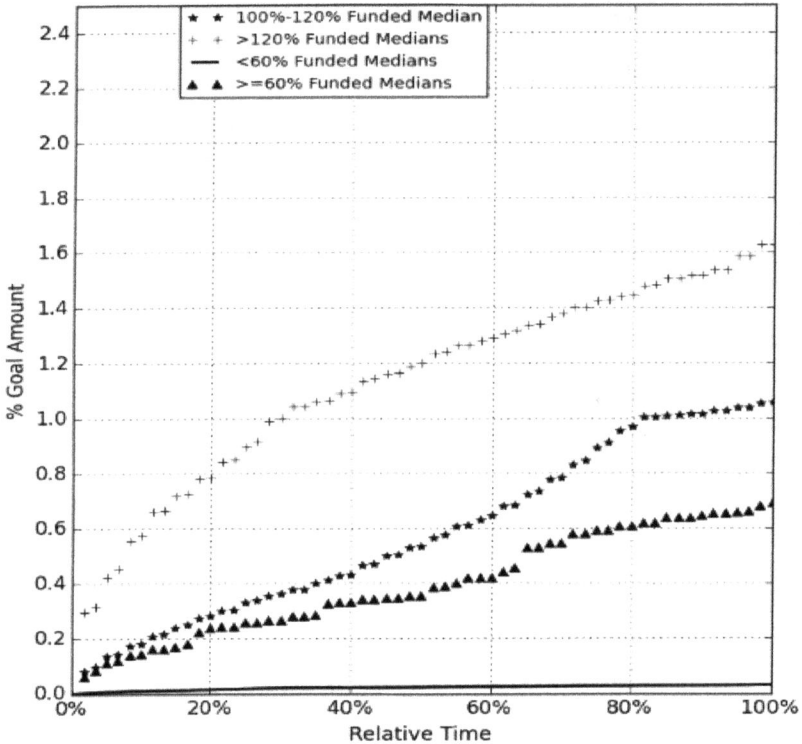

Figure 3. Pledge Analysis (source: Ref. [19])

count 3 on average generated 314 rules. When support count was set to 4, only 59 rules were generated and for support count of 5, only 23 rules were generated. No rules were generated when the support count was above 5. Although, support count is low, but rules generated are important because they could find good number of combination of similar projects funded together. This rule set can be enhanced if backing frequency is increased. Backers may have backed more projects (may be not part of this dataset) but backing history includes only projects that occurred in this dataset.

Recommender: Recommender module is executed for each fold of cross validation. Recommendation model recommends projects that are popular among backers, have sound quality and have potential to reach goal amount. Recommender module has two sub phases:

1   Generating recommendation as directed by association rules.

2   By matching all category and subcategory of projects, backer had funded in past, module selects projects from project dataset having probability 0.5 or more and following funding pattern of potential projects. Projects that are having funding pattern or lie in between > 60% funded and 100-120% funded graph (shown in Fig. 3) are potential projects. We observed current pledged status w.r.t their current funding time passed of all projects having predicted success probability 0.5 or more. All live projects who have met their goal are removed. Rest are analysed for pledged status. All projects that could raise 10-20% within first 20% of funding time or 20-40% within next 20-40% of funding time or 35-60% in 40-60% of funding time or raised 40% or above in 60-80% of time or 60% or above are considered for recommendations.

Then, these two results are integrated and ranked on basis of probability high and level of funding low. This set will consist of all projects popular in the network and projects that have potential to succeed. Popular projects are identified using association rule mining. Promising projects are identified by predictor and trend monitor.

Table 4 shows performance analysis of phase 1 of recommender module i.e. recommendations generated using association rules. Two parameters - coverage and accuracy are used to measure performance of this phase. Accuracy tells how many recommendations are correct. Coverage measures how many are likely to be backed by backer. The average coverage of 41.94% and average accuracy of 26.56% was obtained on experimentation. Accuracy and coverage are measured at category level i.e. by considering category of projects backed and recommended by the model on the basis of information of one project (chosen randomly) backed by backer. Recommendations generated using rules had limitations of covering less number of projects. Recommendations are generated by matching subsets of projects backed by backer with left hand side of rules. Right hand side of rules that match with subsets are recommended. As, rules had covered limited number of projects, most of the backer's backing list and their subsets did not match with any of the rules, as a result recommendations are not generated for these backers in the test set. So, to improve this generated list of recommendations and to provide recommendations for all backers of test set, we directly matched projects with backer's profile. For example, if a backer has backed project(s) of category- Film & video and subcategory – Documentary earlier, then recommender selects top 10 ranked live projects of this category and subcategory and recommends them to the backer. So, the accuracy of these recommendations is 100% as they match with the category of projects backed by backer. If a backer has backed projects of different subcategories of a category or different categories, recommender generates maximum 10 top ranked projects of these categories. Suppose, a backer has backed projects of category Food and Games both, then recommender selects top 5 ranked projects of each category.

Table 4. Analysis of Recommendation using Rules

Cross fold Iteration No.	No. of Rules Generated	Average Accuracy per backer	Average Coverage per backer
I	354	21.63%	38.80%
II	396	24.39%	42.98%
III	350	21.53%	37.08%
IV	399	22.52%	43.86%
V	73	42.72%	46.98%

4.3 Summary

This work follows the thought of Ref. [20] that Capturing signs of success in early stage of a project and taking action accordingly helps in meeting goals. Probability of acceleration in contribution by funders also increases with the increase in funding level [6]. Research also suggests that initial funding comes from family and friends and support from family and friend also signals sound project quality [13], If a project starts well, later slows down, it may be because it could be publicized among interested funders. With this perception; we identified projects to be promoted in early stage by just observing projects performance within 20% of funding period of time. If a projects is able to achieve 10% to 20% within 20% of time and signals good quality are promoted. This work not only monitors and promotes projects soon after their launch but also keeps track of projects in later stages of funding cycle and promotes those having slow growth to improve overall success rate of crowdfunding platforms.

This model generates recommendations by analysing projects' and backers' characteristics and behaviour. This model integrates the recommendations generated by association rules and recommendations generated by matching backers' profiles and projects' features. The recommendations generated using association rule mining is though very less, but accurate. Observations revealed that projects recommended by association rule match with projects backed by funder in many dimensions such as category, subcategory, status. And projects generated by profile matching find projects interesting to backers.

Our dataset has 1,720 projects that have predicted success probability of 0.5 or more. Out of 1,720 projects, 312 projects are the ones that failed. Had these projects been tracked and promoted timely, success rate would have improved to 53.65% from 46% i.e. an increase of approximately 7%.

Conclusions and Future work

In this paper, a recommendation model based on Association rule mining technique and past backing behaviour is presented. As recommendations, generated using only association rules, cover only limited number of backers and projects, the model is refined and projects matching with the backers' profile and lacking by less margin from their target amount are identified and recommended. This method helps in improving accuracy of the model at category and subcategory level. Model helps in promoting new projects soon after their launch if required.

Projects launched over crowdfunding platform remain live for short period of time. Everyday numerous projects are launched on the platforms and new/old backers keep on backing them. So, project dataset and backing history are dynamic. This requires rules set should be evolved to accommodate new projects and their backing pattern. To upgrade rule set, a rule generating recommendations of non-live projects is updated / removed and rules referring new launched projects are added to the rule base. This rule set can be improved if more number of backers has backed multiple projects.

Accuracy and coverage of rules are evaluated on category and subcategory level both. The average coverage and accuracy at category level is 41.94% and 26.56% respectively, which is quite satisfactory. The average coverage and accuracy at subcategory level is 32.46% and 20.78% respectively. Accuracy and coverage is evaluated on basis of recommendations provided by model when information of only one project backed by backer is given. System is required to be evaluated for different support and confidence threshold value and for subset of projects backed by backer. More experiments are required to be performed to compare with other recommendations model and other models of crowdfunding.

This model can be improved, if crowdfunding platforms provide additional information such as amount and time of contribution for each backer, then key backers (backers that have potential influence on social network and can help in promotion of projects) can be identified to promote these projects. This work cannot generate recommendations for backers who are new to the system and have not backed any project. To generate recommendations, backer must have backed at least one project. As this work has limitation of promoting projects on crowdfunding platforms, we would like to further enhance its capability to promote them on social media such as Twitter, Facebook etc., as projects promoted over social media have great chance of success.

This work emphasises on project quality and backers' interest. Crowdfunding is vulnerable to fraud. With the increase in volume of crowdfunding, chances of frauds are also increasing. So, we would like to extend this model so that promising projects of reliable and reputed creators only are recommended among potential backers.

Acknowledgement

This analysis is based on dataset downloaded from web site: http://www.kickspy.com/. This is released by owner of this website. This dataset is rich in Kickstarter projects features and contained each day pledged status of these projects during funding cycles that enabled us to analyze the pledge characteristics of projects and promote them when required. We would like to thank owner of this website for uploading this dataset and allowing others to use it for research work.

References

1   J.S Hui, M. Greenberg, E Gerber, "Understanding the role of community in Crowdfunding work", CSCW 2014, ACM, Baltimore, MD, USA, 2014.
2   J. Hui, E. M. Gerber, D. Gergle, " Understanding and Leveraging social networks for Crowdfunding: opportunities and challenges", in proceedings of the 2014 conference on Designing interactive systems, ACM, New York, USA,2014, 677-680.
3   A. Ordanini, L. Miceli, M. Pizzetti, A. Parasuraman, "Crowdfunding: transforming Customers into Investors through innovative service platforms", Journal of Service Management, Vol. 22, Issue 4, 2011, 443 – 470.
4   P. Belleflamme, T. Lambert, A. Schwienbacher. 2011, "Crowdfunding: Tapping the Right Crowd", Journal of Business Venturing, vol. 29, 5, 2014, 585-609.

5   Philipp Haas, Ivo Blohm, Jan Marco Leimeister, "An Empirical Taxonomy of Crowdfunding Intermediaries", in proceedings of 35th International Conference on Information Systems (ICIS), Auckland, New Zealand.

6   Ajay K. Agrawal, Christian Catalini, Avi Goldfarb, "Some simple Economics of Crowdfunding", National Bureau of Economic Research, Working Paper 19133,available at: http://www.nber.org/papers/w19133, 2013.

7   Jacob Solomon, Wenjuan Ma, Rick Wash, "Don't Wait!: How Timing Affects Coordination of Crowdfunding Donations", CSCW '15, in proceedings of the 18th Conference on Computer Supported Cooperative Work & Social Computing, ACM New York, USA,2015, 547-556.

8   J. Hemer, "A Snapshot on Crowdfunding", Working Papers Firms and Region, No. R2/2011, 2011.

9   J. An, D. Quercia, J. Crowcroft, "Recommending Investors for Crowdfunding Projects", in Proceedings of the 23rd international conference on World wide web, ACM New York, USA,2014 , 261-270.

10 V. Rakesh, J. Choo, C. K. Reddy, "What motivates people to invest in Crowdfunding projects? Recommendation using heterogeneous traits in Kickstarter", Association for the Advancement of Artificial Intelligence, 2015.

11 V. Kuppuswamy and B. L. Bayus. "Crowdfunding Creative Ideas: The Dynamics of Project Backers in Kickstarter", Social Science Research Network, Working Paper Series, March 2013.

12 Chun-Ta Lu , Sihong Xie , Xiangnan Kong , Philip S. Yu, "Inferring the impacts of social media on Crowdfunding", in proceedings of the 7th ACM international conference on Web search and data mining, New York, USA, 2014.

13 Ajay K. Agrawal, Christian Catalini, Avi Goldfarb, "The Geography of Crowdfunding", National Bureau of Economic Research, Working Paper 16820, available at: www.nber.org/papers/w16820, 2011.

14 Rick Wash, J. Solomon, "Coordinating Donors on Crowdfunding Websites", in CSCW'14, ACM, Baltimore, USA, 2014.

15 E. Mollick, "The dynamics of Crowdfunding: An exploratory study", Journal of Business Venturing, 29, 2014, 1–16.

16 Jaya Gera, Harmeet Kaur, "Identifying Significant Features to Improve Crowd Funded Projects' Success", in press.

17 Q. Zhao, Sourav S. Bhowmick, "Association Rule Mining: A Survey", Technical Report, CAIS, Nanyang Technological University, Singapore, No. 2003116, 2003

18 N. Bendakir and E. Aimeur, "Using Association Rules for Course recommendation", In Proceedings of the AAAI Workshop on Educational Data Mining, 2006, pages 31-40.

19 Jaya Gera, Harmeet Kaur, "Crowd Funding: Refining Accomplishment", unpublished.

20 G. Burtch, A.Ghose, S. Wattal, "An Empirical Examination of the Antecedents and Consequences of Contribution Patterns in Crowd-Funded Markets", SSRN Working Paper, http://papers.ssrn.com/sol3/papers.cfm?abstract id=1928168.

P. S. Hiremath[1] and Rohini A. Bhusnurmath[2]

Colour Texture Classification Using Anisotropic Diffusion and Wavelet Transform

Abstract: In the present paper, a novel method of colour texture classification using anisotropic diffusion and wavelet transform is experimented on different colour spaces. The aim of the proposed method is to investigate the suitability of colour space for texture description. The directional subbands of the image are obtained using wavelet transform. Texture component of the directional information is obtained using anisotropic diffusion. Further, various statistical features are obtained from the texture approximation. The class separability is boosted using LDA. The proposed method is evaluated on Oulu colour dataset. The k-NN classifier is used for texture classification. The proposed approach has been effectively tested on RGB, HSV, YCbCr and Lab colour spaces. The experimental results are promising and show the efficiency in terms of reduced time complexity and classification accuracy as compared to the other methods in the literature.

Keywords: Colour texture classification, Anisotropic diffusion, Wavelet transform, Partial differential equation (PDE), time complexity

1 Introduction

Texture can be defined as a local statistical pattern of pixels. Texture classification assigns texture labels to unknown samples, according to classification rules learnt from training samples. In the presents work combined texture and colour analysis technique is used for texture image classification. There are large numbers of colour spaces with respect to different properties. The colour texture classification is influenced by choice of the colour space. Many authors

1 Department of Computer Science (MCA), KLE Technological University, BVBCET Campus, Hubli-580031, Karnataka, India.
hiremathps53@yahoo.com
2 Department of P.G. Studies and Research in Computer Science, Gulbarga University, Kalaburgi-585106, Karnataka, India.
rohiniabmath@gmail.com

have compared the classification performances due to different colour spaces in order to determine the suitability of colour space for texture analysis [1- 4].

In [5], the multiscale technique for texture classification in gray scale is extended to colour space using genetic algorithm and wavelet transform. The colour texture image classification based on HSV colour space, wavelet transform and motif patterns is implemented and support vector machine is used to classify texture classes [6]. Neural networks and machine learning are also employed to learn and classify texture classes using wavelet and information theory features [7]. Selvan et al. [8] modelled probability density function by singular value decomposition on wavelet transformed image for texture classification. The colour texture classification using wavelet based features obtained from an image and its complement is implemented in [9]. Karkanis et al. [10] proposed a new approach based on wavelet transform for the detection of tumors in colonoscopic video, in which a colour feature extraction scheme is designed to represent the different regions in the frame sequence. The features called colour wavelet covariance are based on the covariance of second order textural measurement. A linear discriminant analysis is used for classification of the image regions. Sengur [11] proposed wavelet transform and ANFIS for colour texture classification. Crouse et al. [12] introduced a framework for statistical signal modeling based on the wavelet domain hidden Markov tree. The algorithm provides an efficient approach to modeling of wavelet coefficients that are often found in real world images. Xu et al. [13] have shown that the wavelet coefficients possess certain inter-dependences between colour planes and have used wavelet domain hidden Markov model for colour texture analysis. In this approach, modeling the dependences between colour planes as well as the interactions across scales is done. The wavelet coefficients at the same location, scale and sub-band, but with different colour planes, are grouped into one vector and a multivariate Gaussian mixture model is employed for approximating the marginal distribution of the wavelet coefficient vectors in one scale.

Many statistical texture descriptors are explored for the extraction of texture features [14 -21]. The combination of gray scale texture features, colour histograms and moments are computed in [22, 23], which are used for colour texture classification. The hybrid texture features extracted from different colour channels are used for texture classification in [24, 25]. More sophisticated techniques use a combination of in-betweens colour bands to obtain texture features [26, 27]. Van de Wouver et al. [26] proposed wavelet energy correlation signatures. The transformation of these signatures is derived upon linear colour space transformation and LBP histogram based method for colour texture classification is developed [27]. The wavelet based co-occurrence method and the second order statistical features for colour texture classification is proposed by Ariva-zhagan et al. [28]. A set of features are derived for different colour models and colour texture classification is done for different combinations of the features. The non-subsampled contourlet transform (NSCT) and local directional binary patterns (LDBP) based texture classification using k-NN classifier is implement-ed in [29]. The dominant LDBPs to characterize image texture are investigated in [30]. In [31], texture classification based on rotation and shift invariant features is proposed.

In image processing, an anisotropic diffusion filter based on partial differ-ential equation (PDE) is used. The PDE techniques are extensively employed in signal and image processing [32]. Most PDE methods aim at smoothing the im-age while sharpening the edges [33]. An effective method using local directional binary pattern co-occurrence matrix and anisotropic diffusion is presented for texture classification in [34]. Anisotropic diffusion in conjunction with LDBP features is used for colour texture image classification on RGB colour space in [35]. The effect of LDBP features using anisotropic diffusion approach for texture classification is studied for different datasets in [36].

In this paper, an effective framework for colour texture classification based on wavelet transform and anisotropic diffusion is proposed. The objective of this paper is to obtain better classification accuracy at reduced computational cost. Wavelet transform is used to obtain multiscale directional information. The anisotropic diffusion is employed to obtain texture approximation from directional information. Further, various statistical features are computed from texture approximation for colour bands. The proposed algorithm has been test-ed on the different colour spaces, namely, RGB, HSV, YCbCr and Lab. The meth-od is evaluated on sixteen texture classes from Oulu colour texture dataset. The classification is performed using k-NN classifier. The experimental results indi-cate the effectiveness of the proposed method in terms of improved classifica-tion at reduced computational cost.

2 Proposed Method

The proposed method extracts features according to relationships between the colour components of neighboring pixels, i.e luminance based texture features, combined with pure chrominance based statistical moment features. The lumi-nance based texture features are obtained as follows.

i.   Apply Haar wavelet transform on luminance component of input image to obtain detail subbands of the mage.

ii. Apply anisotropic diffusion on the detail component up to n diffusion steps and obtain texture approximation.

iii. Extract statistical features from the texture approximation image.

The k-NN classifier is used for the classification of texture features. These methods are described briefly as given below.

2.1 Wavelet transform

Wavelets represent signals which are local in scale and time, that have generally irregular shape. A wavelet is a waveform of limited duration that has zero average value. Many wavelets including Haar possess the property of orthogonality to display signal compactly. This property ensures the non redundancy of data. Discrete wavelet transform expands the signals using scaling function and base function, which results in decomposition of input image into approximation (A) and detail subbands (H, V, D). The A subband represents smoothness or low pass band. The D subband gives diagonal details, H represents horizontal high frequency, V gives vertical high frequency details in the image. Another important characteristic of the Haar wavelet transform is its low computing requirements for image processing and pattern recognition. The implementation and theory of wavelet based algorithms are presented in [37, 38]. The method is described in brief in Appendix.

2.2 Anisotropic diffusion

Witkin [33] introduced the scale-space representation of image. A new definition of scale space pioneered by Perona and Malik is anisotropic diffusion (AD). It is a non-linear diffusion process based on partial differential equation (PDE) [32]. This process overcomes the unwanted effects such as blurring and edge smoothing, which are the result of linear smoothing filtering. AD has been effectively used in edge detection, image smoothing, image enhancement and image segmentation. AD filtering successfully smoothes image preserving image boundaries sharp. The basic equation of anisotropic diffusion [32] is represented by the Eq. (1):

$$\frac{\partial I(x,y,t)}{\partial t} = div\big[g\big(\|\nabla I(x,y,t)\|\big)\nabla I(x,y,t)\big] \qquad (1)$$

where $I(x,y,0)$ is the original image, t is the time parameter, $\nabla I(x,y,t)$ is the gradient of the image at time t and g(.) is the called conductance function. The diffusion is maximum within uniform regions, where as it is minimum across the edges. The edge stopping function proposed by Perona and Malik is given by the Eq. (2):

$$g(x) = \exp\left[-\left(\frac{x}{K}\right)^2\right] \tag{2}$$

where K is the gradient magnitude that controls the rate of the diffusion. The exponential form favours high-contrast edges over low-contrast ones. A discrete form of the Eq.(1) is given by the Eq. (3):

$$I_s^{t+\Delta t} = I_s^t + \frac{\lambda}{\left|\overline{\eta}_s\right|} \sum_{p \in \eta_s} c\left(\nabla I_{s,p}^t\right) \nabla I_{s,p}^t \tag{3}$$

where I_s^t is the discretely sampled image, s denotes the pixel position in a discrete two-dimensional (2-D) grid, and $0 \le \lambda \le 1/4$ is a scalar that controls the numerical stability, $\overline{\eta_s}$ is the number of pixels in the window (usually four, except at the image boundaries), and $\nabla I_{s,p}^t = I_p^t - I_s^t, \forall p \in \overline{\eta_s}$.

2.3 Statistical features

The two types of texture feature measures are first order and second order statistics. The first order texture feature statistics do not consider neighboring pixel relationships. The second order statistics computation uses the relationship between neighboring pixels [39]. The different feature sets based on first order statistics (F1) and second order statistics (F2-F9) considered in the present study are given in the Table 1.

3 Texture Training and Texture Classification

In texture training phase, colour and texture features are extracted from colour image for texture analysis. The different colour spaces, namely, RGB, HSV,

Table 1. The different feature sets based on first order statistics (F1) and second order statistics (F2-F9)

Feature set	Description	Features extracted	No. of features
F1	First order statistics	median, mean, standard deviation, skewness and kurtosis	5
F2	Haralick features [14]	entropy, homogeneity, contrast, energy, maximum probability, cluster shade and cluster prominence	7 features x 4 angles = 28 number
F3	Gray level difference statistics [15]	contrast, homogeneity, energy, entropy and mean	5
F4	Neighborhood gray tone difference matrix [16]	busyness, complexity, coarseness, contrast and texture strength	5
F5	Statistical feature matrix [17]	coarseness, contrast, period and roughness	4
F6	Law's texture energy measures [18, 19]	Six texture energy measures	6
F7	Fractal dimension texture analysis [20]	roughness of a surface	4
F8	Fourier power spectrum [21]	radial sum and angular sum	2
F9	Shape	size(x,y), area, perimeter and perimeter^2 /area	5

YCbCr and Lab are used. These colour spaces are explained in Appendix. In HSV, YCbCr and Lab colour spaces, V, Y, L channels contain the luminance information, respectively, while the remaining two channels in each colour space contain chrominance information. Texture features are computed from the luminance channel, while the first order statistical features namely, mean and standard deviation, are computed from the chrominance channels.

3.1 Texture training

The texture training algorithm is given in the Algorithm 1.

Algorithm 1: Training Algorithm

Step 1 : Input the training color image block I (in RGB).

Step 2 : Convert RGB color image to HSV (or YCbCr or Lab) space, where V (or

Y or L) is the luminance component I_{lmn}.

Step 3 : Extract the luminance (I_{lmn}) and chrominance components of the color image I.

Step 4 : Using Haar wavelet transform, decompose I_{lmn} into horizontal (H), vertical (V) and diagonal (D) components.

Step 5 : Subject the H, V and D components to anisotropic diffusion up to t steps and obtain texture approximations $I_{Htxr.}$, $I_{Vtxr.}$, I_{Dtxr} for H, V and D components, respectively

Step 6 : Compute statistical features F1-F9 (as listed in the Table 1) for images $I_{Htxr.}$, $I_{Vtxr.}$, I_{Dtxr} (obtained in Step 5).

Step 7: For chrominance components, compute statistical moment features (mean and SD).

Step 8 : Form feature vector F containing features computed from luminance (Step 6) and chrominance components (Step 7) and store F in the feature database, with class label..

Step 9 : The Steps 1 – 8 are repeated for all the training image blocks of all the texture class images and the training feature set (TF) is obtained.

Step 10 : LDA is applied on training feature set (TF) of Step 9 and the discriminant feature set (TFLDA) is obtained, which is then used for texture classification.

Step 11 : Stop.

The Algorithm 1 is executed up to different numbers of diffusion steps (t) and extracted various features as listed in the Table 1.

3.2 Texture classification

The texture classification is performed using k-NN classifier with ten-fold experimentation, based on Euclidean distance [40]. The testing algorithm is given in the Algorithm 2.

Algorithm 2 : Testing Algorithm (Classification of test images)

Step 1 : Input the testing colour image block I_{test} (in RGB).

Step 2 : Convert RGB colour image to HSV (or YCbCr or Lab) space, where V (or Y or L) is the luminance component $I_{testlmn}$.

Step 3 : Extract the luminance ($I_{testlmn}$) and chrominance components of the colour image I_{test}.

Step4 : Using Haar wavelet transform, decompose $I_{testImn}$ into horizontal (H), vertical (V) and diagonal (D) components.

Step 5 : Subject the H, V and D components to anisotropic diffusion up to t steps and obtain texture approximations $I_{testHtxr.}$, $I_{testVtxr.}$, $I_{testDtxr}$ for H, V and D components, respectively

Step6 : Compute statistical features F1-F9 (as listed in the Table 1) for images $I_{testHtxr.}$, $I_{testVtxr.}$, $I_{testDtxr}$ (obtained in Step 5).

Step 7: For chrominance components, compute statistical moment features (mean and SD).

Step8 : Form feature vector F_{test} containing features computed from luminance (Step 6) and chrominance components (Step 7).

Step9 : Project F_{test} on LDA components stored in TFLDA and obtain the weights which constitute test image feature vector $F_{testLDA}$

Step10 : (Classification)

Apply k-NN classifier (k = 3), based on Euclidean distance between $F_{testLDA}$ and TFLDA vectors, to determine the class of the test image block I_{test} .

Step 11: Repeat Steps 1 to 10 for all the test image blocks.

Step 12: Stop.

4 Experimental Results and Discussion

4.1 Dataset

The sixteen texture images from Oulu colour dataset [41] are considered for experimentation and are shown in the Fig. 1. Each Oulu texture sample represents one class. Each texture image is of 512x512 pixels. Each texture image is divided into 16 equal sized non overlapping blocks of size 128x128. Thus, totally 256 blocks are considered. The half of the randomly chosen blocks are used as the training samples and the remaining blocks are considered as test samples for each texture class.

4.2 Experimental results

The experimentation of the proposed method is carried out on Intel® Core™ i3-2330M @ 2.20GHz with 4 GB RAM using MATLAB 7.9 software. The Haar wavelet transform is employed to decompose the luminance component of the image, which results in average (A), horizontal (H), vertical (V) and diagonal (D) components. The H, V and D components of the image are then subjected to aniso-tropic diffusion to find texture approximation. Further, different statistical features (F1-F9) given in the Table 1 and their combinations (F10-F30), as defined in the Table 2 for texture description, are computed from the texture approxima-tion image. Thus, there are thirty feature sets considered for experimentation.

Table 2. The combinations (F10-F30) of different feature sets (F1-F9) given in the Table 1

Feature set name	Feature set combination	Feature set name	Feature set combination
F10	F1+F3	F21	F1+F3+F5
F11	F1+F3+F4	F22	F1+F3+F6
F12	F1+F3+F4+F5	F23	F1+F4+F5
F13	F1+F3+F4+F5+F6	F24	F3+F4+F5
F14	F1+F3+F4+F5+F6+F7	F25	F6+F7
F15	F1+F3+F4+F5+F6+F7+F8	F26	F4+F5
F16	F1+F3+F4+F5+F6+F7+F8+F9	F27	F3+F4
F17	F1+F3+F5+F6	F28	F5+F6
F18	F3+F4+F5+F6	F29	F8+F9
F19	F1+F4+F5+F6	F30	F2+F4
F20	F6+F7+F8+F9		

The LDA is used to enhance the class separability. The k-NN classifier is used for classification. The optimal values of parameters of anisotropic diffusion are: lambda = 0.25 and conduction coefficient = 60. The experimentation for each feature set is executed up to 10 diffusion steps (t).

The average classification accuracy is computed for the sixteen class prob-lem, where 16 texture classes are considered for the experimentation from Oulu texture dataset [41]. The proposed method is experimented on different colour spaces, namely RGB, HSV, YCbCr and Lab to determine the suitability of the colour space for texture classification. The Table 3 shows the comparison of

Figure 1. Colour texture images from Oulu dataset. From left to right and top to bottom: Grass, Flowers1, Flowers2, Bark1, Clouds, Fabric7, Leaves, Metal, Misc, Tile, Bark2, Fabric2, Fabric3, Food1, Water, and Food2 [5, 6, 8, 11, 35].

optimal average classification accuracy obtained among the thirty features sets that are experimented. The corresponding average training time and average testing time for different feature sets using the optimal number of diffusion steps for Oulu dataset is recorded for different colour spaces. The optimal number of diffusion steps is the diffusion step at which the best classification results are obtained for a given feature set. It is observed from the Table 3 that the proposed method performs better for HSV colour space as compared to RGB, YCbCr and Lab. Further, it is observed that the combination of feature sets gives improved classification accuracy when compared to single feature set. Hence, the accuracy of classification depends on the type of features used.

It is also observed from the Table 3 that the feature sets F6, F5, F4, F3, and F1 are dominant that result in improved classification accuracy when taken in combinations. The optimal average classification accuracy of 99.45% is observed for HSV colour space for the features set F17 which consumes training time of 24.28 sec. and testing time of 1.52 sec. The average classification accuracy of 97.11%, 98.36% and 98.44% is obtained for RGB, YCbCr and Lab colour spaces respectively. The training and testing time taken by these colour spaces are 25.73 sec and 1.61 sec, 8.60 sec. and 0.54 sec., 21.86 sec. and 1.37 sec.

Table 3. Comparison of average classification accuracy (%), average training time and average testing time of the proposed method on different colour spaces and the corresponding optimal feature set and diffusion step

Colour space	RGB	HSV	YCbCr	Lab
Average training time (sec.)	25.73	24.28	8.60	21.86
Average testing time (sec.)	1.61	1.52	0.54	1.37
Average classification accuracy (%)	97.11	99.45	98.36	98.44
Optimal feature set	F20	F17	F22	F26
Optimal no. of diffusion step	2	8	6	8

The proposed method is experimented on the same Oulu dataset [41] as used in [35, 5, 6, 11, 8], so that the results can be compared. The Table 4 shows the comparison of optimal average classification accuracy obtained by the proposed method and other methods in the literature on the Oulu dataset.

Table 4. Comparison of average classification accuracy (%) obtained by the proposed method and other methods in the literature [35, 5, 6, 11, 8] on the Oulu dataset

Sl. No.	Image name	Proposed method	Hiremath and Rohini [35]	Abdulmunim [5]	Chang et al. [6]	Sengur [11]	Selvan and Ramakrishnan [8]
1	Grass	100	100	100	96	95	86
2	Flowers1	100	96.87	100	99	100	90
3	Flowers2	100	100	100	96	95	89
4	Bark1	98.75	100	97	98	99	90
5	Clouds	100	100	98	100	100	97
6	Fabric7	100	100	100	97	96	92
7	Leaves	100	100	100	100	100	92
8	Metal	100	100	100	93	93	90
9	Misc	100	100	100	99	100	93
10	Tile	100	100	99	97	96	95
11	Bark2	96.25	100	100	97	96	90
12	Fabric2	100	100	95	100	100	97
13	Fabric3	100	100	98	100	100	99
14	Food1	100	100	95	99	99	95
15	Water	96.25	100	99	100	100	100
16	Food2	100	100	100	95	93	90

Average classification accuracy (%)	99.45	99.80	98.81	97.87	97.63	92.81
Average training time (sec.)	24.28	497.68	-	-	-	-
Average testing time (sec.)	1.52	26.11	-	-	-	-

It is observed from the Table 4 that the average classification accuracy of the proposed method is improved as compared to the methods in [5, 6, 11, 8]. The average training time and average testing time of the proposed method is significantly reduced by 95.12% and 94.18% respectively, as compared to the method in [35] yielding comparable average classification accuracy. Thus, the proposed method is computationally less expensive and more effective.

Conclusions

In this paper, a novel method of colour texture classification using partial differential equation for diffusion in colour spaces is proposed. The proposed method is tested on sixteen Oulu colour textures.The experimental results obtained in different colour spaces, namely, RGB, HSV, YCbCr and Lab, are compared with other methods in the literature. Following conclusions can be made from the experimentation:
- The HSV colour space yields better classification results.
- The computational cost is reduced significantly up to 95% as compared to the method in [35].
- Better classification accuracy is attained as compared to the other methods in literature
- Experimental results demonstrate the effectiveness of the proposed method in terms of classification accuracy and reduced time complexity.

Acknowledgment

The authors are grateful to the reviewers for critical comments and suggestions, which improved the quality of the paper to greater extent.

References

1   Drimbarean, A. & Whelan, P.F., "Experiments in Colour Texture Analysis", Pattern Recognition Letters, No. 22, 2001, pp 1161-1167.
2   Maenpaa, T. & Pietikainen, M., "Classification With Color and Texture: jointly or separately?", Pattern Recognition, No. 37, 2004, pp 1629-1640.
3   Palm, C., "Color Texture Classification by Integrative Co-Occurrence Matrices", Pattern Recognition, No. 37, 2004, pp 965-976.
4   Xu, Q., Yang, J., Ding, S., "Color Texture Analysis Using the Wavelet Based Hidden Markov Model", Pattern Recognition Letters, No. 26, 2005, pp 1710-1719.
5   Abdulmunim Matheel E., "Color Texture Classification Using Adaptive Discrete Multiwavelets Transform", Eng. & Tech. Journal, Vol. 30, No. 4, 2012, pp 615-627.
6   Chang, Jun-Dong , Yu, Shyr-Shen, Chen, Hong-Hao & Tsai, Chwei-Shyong, "HSV-Based Color Texture Image Classification Using Wavelet Transform and Motif Patterns", Journal of Computers, Vol. 20, No. 4, 2010, pp 63-69.
7   Sengur, A., Turkoglu, I. & Ince, M. C., "Wavelet Packet Neural Networks for Texture Classification", Expert Systems with Applications, Vol. 32, No. 2, 2007, pp 527-533.
8   Selvan, S., & Ramakrishnan, S., "SVD-Based Modeling for Image Texture Classification Using Wavelet Transformation", IEEE Transactions on Image Processing, Vol. 16, No. 11, 2007, pp 2688-2696.
9   Hiremath, P. S., & Shivshankar, S., "Wavelet based features for color texture classification with application to CBIR", International Journal of Computer Science and Network Security, Vol. 6, No. 9, 2006, pp.124-133.
10 Karkanis, S. A., Iakovidis, D. K., Maroulis, D. E., Karras, D. A., & Tzivras, M. , "Computer-Aided Tumor Detection in Endoscopic Video Using Color Wavelet Features", Information Tech. in Biomedicine, IEEE Transactions, Vol. 7, No. 3, 2003, pp. 141–152.
11 Sengur, A., "Wavelet Transform and Adaptive Neuro-Fuzzy Inference System for Color Texture Classification", Expert Systems With Applications, Vol. 34, No. 3, 2008, pp. 2120–2128.
12 Crouse, M. S., Nowak, R. D., & Baraniuk, R. G., "Wavelet Based Statistical Signal processing Using Hidden Markov Model", IEEE Trans. Signal Process, Vol. 46, No. 4, 1998, pp. 886–902.
13 Qing, X., Jie, Y., & Siyi, D., "Color Texture Analysis Using the Wavelet-Based Hidden Markov Model", Pattern Recognition Letters, Vol. 26, No. 11, 2005, pp. 1710–1719.
14 Haralick, R.M., Shanmuga, K. & Dinstein I., "Textural Features for Image Classification", IEEE Transactions on Systems, Man and Cybernetics, Vol. 3, 1973, pp 610-621.
15 Weszka, J. S., Dyer, C. R., Rosenfield, A., "A Comparative Study of Texture Measures for Terrain Classification", IEEE Transactions on Systems, Man. & Cybernetics, Vol. 6. 1976, pp 269-285.
16 Amadasun, M. & King, R., "Texural Features Corresponding to Texural Properties", IEEE Transactions on Systems, Man, and Cybernetics, Vol. 19, No. 5, 1989, pp 1264-1274.
17 Chung-Ming, Wu., & Yung-Chang, Chen, "Statistical Feature Matrix for Texture Analysis", CVGIP: Graphical Models and Image Processing, Vol. 54, No. 5, 1992, pp 407-419.
18 Laws, K. I., "Rapid Texture Identification", SPIE, Vol. 238, 1980, pp 376-380.
19 Haralick, R. M., & Shapiro, L. G., Computer and Robot Vision Vol. 1, 1992, Addison-Wesley.
20 Mandelbrot, B. B., The Fractal Geometry of Nature, San Francisco, CA, 1982, Freeman.

21 Rosenfeld, A., & Weszka, J., "Picture Recognition", in Digital Pattern Recognition, K. Fu (Ed.), Springer-Verlag, 1980, pp 135-166.

22 Mirmehdi, M. & Petrou, M., "Segmentation of Color Textures", IEEE Trans. on Pattern Analysis and Machine Intelligence, Vol. 22, No. 2, 2000, pp142-159.

23 Dimbarean, A. & Whelan, P., "Experiments in Color Texture Analysis". Pattern Recognition Letters No. 22, 2001.

24 Chang, C.C. & Wang, L.L., "Color Texture Segmentation for Clothing in a Computer-Aided Fashion Design System", Image and Vision Compuing, Vol. 14, No. 9, 1996, pp 685-702.

25 Hauta-Kasari, M., Parkkinen, J., Jääskeläinen, T. & Lenz, R., "Generalized Co-Occurrence Matrix for Multispectral Texture Analysis", 13th International Conference on Pattern Recognition, 1996, pp 785-789.

26 Wouwer, G. Van de, Scheunders, P., Livens, S. & Dyck. Van D., "Wavelet Correlation Signatures for Color Texture Characterization", Pattern Recognition Letters, Vol. 32, No. 3, 1999, pp 443-451.

27 Arvis, V., Debain, C., Berducatand, M. & Benassi. A., "Generalization of the Cooccurrence Matrix for Colour Images : Application to Colour Texture Classification". Journal of Image Analysis and Stereology, No. 23, 2004, pp. 63-72.

28 S. Arivazhagan, L. Ganesan, and V. Angayarkanni, "Color Texture Classification Using Wavelet Transform". In proceedings of the Sixth International Conference on Computational Intelligence and Multimedia Applications, ICCIMA 2005.

29 Hiremath, P. S., & Bhusnurmath, Rohini A., "Texture Image Classification Using Nonsubsampled Contourlet Transform and Local Directional Binary Patterns". Int. Journal of Advanced Research in Computer Science and Software Engineering, Vol. 3, No. 7, 2013, pp 819-827.

30 Hiremath, P. S., & Bhusnurmath, Rohini A., "A Novel Approach to Texture Classification using NSCT and LDBP". IJCA Special Issue on Recent Advances in Information Technology, No. 3, 2014, pp 36-42 (ISBN-973-93-80880-08-3).

31 Hiremath, P. S., & Bhusnurmath, Rohini A., "Nonsubsampled Contourlet Transform and Local Directional Binary Patterns for Texture Image Classification Using Support Vector Machine", Int. Journal of Engineering Research and Technology, Vol. 2, No. 10, 2013, pp 3881-3890.

32 Perona, P., & Malik, J., "Scale-Space and Edge Detection Using Anisotropic Diffusion", IEEE transaction on pattern analysis and machine intelligence. Vol.12, No.7, 1990, pp 629-639.

33 Witkin, A. P., "Scale Space Filtering", Proc. Int. Joint Conf. Artificial Intelligence, 1983, pp 1019-1023.

34 Hiremath, P. S., & Bhusnurmath, Rohini A., "Texture Classification Using Anisotropic Diffusion and Local Directional Binary Pattern Co-Occurrence Matrix", Proceedings of the Second International conference on Emerging Research in Computing, Information, Communication and Applications (ERCICA 2014), No. 2, 2014, pp 763-769, Aug. 2014. ISBN: 9789351072621, Elsevier Publications 2014.

35 Hiremath, P. S., & Bhusnurmath, Rohini A., "RGB – Based Color Texture Image Classification Using Anisotropic Diffusion and LDBP", Multi-disciplinary Trends in Artificial Intelligence, 8th International Workshop, MIWAI 2014, M.N. Murty et al. (Eds.) LNAI 8875, 2014 pp 101–111, DOI 10.1007/978-3-319-13365-2_10, Springer International Publishing Switzerland 2014.

36 Hiremath, P. S., & Bhusnurmath, Rohini A., "Diffusion Approach For Texture Analysis Based On LDBP", Int. Journal of Computer Engineering and Applications, Vol. 9, No. 7, Part I, 2015, pp 108-121.
37 Daubechies, I., Ten Lectures on Wavelets, 1992, SIAM, Philadelphia, PA.
38 Mallat, S. G., "A Theory of Multiresolution Signal Decomposition: The Wavelet Representation", IEEE transactions on pattern analysis and Machine intelligence, No. 11, 1989, pp 674-693.
39 Ojala, T., & Pietikäinen, M, Texture Classification, Machine Vision and Media Processing Unit, University of Oulu, Finland, 2004.
40 Duda, R. O., Hart, P. E., & Stork, Pattern Classification, Wiley publication, New York, 2001.
41 Internet: University of Oulu texture database, 2005, http://www.outex.oulu.fi/outex.php.
42 Wyszecky, G. W., & Stiles, S. W., "Color Science: Concepts and Methods, Quantitative Data and Formulas", Wiley, 1982.
43 Olga, Rajadell, & Pedro, García-Sevilla, "Influence of color spaces over texture characterization", Advances in Intelligent and Information Technologies, Research in Computing Science 38, 2008, pp. 273-281. © M.G. Medina Barrera, J.F. Ramírez Cruz, J. H. Sossa Azuela (Eds.)
44 Hunter, Richardsewall, "photoelectric color-difference meter", Josa, Vol. 38, No. 7:661, 1948. (Proceedings of the winter meeting of the optical society of America).
45 Hunter, Richardsewall, "Accuracy, precision, and stability of new photo-electric color-difference meter", Josa 38 (12): 1094, December 1948. (Proceedings of the thirty-third annual meeting of the optical society of America).

APPENDIX

Wavelet Transform

The theoretical aspects of wavelet transform are discussed in [37, 38]. The continuous wavelet transform of a 1-D signal $f(x)$ is defined as in Eq. (1) and Eq. (2):

$$(W_\psi f)(a,b) = \langle f, \psi(a,b) \rangle = \int f(x)\psi^*_{(a,b)}(x)dx \tag{1}$$

$$\psi_{a,b} = a^{-1/2}\psi((x-a)/b) \tag{2}$$

where a is the scaling factor, b is the translation parameter related to the location of the window, and $\psi^*(x)$ is the transforming function. An image is a 2-D signal. A 2-D DWT can be seen as a 1-D wavelet scheme which transform along the rows and then a 1-D wavelet transform along the columns.

Haar transform

Haar functions form an orthonormal system for the space of square integrable function on the unit interval [0, 1] and the representation of Haar wavelet is shown in the Fig. A1. The Haar transform serves as a prototype for all other wavelet transforms. Like all wavelet transforms, the Haar transform decomposes a discrete signal into two sub-signals of half its length.

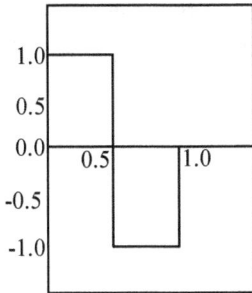

Figure A1. Representation of Haar wavelet

Colour Spaces

For a RGB colour image, the different texture features are computed on each of the channels red (R), green (G) and blue (B) of the input RGB image.

The theory of different colour spaces is explained in [42]. The proposed method is experimented on HSV, YCbCr and Lab color spaces. From each colour space, one channel containing the luminance information and two others are containing chrominance information is obtained. The luminance channel is then used to compute the texture features. The first order statistical features namely, mean and standard deviation, are computed from the chrominance channel as given in the Eq. (3) and Eq. (4):

$$Mean(m) = \frac{1}{N^2} \sum_{i,j=1}^{N} p(i,j) \tag{3}$$

$$StandardDe\,viation\,(sd) = \sqrt{\frac{1}{N^2} \sum_{i,j=1}^{N} [p(i,j)]^2} \tag{4}$$

HSV colour space

The HSV (hue, saturation, value) colour space corresponds to how people perceive the colour than the RGB colour space does. The wavelength of a colour is represented by Hue (H). The value of Hue changes from 0 to 1 when colour goes from red to green then to blue and back to red. As colour is rarely monochromatic, saturation (S) represents the amount of white colour mixed with the monochromatic colour. The Value (V) represents the brightness and does not depend on the colour. So, V is intensity whereas H and S represent chrominance. The following equations Eq. (5) to Eq. (9) represent transformation of RGB in [0,1] to HSV in [0,1]:

$$V = \max(R, G, B) \tag{5}$$

$$S = \frac{V - \min(R, G, B)}{V} \tag{6}$$

$$H = \frac{G - B}{6S}, \quad if \ V = R \tag{7}$$

$$H = \frac{1}{3} + \frac{B - R}{6S}, \quad if \ V = G \tag{8}$$

$$H = \frac{2}{3} + \frac{R - G}{6S}, \quad if \ V = B \tag{9}$$

YCbCr colour space

The YCbCr colour space is been widely used for digital video. In this colour space, luminance information is represented by Y component and chrominance information is represented by Cb and Cr components. Cb is the difference between the blue component and a reference value. Cr is the difference between the red component and a reference value. These features are defined for video processing purposes and so are not meaningful for human perception. The following equations Eq. (10) to Eq. (12) show the transformation of RGB in [0,1] to YCbCr in [0,255].

$$Y = 16 + 65.481R + 128.553G + 24.966B \tag{10}$$

$$Cb = 128 - 37.797R - 74.203G + 112B \tag{11}$$

$$Cr = 128 + 112R - 93.786G - 18.214B \tag{12}$$

Lab colour space

The Lab colour space is discussed in [43]. The three standard primaries (X, Y and Z) to replace red, green, and blue is defined by the International Commission on Illumination (CIE). All visible colours could not be specified with positive values of red, green and blue components [44, 45]. However, XYZ is not perceptually uniform. Perceptually uniform means that a change of the same amount in a colour value should produce a change of about the same visual importance [42]. Lab space is derived from the master colour space CIE XYZ. The idea of Lab colour space is to invent a space which can be computed from the XYZ space, but being perceptually uniform. Lab colour space is a colour-opponent space with dimension L for lightness and a, b for the colour opponent dimensions [42]. Therefore, this colour space represent the chromaticity with components a and b and the luminance L separately.

This space is transferred from CIE lab trisimulus values into an achromatic lightness value L and two chromatic values a and b using the transformation equations Eq. (13) to Eq. (18). The Xn, Yn and Zn are the trisimulus values of the reference (neutral) white point and f(t) is defined in Eq. (19). Texture features are computed from luminance channel L and chrominance features are computed from channels a and b.

$$X = 0.412453R + 0.357580G + 0.180423B \tag{13}$$

$$Y = 0.212671R + 0.715160G + 0.072169B \tag{14}$$

$$Z = 0.019334R + 0.119193G + 0.950227B \tag{15}$$

and

$$L = 116 \ f\left(Y/Y_n\right) - 16 \tag{16}$$

$$a = 500\left[f\left(X/X_n\right) - f\left(Y/Y_n\right)\right] \tag{17}$$

$$b = 200\left[f\left(Y/Y_n\right) - f\left(Z/Z_n\right)\right] \tag{18}$$

where

$$f(t) = \begin{cases} t^{1/3} & , \ t > 0.008856 \\ 7.787t + \dfrac{16}{116} & , \ t \leq 0.008856 \end{cases} \tag{19}$$

I.Thamarai[1] and S. Murugavalli[2]

Competitive Advantage of using Differential Evolution Algorithm for Software Effort Estimation

Abstract: Software effort estimation is the process of calculating the effort required to develop a software product based on the input parameters that are usually partial in nature. It is an important task but most difficult and complicated step in the software product development. Estimation requires detailed information about project scope, process requirements and resources available. Inaccurate estimation leads to financial loss and delay in the projects. Due to the intangible nature of software, most of the software estimation process is unreliable. But there is a strong relationship between effort estimation and project management activities. Various methodologies have been employed to improve the procedure of software estimation. This paper reviews journal articles on software development to get the direction in the future estimation research. Several methods for software effort estimation are discussed in this paper including the data sets widely used and metrics used for evaluation. The use of evolutionary computational tools in the estimation is dealt in detail. A new model for estimation using Differential Evolution Algorithm called DEAPS is proposed and its advantages discussed.

Keywords: Software Effort Estimation Methods, Algorithmic and Non-Algorithmic Models, Evolutionary Computation, Differential Evolution.

1 Introduction

Planning a software project is one of the most important activities in any software development process. Many factors are to be considered to estimate the software cost and effort. The most important factors are size of the project, number of persons involved and schedule. Prediction of software effort is a

1 Research Scholar, Sathyabama University, Chennai, India
ilango.thamarai@gmail.com
2 Research Supervisor, Sathyabama University, Chennai, India
murugavalli26@rediffmail.com

difficult and complicated task. Software is intangible in nature. So the measurement of progress is the software process is very difficult to access. Also the requirements of software project change continually which causes the change in estimation. Inaccurate estimation of effort is the usual cause of software project failures. In [1], Magne Jorgensen and Tanja M. Grushke suggested that the type of individual lesson learned processes may have effect on the accuracy of estimates. Budget and schedule pressure also plays an important role in effort calculation according to Ning Nan and Donald E.Harter [2]. In [3], Linda M.Laird lists out the reason for the inaccurate software project estimation. They are:

a)   Lack of education training
b)   Confusion in the schedule
c)   Inability of team members
d)   Incomplete and changing requirement
e)   Hope based planning

In [4], Magne Jorgenson and Martin Sheppard present a systematic review of various journals. According to them, the properties of dataset impact the result when evaluating the estimation. In [5], Karel Dejaegar et al present an overview of the literature related to software effort estimation. The parameters used were project size development and environmental related attributes. The paper gives detailed study with different processing steps and addresses many issues like data quality, missing values etc. The authors confirm that the CO-COMO model performed equally well as the non-linear techniques. In [6], Tim Menzies, Andrew Butcher et al evaluated the lessons that are global to multiple projects and local to particular projects in software effort estimation. They used Promise repository and concluded that software effort estimation team should build clusters from the available data and the best cluster is the one that is near the source data that is not from the same source as the test data.

In [7], Ekrem Kocaguneli et al explored the general principles of effort estimation that can guide the design of effort estimation. The author concludes that, the effort estimation shall improve significantly, if the situations when the assumptions are violated are identified and removed. Also it is said in the paper that estimation can be improved by dynamic selection of nearest neighbor with small variance.   Nicolaos Mittas and Lefteris Angelis proposed a statistical framework based on multiple comparison algorithms to rank several cost estimation models [8]. In [9], Mark Harman and Afshin Mansouri propose the application of search based optimization for the software effort estimation. They list out the advantages of using search based optimization as robustness, scalability and powerful. In [10], Ekrem kocaguneli et al propose a tool called QUICK TOOL to reduce the number of features and instances required to capture the information for software effort estimation. This reduces the complexity in data interpretation. The distance between features is taken for analysis. The tool is suitable for small data sets. In [11], Ray Ashman suggested a simple Use Case based model. In this, the relationship between estimated and actual data is used to improve the future estimates. This model works best in an interactive development process. The aim of the Use Case based model is to capture the experience of specialists and enable a consistency on the timescale.

As today's software products are largely component based, the prediction is also done on the basis of individual components of the software. The performance of a system can be predicted by creating models to calculate the performance of every single component that comprises the full system.To improve the uncertainty in cost assessments, Magne Jorgensen provide evidence based guidelines in [12]. The methods for assessing software development cost uncertainty are based on results from empirical studies. Some of the important guidelines provided are, not to rely solely on unaided, intuition based uncertainty assessment process and to apply structured and explicit judgment based process. His advice is to combine uncertainty assessments from different sources through group work and not through mechanical combination. Thus it can be seen that various aspects has to be considered for Software Effort Estimation Models. The main aim of this paper is to give a detailed description about various software effort estimation methods. The paper is organized as follows: Section 2 consists of discussion on various traditional software estimation methods and Algorithmic Models. In Section 3, we discuss about the recent trend of using evolutionary computation methods in software effort estimation. In Section 4, we propose a new model called DEAPS for the selection of most relevant project in Analogy for Effort Estimation and its experimental results. Section 5 is the conclusion and recommendation for future work.

2 Software Estimation Methods

There are many traditional methods such as Expert Judgment method, Function Point method, COCOMO, SLIM Model, Case Based Reasoning Model to the recent methods that uses Neural Networks, Fuzzy Logic, Genetic Algorithm, Genetic Programming, Particle Swarm Optimization etc. The Software Effort Estimation models are primarily divided into four main categories. The first and foremost is the Expert Judgment Method, where the effort is estimated by experts in the field. Algorithmic models are based on the mathematical formulas. Some of the

Algorithmic Models are FP (Function Point), COCOMO (Constructive Cost Model) and SLIM (Software Life cycle Management model). These models depends on various parameters like LOC (Lines of code), Complexity, Number of Interfaces etc. The limitation of Algorithmic models such as inflexibility led to the Non-algorithmic models. Case Based Reasoning (CBR) is a popular Non-algorithmic method. Analogy is a CBR methodology. Also with the advent of soft computing techniques, new methodology of evolutionary computation came into existence. In this, many Machine Learning methods are used including Neural Networks, GA, GP and DE. These methods are discussed in the following sections.

All the approaches have their own advantages and disadvantages. It should be noted that there is not a single method which can be said to be best for all situations. Some of the research favors the combination of methods that has been proved successful. In [13], Chao-Jung Hsu et al, integrated several software estimation methods and assigned linear weights for combinations. They proved that their model is very useful in improving estimation accuracy. In [14], Magne Jorgensen and Stein Grimstad had made a detailed study on how misleading and irrelevant information can affect software estimation. They have presented the research questions and hypothesis. Based on the answers, they concluded that the field settings that led to irrelevant information have very small impact on the effort estimation than the artificial experimental settings. They also summarize that the researchers should be more aware of the different role of lab and field experiments.

3 Evolutionary Computation Models

The use of Evolutionary Computation Model is suggested recently to estimate the software projects. They have the advantage of handling large search spaces. The basic idea is the Darwin's theory of evolution according to which the genetic operations between chromosomes lead to the survival of the fittest individuals. These methods are the extension of machine learning algorithm such as ANN.

3.1 ANN (Artificial Neural Network)

ANN Models are inspired by human neural system to solve problems. An ANN is an information processing system that has certain performance characteristics in common with the biological neural network. This type of network has two layers, namely input layer and output layer. There are links between the layers.

Each link carries weights. There can be hidden layers in between these layers. Back propagation Algorithm is the most popular method for training. In this, there are two passes, a forward pass and a backward pass. In the forward pass, the weights are all fixed. During the backward pass, the weights are adjusted according to the error correction rules. The weight adjustment is based on the error produced between the desired output and actual output.

Mair et al in [15] conclude that Artificial Neural Network offer accurate effort prediction but their configuration and interpretation is difficult to achieve. In [16], Gavin R.Finnie and Gerhard E.Wittig examined the performance of ANN and CBR tools for software effort estimation. The concept of back propagation neural network on Desharnais dataset and ASMA (Australian Software Metric Association) dataset were explored by them. They proved that ANN results in high level of accuracy. They concluded that ANN models are capable of providing adequate estimation models. They also said that the performance depends on the training data. In [17], Ruchika Malhotra and Ankita Jain evaluated the use of machine learning methods such as ANN, Decision tree, Support Vector Machine in effort prediction. According to their research, it is proved that decision tree method is the best method among the three. The advantage of Artificial Neural Network is that they can handle heterogeneous database but there are no guidelines for design. Accuracy largely depends on training data set.

3.2 GA (Genetic Algorithm)

GA is a search based algorithm to get an optimal solution. It is a evolutionary computation method. Genetic Algorithm creates consecutive population of individuals due to which, we get optimal solution for the given problem. The search process is influenced by the following components:

a)   An encoding of solutions to the problem known as chromosome
b)   A function to evaluate the fitness
c)   Initialization of initial population
d)   Selection operator
e)   Reproduction operator

The important issues related with GA are the representation of solution, selection of genetic operators and choosing the best fitness solution. GA can efficiently search through the solution space of complex problem. In [18], GA is used for project selection. The steps involved are encoding, population generation, fitness function evaluation, cross over, mutation, elitism and stopping criteria.

The two real world data sets Desharnais data set and Albrecht data set are used for the experiments. The authors applied two ABE based models on these data sets. The first model uses GA to select appropriate projects subsets. This is named as PSABE (Project Selection in Analogy Based Estimation). The second model FWABE (Feature Weighting ABE) assign relevant feature weights by GA. The results are better than the other software estimation methods. In [19], Klaus Krogmaun et al presented a reverse engineering approach that combine genetic search, static and dynamic analysis to predict the performance of software application. The performance of file sharing approach is predicted using runtime byte code count. The average accuracy is proved to be better than other methods. The main disadvantage is that this approach supports only component for which java byte code is available. This method is useful for component based applications.

3.3 GP (Genetic Programming)

GP is a field of Evolutionary Computation that works on tree data structure. The Non continuous functions are very common in software engineering applications due to the use of branching statements. Genetic Programming can be effectively used in such situations. Using a tree based representation in Genetic Programming requires adaptive individuals and domain specific grammar GP begin with a population of randomly created programs. The programs consist of functions corresponding to the problem domain. Each program is evaluated based on fitness function. Unlike GA, mutation operation is usually not needed in GP because the crossover operation can provide for point mutation at nodes. The process of selection and crossover of individual continues till the termination criteria are satisfied. Colin J.Burgess and Martin Lefley analyzed the potential of G.P in Software Effort Estimation in terms of accuracy and ease of use [20]. The research was based on Desharnais data set of 81 software projects. The authors prove that the use of GP offer improvement in accuracy but this improvement depends on the measure and interpretation of data used.

3.4 DE (Differential Evolution)

Differential Evolution is an important evolutionary computation method in recent days that can be used to improve the exploration ability. Differential Evolution is similar to Genetic Algorithm, but it differs in the sense that distance and direction information from the current population is used to guide the search process. Differential evolution (DE) is a method that optimizes a problem, iteratively to improve a solution . There are many types of DE such as Simple DE, Population based DE, Compact DE, etc. DE performs well than any other contemporary algorithm and it is proved that it offers good optimization due to higher number of local optima and higher dimensionality. In a simple DE algorithm an initial population is created by random set of individuals. For each generation, three individuals say x_1, x_2 and x_3 are selected. An off spring x'_{off} is generated by mutation as

$$x'_{off} = x_1 + F(x_2 - x_3)$$

Here, F is a scale factor. Then crossover is done based on some condition. In [21], we have proposed a new method of using Differential Evolution Algorithm for the selection of similar projects in analogy method. In the proposed algorithm which is based on population based DE, the Primary population (Pp) set consists of selected individuals. The secondary population (Ps) serves as an archive of those offspring rejected by the selection operator.

4 Selection of most relevant project using Differential Evolution Algorithm

The main advantages of DE are its ability to provide multiple solutions. It can be easily applied to real problems despite noisy and multidimensional space. It is simple but has effective mutation process that ensures search diversity. Here, we propose a new model called DEAPS (Differential Evolution in Analogy for Project Selection). The following Figure gives the framework for using Differential Evolution Algorithm to select the relevant project from the available set of historical projects.

The proposed method combines Analogy concept with Differential Evolution Algorithm, The retrieval of most similar project is done in two stages. In the 1st stage, there is a reduction of historical database to a set of most similar projects using Similarity Measure. In the 2nd stage, DE is applied to retrieve the most relevant project.

4.1 Performance Evaluation Metrics

Evaluation criteria are essential for the validation of Effort Estimation Models. Metrics are used for this purpose. The most commonly used metrics are given

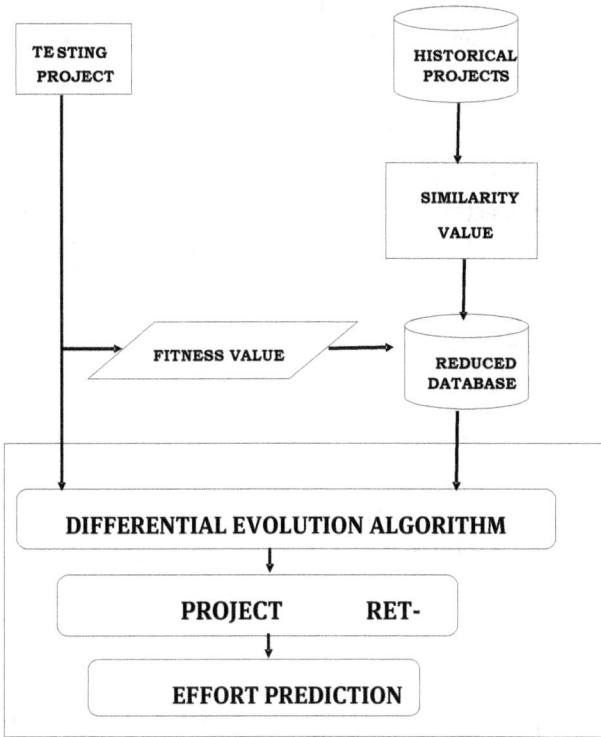

Figure 1: Differential Evolution in the selection of relevant project

below:

MRE (Magnitude of Relative Error): Relative Error is the difference between the actual value and estimated value. MRE is the absolute value of the relative error.

$$\text{MRE} = \frac{|A-E|}{|A|}$$

Where A is the Actual Effort Value and E is the Estimated Effort Value.

MMRE (Mean Magnitude of Relative Error): MMRE is the average percentage of the MRE over an entire dataset

$$\text{MMRE} = \sum_{i=1}^{i=n} \left| \frac{A_i - E_i}{A_i} \right| * \frac{100}{n}$$

Where A_i is the Actual Effort and E_i is the Estimated Effort of the ith project, n is the number of projects.

Pred(q): The prediction level pred(q), is the average percentage of prediction that falls within a specified percentage (q%) of the actual value. If the value of pred(q) is high, then the estimation is good

$$\text{pred(q)}= \frac{p}{n}$$

Where, p is the number of projects whose MRE is less than or equal to q. The commonly used metric is pred(.25) which is the percentage of predictions that is less than 25% of the actual value

$$\text{Pred (.25)} = \frac{1}{n}\sum_{i=1}^{n}\left(\frac{|A_i - E_i|}{|A_i|}\right) <= 0.25$$

4.2 Experimental Results

The input is the project parameters from the Albrechdt dataset whose values are slightly changed. It is found that, by using the DE Algorithm, the most relevant project is retrieved. The results are given in the following table:

Table 1: Results of DEAPS Model on Albrechdt Dataset

Project Id	Estimated Effort	Actual Effort	MRE	Pred(0.25)
1	100	102.4	0.0234	1
2	94	105.2	0.1065	1
3	25	11.1	1.2523	0
4	15	21.1	0.2891	0
5	35	28.8	0.2153	1
6	6	10	0.4000	0
7	10	8	0.2500	1
8	1	4.9	0.7959	0
9	20	12.9	0.5504	0
10	20	19	0.0526	1
11	10	10.8	0.0741	1
12	10	8	0.2500	1
13	9	7.5	0.2000	1
14	10	12	0.1667	1
15	1	0.5	1.0000	0
16	12	15.8	0.2405	1

17	15	18.3	0.1803	1
18	6	8.9	0.3258	0
19	30	38.1	0.2126	1
20	51	38.1	0.3386	0
21	5	3.6	0.3889	0
22	11	11.8	0.0678	1
23	1	0.5	1.0000	0
24	2	6.1	0.6721	0

Table 2 summarizes the results of various effort estimation methods on Albrechdt dataset. The corresponding test results are compared with the previous research results [22] and the comparison is also shown in diagrammatically in Figure 2, 3 and 4.

Table 2: Comparison of result with previous Models

S.No	Methods	MMRE	PRED(0.25)	MdMRE
1	ABE	0.49	0.13	0.49
2	FWABE	0.42	0.25	0.46
3	PSABE	0.39	0.38	0.45
4	ANN	0.49	0.25	0.51
5	DEAPS	0.38	0.54	0.36

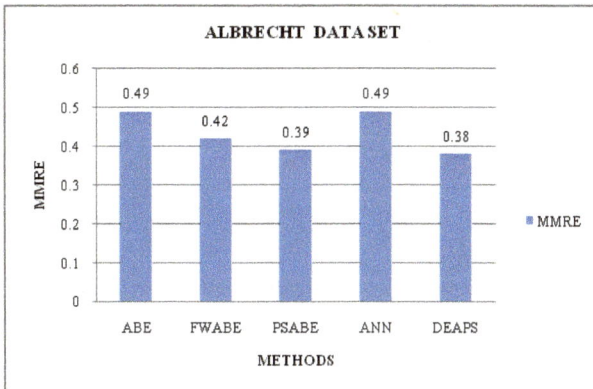

Figure 2 : Comparison of MMRE values

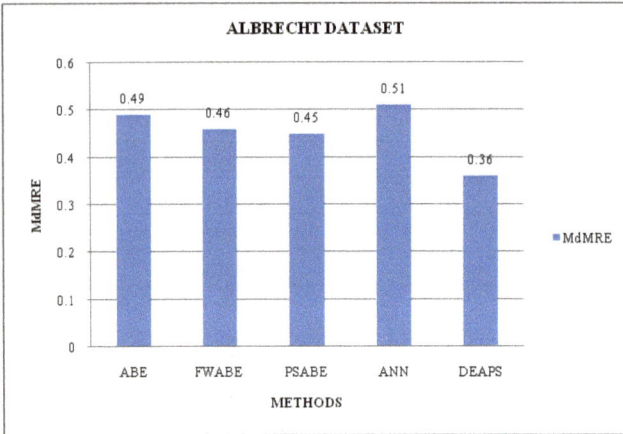

Figure 3 : Comparison of MdMRE values

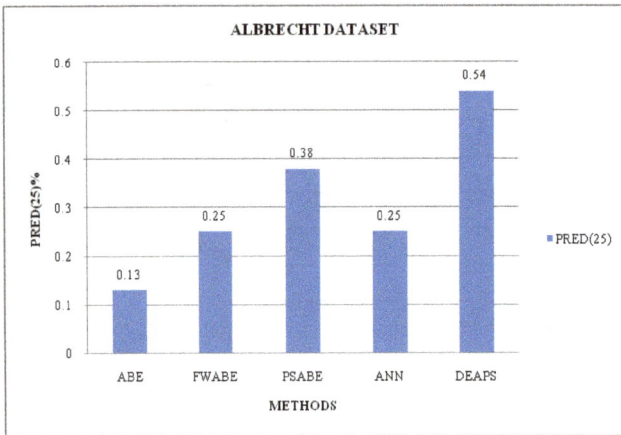

Figure 4. Comparison of Pred(0.25) values

The following Figure 5 shows a combined illustration of the test results. The results shows that the application of the proposed Model DEAPS for the selection of relevant project has the best performance among all methods (0.26 for MMRE, 57 % for Pred (0.25)and 0.22 for MdMRE). The MMRE and MdMRE are lesser than the other methods and also the probability of a project having MRE<=0.25 is also very high when compared with other models.

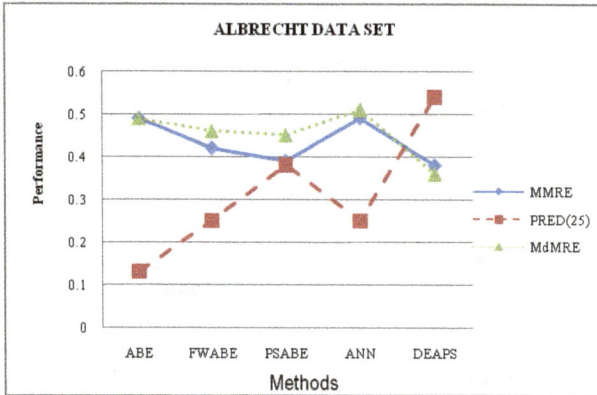

Fig. 5. Results of Proposed model DEAPS on Albrechdt Dataset

Conclusion and Future work

This paper gives a detailed study of how Evolutionary Computation Algorithm has been used in the Software Effort Estimation models. Also a new approach has been proposed to simplify the Analogy based estimation. In our proposed model DEAPS, Differential Evolution Algorithm is used to select the most relevant project from set of historical projects that matches with the new project. The proposed method is implemented in JAVA platform. The experimental results are given and the observation of results clearly indicates that this model is better than existing methods. The metrics used are MMRE, MdMRE and pred(25%). As the search space is big, the Evolutionary Computation method is used which has been proved to be useful. Future work is to analyzes the performance of the model with few more real datasets and to prove efficiency of this method.

References

1   Magne Jorgensen, Tanja M.Grusehke and R. Gupta, "The Impact of Lessons- Learned sessions on Effort Estimation and Uncertainity Assesments", IEEE Trans. on Software Engg., pp. 368-383, 2009.
2   Ning Nan and Donald E.Harter, "Impact of Budget and Schedule Pressure on Software Development Cycle time and Effort", IEEE Trans. on Software Engg., pp. 624-637 , 2009
3   Linda M Laird, "The Limitations of Estimation", IT Pro, pp. 40-45, 2006.

4   Magne Jorgensen and Martin Sheppard, "A Systematic review of Software Development Cost Estimation Studies", IEEE Trans. on Software Engg., pp. 33-53, 2007
5   Karel Dejaeger, Wouter Verbeke, David Martens, Bart Baesens, "Data mining techniques for Software Effort Estimation : A Comparative study", IEEE Trans. on Software Engg., vol. 38, pp. 375-397, 2012
6   Tim Menzies, Andrew Butcher, David Cok, Lucas layman, Forrest Shull, Burak Turhan, "Local vs Global lessons for defect Prediction and Effort Estimation", IEEE Transactions on Software Engg., Vol.39, pp. 822-834, 2013
7   Ekram kocaguneli , Tim Menzies, Ayse Basar Bener and Jacky W Keung, "Exploiting the essential assumptions of Analogy based Effort Estimation", IEEE Transactions on Software Engg., pp. 425-437, 2012
8   Nikolos Mittas and Lefteris Angelis, "Ranking and clustering Software Cost Estimation Model through a multiple Comparison Algorithm", IEEE Transactions on Software Engg., pp. 537-551, 2013
9   Mark Harman, Afshin Mausouri, "Search based Software Engineering : Introduction to special issue of IEEE Trans. on Software Engg.", IEEE Transactions on Software Engg., pp. 737-741, 2010
10 Ekrem Kocaguneli, Tim Menzies, Jacky Keung, David Cok, Ray Madachy, "Active Learning and Effort Estimation: finding the essential content of Software Effort Estimation data", IEEE Transactions on Software Engg., pp. 1039-1053, 2013
11 Ray Ashman, "Project Estimation: A Simple Use Case based Model", IT Pro, pp. 40-44, 2004
12 Magne Jorgensen, "Evidence based guidelines for assessment of Software development Cost Uncertainty", IEEE Transactions on Software Engg., pp. 942-954, 2005
13 Chao Jung Hsu , Nancy Urbina Rodas, Chin Yu Huang and Kuan- Li Peng, "A Study of im-proving the Accuracy of Software Effort Estimation using Linear Weighted Combination", Proc. of Annual IEEE Comp. Software Applications Conference, pp. 98-103, 2010
14 Magne Jorgensen and Stein Grimstad, "The impact of irrelevant and misleading Info. on Software development Effort Estimates : A Randomized Controlled Field Experiment", IEEE Transactions On Software Engg., pp. 695-707, 2011
15 C. Mair, G. Kadoda, M. Lefley, K. Phalp, C. Schofield, S, Shepperd, S. Webster, "An Investi-gation of Machine Learning based Prediction Systems", Journal of Software Systems, pp. 23-29, 2000
16 Gavin R. Finnie, Gerhard E.Wittig, "AI tools for Software development Effort Estimation", Proc. of International Conference on Software Engg. education and practice, pp. 83-92, 1996
17 Ruchika Malhotra, Ankita Jain, "Software Effort Prediction using statistical and Machine Learning Methods", International Journal of Adv. Comp. science application, pp. 45-52, 2011
18 Y.F.LI, M.Xie, T.N.Goh, "A Study of Genetic Algorithm for Project Selection for Analogy based Software Cost Estimation", Proc. of IEEE IEEM, pp. 1256-1260, 2010
19 Klaus Krogmaun, Michael Kuperberg, Ralf Reussner, "Using Genetic Search for Reverse Engeering of Parametric Behavior Models for Performance Prediction", IEEE Transactions On Software Engg. pp. 865-877, 2007
20 Colin J.Burgess and Martin Lefley, "Can Genetic Programming improve Software Effort Estimation? A Comparative Evaluation", Elsevier, pp. 863-873, 2001

21 I.Thamarai, Dr.S. Murugavalli, "Using Differential Evolution in the Prediction of Software Effort" , Proc. of Fourth International Conference on Advanced Computing, pp. 1-3, 2012

22 Y.F. Li , M. Xie, T.N. Goh, "A study of project selection and feature weighting for analogy based software cost estimation", The Journal of Systems and Software –Elsevier, pp. 241-252, 2009

Shilpa Gopal[1] and Dr. Padmavathi.S[2]

Comparative Analysis of Cepstral analysis and Autocorrelation Method for Gender Classification

Abstract: Gender classification is one of the initial steps in any of the speaker recognition system. This paper deals with the comparative analysis of two of the important gender classification methods, namely Cepstral analysis and Short time Autocorrelation Method, experimented on both natural and synthetic voices. From the experimental results, it is concluded that autocorrelation method performs better gender classification.

Keywords: Autocorrelation, Biometrics, Cepstral Analysis, Gender Classification, Speaker Verification

1 Introduction

Biometric has gained its importance in the field of security years ago. Biometric is the way of using biological characteristics, such as fingerprint, palm geometry, iris, retina, sclera, handwriting, face, tooth, voice etc, for uniquely identifying and thereby authenticating individuals. Speaker recognition, also known as voice recognition, is one of the most cost effective and user friendly biometric methods. Speaker recognition is the process of using characteristic of voices to uniquely identify a person. Speaker Recognition is classified into two categories based on the application, namely Speaker Verification and Speaker Identification [8].In speaker identification, voices of persons are stored in the database. When an unknown voice is given as input, then this voice is compared with those in the database and thus identifying the speaker. Speaker verification, on the other hand, is used to prove that the input voice is belonging to a particular person or not. In short, in speaker Identification only the voice is known from which the

---

1 Department of Computer Science and Engineering, Amrita School of Engineering, Amrita Vishwa Vidyapeetham (University), Coimbatore-641 112
shilpag32@gmail.com
2 Department of Computer Science and Engineering, Amrita School of Engineering, Amrita Vishwa Vidyapeetham (University), Coimbatore-641 112
s_padmavathi@cb.amrita.edu

speaker has to be identified whereas in case of Speaker Verification, both the speaker and voice are known with which we have to check if he/she is the correct speaker. Speaker recognition is one of the few biometric methods which can be used for onsite applications, such as accessing facilities or objects, remote applications like remote authentication via telephone and also for interactive gaming.

Characteristics of voice signal differ largely for male and female and hence gender classification is considered to be one of the initial steps in the process of voice biometrics. Some of the existing gender classification methods use short time average magnitude, short time energy, short time zero crossing rate , neural network and so on. In this paper, gender classification is done using cepstral analysis and autocorrelation method and a comparative analysis of the two are summarized.

This paper is motivated based on the need of gender identification in a real time environment where the words spoken are not common among people. Most of the existing methods may fail in such circumstances. Since 'a' and 'e' are the most commonly used vowels in English language, the presence of their sounds 'aah' and 'eee' is common to utter by the people in the natural environment. Hence this paper concentrates on gender classification based on these two sounds.

In chapter 2, a literature survey on the existing gender classification methods is discussed. In chapter 3, brief description about the methods considered in this paper, is made. Chapter 4 deals with the experiment and analysis results. Conclusion and References are respectively in chapter 5 and 6.

2 Literature Survey

In [1], short time analysis of the speech signal is taken into consideration. The speech signal is quasi stationary and hence the signal is broken into sub signals called windows for analysis. Short-time average magnitude, short time energy, short-time zero crossing rate, short time auto-correlation are calculated on voice signals and observations are noted. 'Oh my God 'is the sentence spoken by male and female speakers which are used for the analysis. From the observations it is concluded that in the case of female speakers, the average short time energy value and average short time zero crossing rate are found to be higher than that of male speakers. Also, significant differences are observed in case short time average magnitude and short time average autocorrelation plots.

In [2], the gender detection is done based on the pitch difference in male and female. Here, zero crossing rate is considered and also confirmed that for female the rate is higher. It is observed that the center of gravity of the spectrum for male voice is closer to low frequencies than those of female frequencies. On considering these observations, a variable, defined by a function of zero crossing rate and the centre of gravity of the acoustic vector, is proposed for performing the gender classification.

Gender classification is done using neural networks in [3], where three features namely energy entropy, short time energy and zero crossing rate are considered and fuzzy logic and neural networks are used for gender classification. Training the fuzzy logic and neural network is done using the dataset generated by considering the three features. In case of testing, for an input speech signal, fuzzy and neural network gives an output. Mean of the output value is used to determine the gender class. This method may show better results but the main disadvantage is that the need of training which is time consuming.

Existing gender classification methods do not test for consistency in different environments. In this paper, the consistency of the proposed method is tested by considering the performance in different environments.

3 Proposed Work

Pitch is a psycho acoustical attribute of sound according to which sound can be scaled from low to high. It can be quantified as frequency and determines how quickly the sound makes air to vibrate. From the common view point, pitch range of male and female differ significantly and this fact is used in the process of gender classification. The pitch range of male generally ranges from 90Hz to 120Hz and that of female ranges from 150Hz to 300Hz. Pitch determination in frequency domain is achieved by cepstral analysis and that in time domain is achieved by autocorrelation method. This paper gives a comparative analysis of cepstral analysis and autocorrelation method used for gender classification.

3.1 Cepstral Analysis

Cepstrum is one of the common transforms to gain information from a speech signal. A speech signal is composed of both excitation source and vocal tract system components [6][7]. The excitation source component lies in the high quefrency region while the vocal tract component lies in the low quefrency region, where quefrency is the time domain obtained when inverse Fourier transform is applied on log power spectral density. Cepstral analysis can be used for separating the components from speech for further processing.

Speech signal can be represented as the convolution of the respective excitation sequence e(n), and the vocal tract filter characteristics h(n) as in equation (1)

$$S(n)=e(n)*h(n) \qquad\qquad (1)$$

In the frequency domain, equation (1) can be written as equation (2)

$$S(w) = E(w).H(w). \qquad\qquad (2)$$

According to convolution theorem, convolution in the time domain is same as multiplication in the frequency domain. Next, the speech sequence has to be deconvolved into excitation and vocal tract components in the time domain. Deconvolution is the process of separating the two components. The equivalent frequency domain action is to convert the multiplication of two components into linear combination of the same. This transformation of the multiplied source and system components in the frequency domain to linear combination of the two components in the cepstral domain can be done using the cepstral analysis. The sequence of steps involved in cepstral analysis is as follows

1. Apply Discrete Fourier Transform to the signal S(n) in equation (1) to obtain equation (2). From equation(2), the magnitude spectrum of the speech can be represented as equation (3)

$$|S(w)| = |E(w)|.|H(w)| \qquad\qquad (3)$$

    Where w is the frequency index.

2. On applying logarithm to equation (3), linear combination of E(w) and H(w) is obtained, as in equation (4)

$$\log_{10}|S(w)|=\log_{10}\{|E(w)|\}+ \log_{10}\{|H(w)|\} \qquad (4)$$

    Thus product of the components is transformed to linear summation of the same.

3. By performing inverse discrete Fourier transform on equation (4), the components can be separated. Applying Inverse Discrete Fourier Transform to the log spectra transforms it to quefrency domain which is similar to time domain.

$$c(n)=IDFT[\ \log_{10}\{|E(w)|\} + \log_{10}\{|H(w)|\}] \qquad (5)$$

The vocal tract components are concentrated at the low quefrency region while the excitation components are concentrated at the high quefrency region. The excitation appears as periodic modes in the high quefrency region and the peaks are located at the period of fundamental frequency. In this paper only the high quefrency region is analyzed as, it is related to the pitch.

Variance is related to pitch of a given voice signal while kurtosis is related to the flatness of a signal. Thus, variance and kurtosis of the cepstrum can also be considered for gender classification. Variance is the measure of how data is distributed itself about the mean or the expected value. The equation of variance [6] is given by the formula (6)

$$V = \frac{1}{N-1} |A_i - \mu^2|$$
(6)

Where A is the random variable vector made up of N scalar observations and μ is the mean of A, given by equation (7)

$$\mu = \frac{1}{N} \sum_{i=1}^{N} A_i$$
(7)

Variance of female voice is much lesser than that of male. This distinction can be used for gender classification.

Kurtosis of a distribution [9] is given by equation (8). It is the measure of flatness of a signal, which is usually higher for male.

$$K = \frac{E(x-\mu)^4}{\sigma^4}$$
(8)

Where μ is the mean, σ is the standard deviation and E(x) is the expected value of x.

3.2 Autocorrelation

The similarity of two waveforms can be measured using the method of correlation. The similarity is calculated at different time in intervals. The result of a correlation is a measure of similarity as a function of time lag between the beginnings of the two waveforms [1]. Autocorrelation is the measure of similarity of a signal to a time delayed version of itself. If the time lag is zero, then there is exact similarity. As the time lag increases, there is increase in dissimilarity.

Pitch can be extracted from the speech signal by computing short time autocorrelation function of the speech signal. The autocorrelation function is the correlation of a waveform with itself. The basic steps involved [4] in finding the autocorrelation corresponding to delay τ are

1. Find the value of the signal at a time t,
2. Find the value of the signal at a time t + τ,
3. Multiply those two values together,
4. Repeat the process for all possible times, t, and then
5. Compute the average of all those products.

A function of τ, called the autocorrelation function, is obtained when the process is applied on all other values of τ.
$R(\tau)$ is calculated using equation (9)

$$R(\tau) = \frac{1}{tmax-tmin} \int_{tmin}^{tmax} s(t)\, s(t+\tau)\, dt \qquad (9)$$

The performance of these methods is observed in different environments namely noise free, hall, space, stadium and telephone. The performance of these methods when whispered voices are given as input is also observed.

4 Experimental Results

Synthesized voices of males and females are analyzed for the two methods. Detailed summary about the input signals considered for experimentation is given in Table1.

Table 1. Details About The Input Speech Signals Considered For The Analysis

Name of the wav file	Nature (Natural / Synthesized)	Gender	Environment	Word / Sentence
m_aah.wav	Synthesized	Male	Noise-free	aah
m_aah_hall.wav	Synthesized	Male	Hall	aah
m_aah_space.wav	Synthesized	Male	Space	aah
m_aah_stadium.wav	Synthesized	Male	Stadium	aah
m_aah_telphone.wav	Synthesized	Male	Telephone	aah

m_aah_whisper.wav	Synthesized	Male	Whisper	aah
f_aah.wav	Synthesized	Female	Noise-free	aah
f_aah_hall.wav	Synthesized	Female	Hall	aah
f_aah_space.wav	Synthesized	Female	Space	aah
f_aah_stadium.wav	Synthesized	Female	Stadium	aah
f_aah_telphone.wav	Synthesized	Female	Telephone	aah
f_aah_whisper.wav	Synthesized	Female	Whisper	aah
m_eee.wav	Synthesized	Male	Noise-free	eee
m_eee.wav	Synthesized	Male	Noise-free	eee
m_eee_hall.wav	Synthesized	Male	Hall	eee
m_eee_space.wav	Synthesized	Male	Space	eee
m_eee_stadium.wav	Synthesized	Male	Stadium	eee
m_eee_telphone.wav	Synthesized	Male	Telephone	eee
m_eee_whisper.wav	Synthesized	Male	Whisper	eee
f_eee.wav	Synthesized	Female	Noise-free	eee
f_eee_hall.wav	Synthesized	Female	Hall	eee
f_eee_space.wav	Synthesized	Female	Space	eee
f_eee_stadium.wav	Synthesized	Female	Stadium	eee
f_eee_telphone.wav	Synthesized	Female	Telephone	eee
f_eee_whisper.wav	Synthesized	Female	Whisper	eee

f_rec_eee.wav	Natural	Female	Noisy	eee
m_rec_eee.wav	Natural	Male	Noisy	eee
m_yes.wav	Synthesized	Male	Noise-free	yes
f_yes.wav	Synthesized	Female	Noise-free	no
m_wow.wav	Synthesized	Male	Noise-free	wow
f_wow.wav	Synthesized	Female	Noise-free	wow
m_hello.wav	Synthesized	Male	Noise-free	hello
f_hello.wav	Synthesized	Female	Noise-free	hello
f_rec_hello.wav	Natural	Female	Noisy	hello
m_rec_hello.wav	Natural	Male	Noisy	hello
f_sp3.wav	Synthesized	Female	Noise-free	It's a beautiful day
m_sp4.wav	Synthesized	Male	Noise-free	It's a beautiful day
f_sp5.wav	Synthesized	Female	Noise-free	eSpeak is a speech synthesizer
m_sp6.wav	Synthesized	Male	Noise-free	eSpeak is a speech synthesizer

Pitch is related to loudness. The pitch of female is much higher than that of male. But the pitch variation cannot be significantly noted when the audio signal is plotted. Figure 1 show the speech signal obtained when the word 'verandah' is spoken by male and female.

Here, the plot shows loudness value for male as 0.6 while that for female as 0.5 and thus the result is misleading.

Cepstral analysis tries to identify the pitch by segregating the vocal tract and excitation source components while the autocorrelation method tries to find the loudness or pitch factor in the time domain.

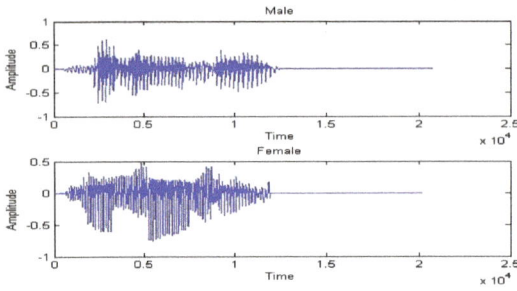

Figure 1. Plot obtained for the word 'verandah' spoken by male and female The top part is the plot obtained for male while the bottom part is that obtained for female

4.1 Cepstral Analysis

When male and female voice inputs, from different environments, are given, the following outputs are obtained.

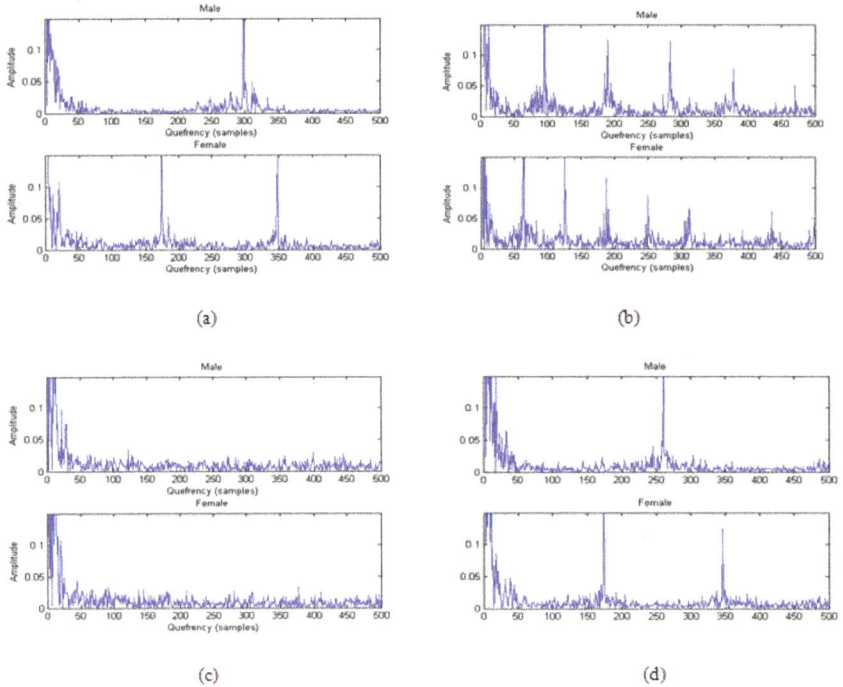

(a)

(b)

(c)

(d)

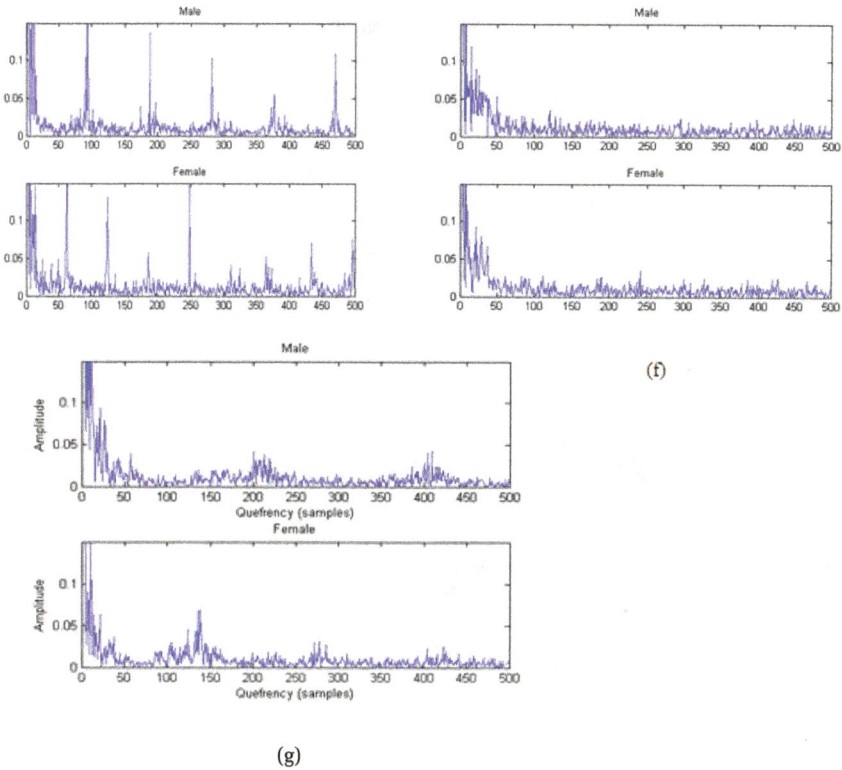

(f)

(g)

Figure 2. Cepstrum of speech segment "aah" spoken by male and female(synthesized) in different environments (a) noise free,(b) telephone, (c)whisper, Cepstrum of speech segment "eee" spoken by male and female(synthesized) in different environments (d) noise free,(e) telephone, (f)whisper (g) Cepstrum of speech segment "wow" spoken by male and female(synthesized voices). The top part is the plot obtained for male while the bottom part is that obtained for female

Figure 2.a shows the graph obtained when noise free speech segment "aah", spoken by male and female are given as input. The excitation appears as periodic modes at the high quefrency region. It can be clearly observed that the number of modes is more in case of females than in case of males. To test the consistency of the method, the voices of male and female from different environments are given as input. Similar results are obtained in case of voices from hall, space and stadium. Voices via telephone also show no difference in the output. Figure 2.b shows the graph obtained when speakers are spoken via telephone. Gender classification using this method from whispered voice is difficult since it does not give any useful pitch information. The plot when whispered voice is given as input is shown in figure 2.c. Similar results are obtained when speech segment "eee" is given as input in different environments. These are shown in figure 2.d, 2.e and 2.f. When speech segment 'wow' is given as input as shown in figure 2.g, the result is similar to that of whispered voice. Since there are no significant modes observed, additional processing like calculating the variance and kurtosis is required for gender classification.

To find the variance, a window size of 50ms quefrency is taken. Sum of variances of each window is calculated. Kurtosis is also calculated similarly. Some of words used for experimentation and the corresponding variance and kurtosis plots are shown in figure 3.

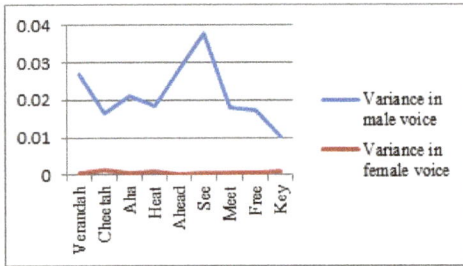

(a)

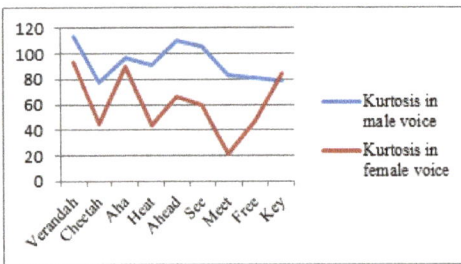

(b)

Figure 3. (a) Variance and (b) Kurtosis plot of male and female for different words

From figure 3.a, it is clearly visible that the variance of male in all cases is much larger than that of female. In general, it can be said that if the value of variance of cepstrum is more than a threshold T, then the voice belongs to male. From the results obtained, it can be concluded that if the variance value is more than 0.01, the input voice belongs to male. Kurtosis plot in figure 3.b does not show much promising result for gender classification.

4.2 Autocorrelation

The autocorrelation plots, obtained when male and female voices from different environments are given as inputs, are as follows.

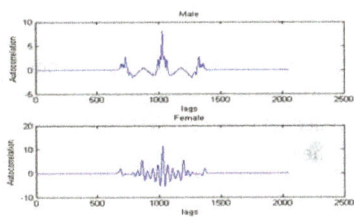

(a)

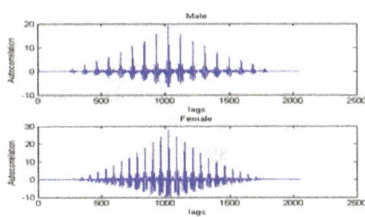

(b)

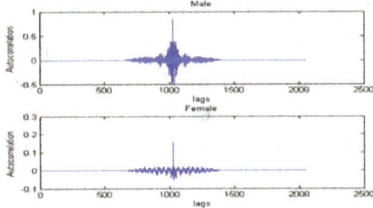

(c)

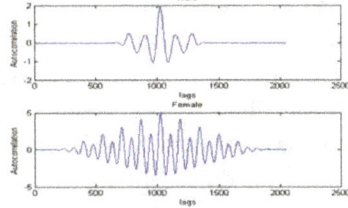

(d)

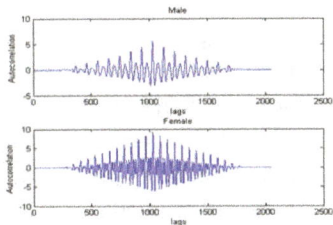

(e)

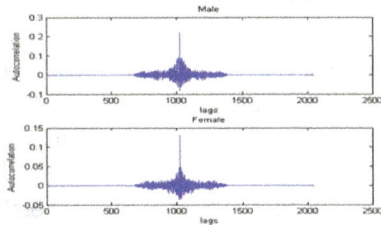

(f)

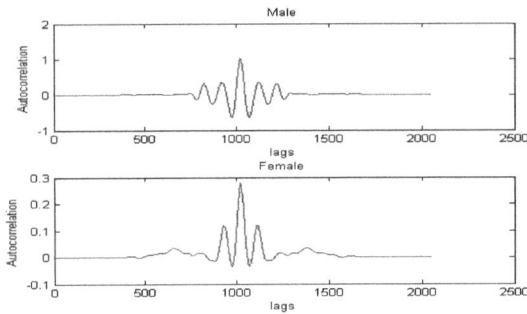

(g)

Figure 4. Autocorrelation plot of speech segment "aah",spoken by male and female (a) noise free,(b) telephone, (c)whisper, Autocorrelation plot of speech segment "eee",spoken by male and female (d) noise free,(e) telephone, (f)whisper (g) Autocorrelation plot of speech segment "wow" spoken by male and female(synthesized voices). The top part is the plot obtained for male while the bottom part is that obtained for female.

Figure 4 shows the time domain plots of speech segments spoken by male and female in different environments. Figure 4.a shows the graph obtained when noise free speech segment "aah" spoken by male and female is given as input. It is observed that the peak value in case of female is higher when compared with that of male. Similar graph is obtained for telephonic voice, which is shown in figure 4.b. As in the case of the first method, to test the consistency, voices of male and female from different environments are given as input. When the whispered voices are given as input the peak value is more in case of male than female. This is because pitch depends also on loudness. The graph is shown in figure 4.c. Similar results are obtained when speech segment "eee" is given as input. Corresponding plots are shown in figure 4.d,4.e and 4.f .Figure 4.g shows the plot obtained when speech segment 'wow' is given as input. In this case, the mode is found to be higher in case of male than female. When a spoken sentence is given as input, the output is as expected, ie, the peak in case of female itself is higher. The reason for the results observed is the fact that the pitch range of female is higher than that of male. In general, it can be concluded that if the autocorrelation value is more than a threshold T, then the voice belongs to female. From the results, it is clear that when the autocorrelation value is more than 5, then the voice belongs to female.

When recorded voices are given as input, the results are not favorable because of the presence of noise and other variable acoustics. The type of microphone used and the transmission channel can also be some of the reasons.

Conclusion

Gender classification is one of the important steps in the process of speech or speaker recognition. It is considered to be one of the initial steps in the process. In this paper, two pitch detection methods, one in frequency domain and other in time domain namely cepstral analysis and autocorrelation method respectively, are considered for the process of gender classification. Pitch difference between male and female is the main criterion used for gender classification. The results of the experiments show that these methods itself are suitable for gender distinction for normally uttered voices. These methods can also be used as the basis for devising a gender classifier.

From the observations, it can be concluded that these methods are not suitable for identifying the gender in case of whispered voice. Also, it is observed that 'aah' and 'eee' can be used to classify gender successfully. So, as a real time system, we can try to extract these segments from the spoken sentences to perform gender classification.

References

1   B.Jena, B.P Panigrahi," Gender Classification by Pitch Analysis" International *Journal on Advanced Computer Theory and Engineering,*Vol. 1, no. 1, pp.2319-2526, 2012.

2   Harb, Hadi, Liming Chen, and Jean-Yves Auloge. "Speech/music/silence and gender detection algorithm." *In Proceedings of the 7th International conference on Distributed Multimedia Systems DMS01.* 2001.

3   Gomathy, M., K. Meena, and K. R. Subramaniam. "Classification of speech signal based on gender: a hybrid approach using neuro-fuzzy systems."*International Journal of Speech Technology* 14.4 (2011): 377-391.

4   Phy.mtu.edu, 'Autocorrelation (for sound signals)', 2015. [Online] Available: http://www.phy.mtu.edu/~suits/autocorrelatiom.html. [Accessed: 15-Jun-2015].

5   Iitg.vlab.co.in, 'Cepstral Analysis of Speech (Theory) : Speech Signal Processing Laboratory : Electronics & Communications : IIT GUWAHATI Virtual Lab', 2015. [Online]. Available: http://iitg.vlab.co.in/?sub=59&brch=164&sim=615&cnt=1. [Accessed: 22-Apr - 2015].

6   In.mathworks.com, 'Variance - MATLAB var', 2015. [Online]. Available: http://in.mathworks.com/help/matlab/ref/var.html?refresh=true. [Accessed: 20- Aug- 2015].

7   v. matlab, 'MATLAB PROJECTS: voice Conversion in matlab', *Matlabsproj.blogspot.in*, 2012. [Online]. Available: http://matlabsproj.blogspot.in/2012/05/voice-conversion-in-matlab.html. [Accessed: 01- Jul- 2015].

8   Zhang, "Speaker Verification: Text-Dependent vs. Text-Independent",*Research.microsoft.com*. [Online] Available: http://research.microsoft.com/en-us/um/people/zhang/Speaker%20Verification/default.htm. [Accessed: 04- Jun- 2015].

9   In.mathworks.com, 'Kurtosis - MATLAB kurtosis', 2015. [Online]. Available: http://in.mathworks.com/help/stats/kurtosis.html. [Accessed: 20- Aug- 2015].

P Ravinder Kumar[1], Dr Sandeep.V.M[2] and
Dr Subhash S Kulkarni[3]

A Simulative Study on Effects of Sensing Parameters on Cognitive Radio's Performance

Abstract: This paper studies experimentally the influence of single Secondary User (SU) on the Primary User (PU) in a single channel Cognitive Radio (CR) system. It studies the effect of sensing time and sensing period of repeatability on the Interference caused to the primary, the opportunity lost by Secondary User and efficiency of the CR system.

Keywords: Cognitive Radio, Transmit (TR) state

1 Introduction

The demand for wireless network is exponentially growing and the spectrum available for wireless applications is band-limited. On, one hand, the research in increasing the spectrum for usage is not very much successful as the need for the spectrum is always outgrowing the increase in spectrum and on the other hand, users owning the spectrum bands, called Primary Users (PU), are under utilizing significant amount of their spectrum [9]. This calls for developing a new avenue in wireless communication to optimally solve the said problems, simultaneously and CR Technology brings a ray of hope in this direction by allowing the SU's to use the underutilized spectrum without disturbing PU's.

This paper deals with simulation to study the effect of Cognitive Radio system, with a single channel, its associated primary user and a secondary user, on the performance of both primary and secondary users. The paper is organised as follows: The next section briefs about the Cognitive Radio system. Details of our

---

1 Jayaprakash Narayan College of Engineering, Mahabubnagar, T.S
ravinderpalem@gmail.com
2 Jayaprakash Narayan College of Engineering, Mahabubnagar, T.S
svmandr@yahoo.com
3 PES Institute of Technology, Bangalore South Campus (Formerly, PES School of Engineering)
subhashsk@gmail.com

model, aim of our experiment and the method to analyse the performance the Cognitive Radio system is discussed in section III. Section IV gives the results of the experiments conducted and the paper is concluded in section V.

2 Background

A CR helps the unlicensed users, called secondary users, to use the channels/spectrum bands without disturbing the PU's activity. It can be attained by allowing the PU and SU work orthogonally either in space, time or frequency domains [1], [2], [6]. This leads to the following basic models of CR.

1. Primary and secondary transmissions are orthogonal in space with no interference to each other: This allows both PU and SU's to transmit simultaneously on same frequency band [1].

2. Primary and secondary transmissions are non-orthogonal in space but a limited interference exists between them: An intelligent receiving technique that neglects the interference can allow simultaneous transmissions of both PU and SU on same frequency bands. Simultaneous transmissions can also be achieved through intelligent transmitters like the ones used with zero forcing technique [2].

3. Primary and secondary transmissions are highly non-orthogonal in space with heavy interference: Here neither intelligent transmission nor intelligent reception will help in avoiding this interference. This force the primary and secondary users transmit orthogonally in time, on the same frequency band. To maintain the right of PU to utilise his licensed frequency band this orthogonal property in time has to be verified only by the SU prior to its transmissions. This forces the SU to continuously track the availability of the channel. The continuous tracking by SU will not provide time to transmit, so the sensing of the channel is done periodically to optimally balance the freedom to PU and efficient SU transmission [3], [10]. This paper aims to study the effect of this model on primary and secondary transmissions in CR.

4. Spatially non-orthogonal Single SU and Multiple PU's with multiple frequency bands: Here the SU opportunistically uses the vacant frequency band. This model assumes that the SU has multiple sensing equipment, one for each band, with transmission capabilities in multiple frequency bands.

5. Multiple SU's and multiple frequency bands: Here adaptive and cooperative schemes are implemented for spectrum and SU scheduling, where the SU's cooperatively sense and schedule to share the available vacant frequency bands for the transmission [7].

The effective utilization of the channel is ensured by periodically sensing the channel. The sensing accuracy is an important factor as it determines the efficiency of the CR system. The literature suggests 3 ways of sensing the activity of the PU. Energy detection, Matched-filter detection and Feature detection. The energy detection is a simple technique where in the PU is considered to be active if the energy measured on the channel is above some threshold. It is a non-coherent technique. In this scheme we can't differentiate between the signal and the noise, making SU not to understand the PU's signal and hence the privacy of the PU is maintained, but is useful only when the noise level is assured to be far below the used threshold.

The Matched-filter technique comes to our rescue when the signal and noise energies are comparable. This is a coherent technique, needs the knowledge transmission properties of PU. This method is optimum when the transmitted signal is known. Advantage of Matched filtering is the short time to achieve a certain probability of false alarm or probability of Miss Detection. Hence, it requires perfect knowledge of PU signalling features such as bandwidth, operating frequency, modulation type and order, pulse shaping and frame format. Since CR needs receivers for all signal types, implementation complexity of sensing unit is impractically large. Hence it requires CR to demodulate received signals. The Matched filter is an optimal linear filter for maximizing the SNR in the presence of additive stochastic noise [11][12].

The Feature detection technique senses the signal over the channel more intelligently by having prior knowledge of the PU activity [4], [5]. Hence, costly equipments as well as algorithm are very costly. Signals have periodic statistic features such as modulation rate and carrier frequency which is usually viewed as cyclostationary characteristics and specific features should associate with the modulated signals transmitted by primary users. In detection the cyclostationary characteristic of a PU's signal can be distinguished from noise in its statistical properties such as its mean and autocorrelation. Compared with energy detection, cyclostationary detection is not sensitive to noise uncertainty, so it has better robustness in low SNR regimes. However this method requires more prior information on the PU signals to decide the occupancy of PU's. As the consequence, feature detection has much greater complexity. Cyclostationary detection can differentiate noise from PU's signals, and can be used to distinguish the different types of transmissions and PU's effectively. The main drawback is its computational complexity for its implementation. Thus the energy detection is the most common method of signal detection, which has low computational and implementation complexities. It doesn't require prior knowledge of the PU's signal, and not need any special designs. It doesn't perform well in low SNR environments [11] [12].

3 Our Work

3.1 System Model

Our system consists of single primary user (PU) that uses Orthogonal Frequency Division Multiplexing (OFDM) technique for its data transmission and its dedicated channel with a single band of spectrum using OFDM technique for data transmission. The PU and Secondary users (SU) transmissions are non-orthogonal in space having higher interference temperature and hence should be orthogonal in time. A single SU is assumed to opportunistically use the channel whenever the channel is vacant without disturbing the PU activity. Here the PU is unaware of the SU's activity, i.e., the primary user is provided with all its right to use its licensed channel without any need to check the availability of the channel. This makes the SU to make sure the channels status for its usage. This vacancy of the channel is detected through the simplest of the available methods by measuring the energy at the channel. All these factors attract one towards Energy Detection method for sensing the PU's presence.

With energy detection, Primary and Secondary signals can't be differentiated, forcing SU's channel sensing and transmission activities to be mutually exclusive in time. First portion of each period of SU activity is used for sensing and the rest for transmission of secondary data if vacancy is detected. To minimise interference to the PU and SU, channel sensing should be done periodically.

3.2 Aim

To conduct experiment to study the effects of spectrum sensing duration (t_s), CR transmission frame length (T), death rate (a), birth rate (b) of the primary transmissions on the interference introduced to the PU, opportunities lost by SU for transmission , the efficiency (η) of the SU to transmit its data and the data throughput of the secondary channel. The outcome of this experiment is expected to help one in deciding the optimum values of t_s , T for a given PU characterised by a, b so that transmission interference (TI), transmission loss (TL) are at minimum and η and throughput are at their maximum.

3.3 Performance Analysis

The SU opportunistically utilizes the channel by sensing it periodically. The state of the PU either ON or OFF is measured through detecting the energy on the channel. The SU checks the channel for sensing the state of the PU. The energy on the channel, if greater than a threshold, is taken as BUSY state and the state of channel is assumed to be FREE otherwise. The CR system will allow the SU to transmit its data on the channel whenever a FREE state of the channel is identified or sensed. If the CR system senses the BUSY state of the channel, the SU is detained to use the channel till the next sense operation. The fading effect of the channel and the noisy environment makes 4 possible sensing situations, as shown in figure 1.

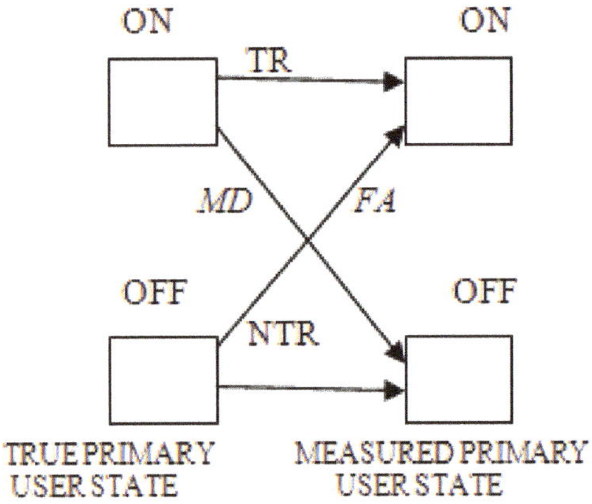

Figure 1. Sensing Activity

When the PU is ON and the SU senses the channel as BUSY, the SU is detained to use the channel and the CR system can said to be in "No Transmission"(NTR) state. When the PU is OFF and the channel is sensed as FREE, the SU is allowed to use the channel and the CR system is said to be in "Transmit" (TR) state. Apart from these two states, the CR system is found to be in 2 more states due to erroneous sensing results. The PU is ON but the channel is miss-detected as FREE, "Miss-Detection" (MD) state. Here the SU starts using the channel and hence produces interference to the PU. The TI is a measure of how many times the PU is interfered by the SU [8].

$$TI(in\ \%) = \frac{Number\ of\ time - slots\ the\ channel\ is\ miss - detected}{Total\ number\ of\ intervels} * 100$$

On the other hand, the channel may produce 'False-Alarm' (FA) by making SU to sense a BUSY State though the PU is OFF. This deters the SU to utilize the channel though it is FREE. This lost opportunity by SU to utilize the channel is measured as 'Transmission Loss' (TL).

$$TL(in\ \%) = \frac{Number\ of\ intervels\ of\ false - alarm}{Total\ number\ of\ intervels} * 100$$

TL is a measure of opportunity lost by SU to use the channel. It is a measure of false-alarm.

The efficiency (η) of the CR system is measured by the time taken to transmit N data as

$$\eta = \frac{N}{Number\ of\ intervels\ that\ the\ SU\ needs\ to\ transmit\ N\ data} * 100$$

The Primary User and Secondary User activities are shown in figure 2. The Primary User is ON and OFF for some duration. The Primary activity represented by continuous line and secondary activity represented by dotted line.

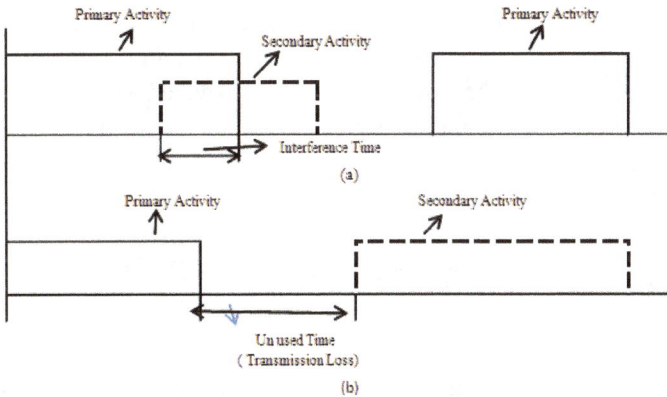

Figure 2. Primary and Secondary user activities

The figure 2(a) shows the Transmission interference between primary user and secondary user. It is due to the Missed Detection by SU. The figure 2(b) shows the data transmission activity by the PU and SU where we come across the unused transmission due to the False-Alarm detected by SU.

Primary User and Secondary User activities for different sensing times and transmission times are shown in figure 3. In its allocated time slot, the secondary user spends ts seconds for spectrum sensing. If the secondary user decides channel is vacant, then spends the remaining time for data transmission. If primary channel is busy it waits for next sensing period. From figure 3 (a) Primary activity is known i.e., when PU is ON and OFF.

Figure 3(b) shows the sensing activity. The duration of sensing is represented by ts. The combination of Transmission period and sensing time gives the total period. Sensing is done periodically. If the PU is ON and same sensed by the secondary user, waits until the next sensing period. When the PU is OFF, SU data transmission starts.

The SU activity is represented during the third and fifth sensing period the SU got the opportunity for transmission. During the second sensing period PU is ON and the rest of time i.e., before the third sensing period starts the PU is OFF, which leads to the transmission loss. At the third sensing period PU is OFF and SU Transmission starts and before the fourth sensing period the PU started transmitting its data which leads to Transmission Interference. This is shown in figure 3(c). Figure 3(d) shows the increased transmission time where the sensing time remains same.

In figure 3(e), due to increase in transmission time, transmission interference increased and transmission loss reduced. In the figure 3(f), sensing period is increased. Due to this the data transmission is reduced which leads in reducing the transmission interference and transmission loss as shown in figure 3(g).

4 Experimentation and Results

The experiment is conducted with PU characterised by its birth rate b and its death rate a .The primary activity is simulated for the combinations of a and b, thereby making the channels availability dependent on both a and b. The SU senses the channel for ts={1,2,3,4,5,6} and the sensing activity is repeated every T={9,18,27,36}. Figure 4 shows the secondary user time slot management.

The experiment is conducted to transmit N data from the SU by opportunistically sensing the channel. This process is repeated 100 times for each combination of a, b, t_s and T and the secondary occupancy, i.e., the intervals wherein the SU is transmitting through the channel, is measured.

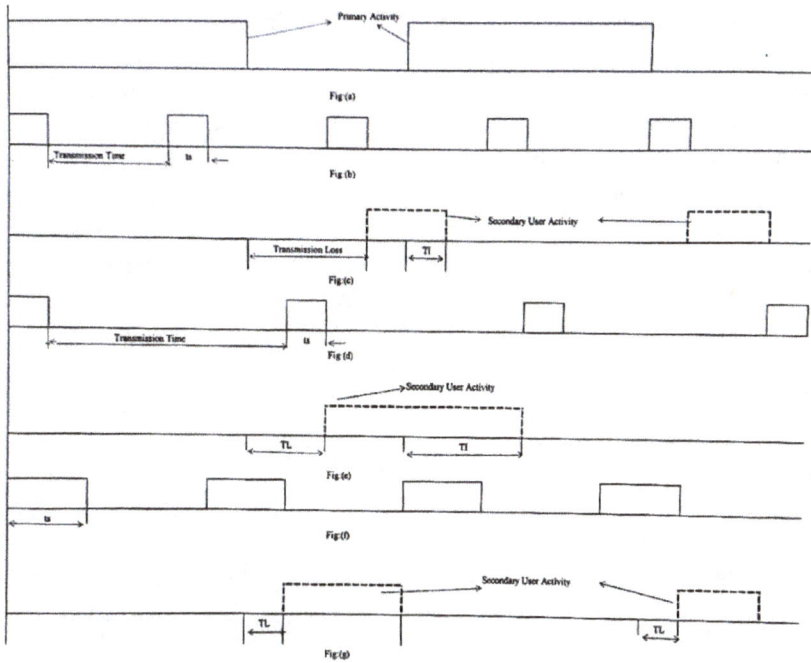

Figure 3: Primary and secondary activities with different sensing times and transmission times

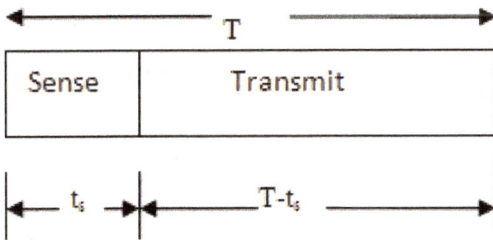

Figure 4: CR time-slot management: sensing duration (t_s) and CR transmission time ($T-t_s$)

Comparing this secondary occupancy with that of the primary, the TI, TL and efficiency are computed from each set of 100 experiments.

The PU on and off states are random. The PU activity is simulated on the basis of birth rate and death rate. Birth rate indicates when the PU is utilizing the channel denoted by 'b'. Death rate indicates when the PU is not utilizing the channel denoted by 'a'. The probability of on rate is given by PRON. Threshold is fixed using exponential distribution based on primary on rate. For channel occupancy simulation, a random number is generated and compared with ON Threshold. If greater a random number of 1's are appended else a random number of 0's are appended. This process is repeated for a random number of times.

The CR user senses the channel periodically, with period T, for occupancy. In a given slot (T), the CR user spends ts seconds for spectrum sensing. If channel is vacant then it spends T-ts seconds to transmit the data else it waits. The secondary occupancy (i.e., the slots when CR user is using channel) is recorded for performance analysis like TI, TL and efficiency.

Transmission Interference is given by number of missed detection states to total number of time periods. Transmission loss is given by number of false alarm states to total number of time periods. Efficiency is the ratio of number of slots CR user awaited to successfully transmit its data to the total number of slots.

Table 1 shows the interference to the primary by the CR system with various sensing time (ts) and sensing period (T). The sensing plays the vital role in minimizing the disturbance to the PU in CR system. The higher the sensing time the lower will be the miss-detection rate and hence the TI reduces. This can be seen clearly in figure 5.

Table 1. Transmission interference in % for various combinations of t_s and T

TI				
ts	T=9	T=18	T=27	T=36
1	5.88	9.45	10.45	10.47
2	3.32	5.25	8.29	5.32
3	3.38	5.868	5.65	9.79
4	1.79	2.999	2.69	6.49
5	0.78	2.135	3.2	1.6.6
6	0.11	0.55	0.52	0.89

Table 2 shows the effect of t_s and T on transmission loss. With increase in t_s the false alarm rate increases hence TL increases. On the other hand with increase in t_s, the time to transmit secondary data reduces and hence for every

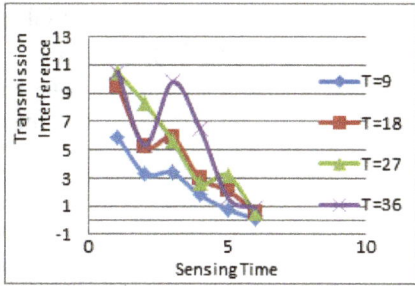

Figure 5: The effect of t_s and T on transmission Interference

false alarm, the opportunity lost reduces. The overall effect is that change in t_s has little effect on TL. This is very easily seen in figure 6.

Table 2. Transmission Loss in % of opportunity lost by the for different combinations of t_s and T

	TL			
ts	T=9	T=18	T=27	T=36
1	7.16	8.8	8.2	7.799
2	7.48	8.71	7.77	8.2
3	9.11	9.2	8.79	8.87
4	8.66	8.61	8.2	9.48
5	9.38	8.97	9.15	8.26
6	8.92	8.78	8.81	8.45

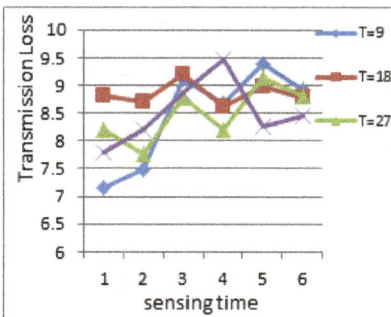

Figure 6. The effect of ts on Transmission Loss for various T values

With increase in T, the time to transmit will increase and for every false alarm TL increases. On the other hand with higher T the sensing frequency reduces and hence rate of false alarm reduces making TL to reduce. The overall effect is that TL is almost independent of T. This can be seen clearly from figure 7.

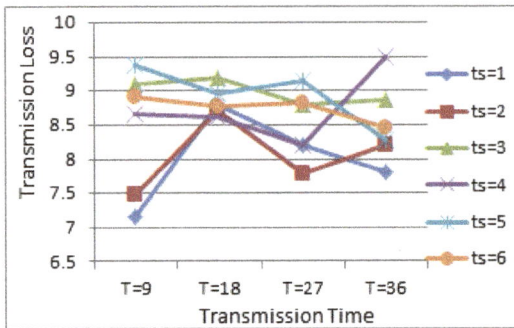

Figure 7. The effect of T on Transmission Loss for various ts values

The efficiency of the CR system is measured through how the opportunities are utilized. Table-3 gives some sample efficiency of our CR system. With increase in t_s time to transmit will reduce and hence efficiency decreases. This is very clearly seen in figure 8.

Table 3. Efficiency of the CR system for different t_s and T

EFFICIENCY				
ts	T=9	T=18	T=27	T=36
1	9.56	11.5	13.54	11.91
2	4.821	6.62	9.998	6.71
3	4.63	6.87	6.701	12.49
4	2.817	3.58	3.27	7.97
5	2.5	2.81	4.098	2.8
6	2.5	2.5	2.5	2.88

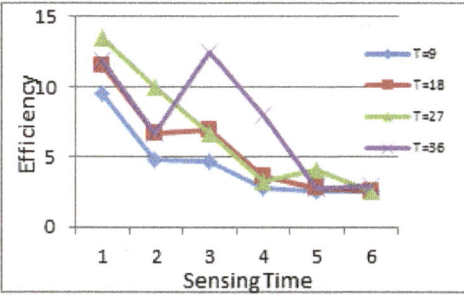

Figure 8. The effect of ts on efficiency of the CR system for various values of ts

Figure 9 shows the effect of T on efficiency of the CR system. It is observed that this variation is unimodel with a peak (Tp). When T is less than Tp, the increase in T increases the opportunities for secondary user to transmit and hence efficiency increases. The other half where T > Tp increase in T will also increase in interference to the PU and hence efficiency reduces.

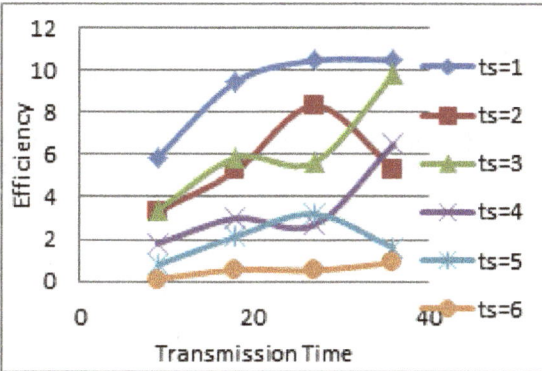

Figure 9. The effect of T on efficiency of the CR system for various values of ts

For an efficient CR system the interference should be least, TL should be minimum and efficiency to be higher. These are dependent on the values of ts and T. The aim of our study is to identify values of ts and T that presents best CR system.

Value of ts should be maintained high in order to minimize the effect on PU but higher ts reduces the efficiency of SU. For better efficiency T=27 is preferred but with T=27, TI is very high. No unique value of ts or T is satisfying our requirements independently. The only option left is to optimally choose the values of ts and T simultaneously for an optimal CR system.

Conclusion

This paper presented the results of experiments conducted to study the CR system with PU and single SU working cooperatively over a single channel. The results confirm that the overall CR system can't be described by a single parameter, either ts or T. The value of ts plays major role in deciding the miss-detection and hence controls the values for TI. The opportunity lost by the SU is almost independent of either ts or T. This encourages one to use higher values of ts as this will reduce TI and no change in TL. But the SU's efficiency gets marginally down with higher ts. In order to attain an optimum performance of CR system these two parameters should jointly be defined. The work is in progress in this direction.

References

1   Sami M. Alamal fouth and Gordon L. Stuber, "Joint Spectrum Sensing and Power Control in Cognitive Radio Networks: A Stochastic Approach," IEEE transactions on wireless communications, 11(12), 2012,4372-4380.
2   S. H. Song and K. B. Letaief, "Prior zero-forcing for relaying primary signals in cognitive network," in Proc. 2011 IEEE Global Communications Conference.
3   W. Zhang, R. K. Mallik, and K. B. Letaief, "Cooperative spectrum sensing optimization in cognitive radio networks," in proc. 2008 IEEE Int.Conf. on commun, 2008, pp.3411-3415.
4   Won-Yeol Lee, , and Ian. F. Akyildiz, "Optimal Spectrum Sensing Framework for Cognitive Radio Networks," IEEE transactions on wireless communications, 2008, 7(10), 3845-3857.
5   S. M. Mishra, A. Sahai, and R. W. Brodersen, "Cooperative sensing among cognitive radios," in proc. IEEE ICC 2006, 2006, 41658-1663.
6   B. Wild and K. Ramchandran, "Detecting primary receivers for cognitive radio applications," in Proc. IEEE DySPAN 2005, 2005, 124-130.
7   S. H. Song, M. O. Hasna, Member, IEEE, and K. B. Letaief, , " Prior Zero Forcing for Cognitive Relaying," IEEE transactions on wireless communications, 2013, 12(2), 938-947
8   Y. C. Liang, Y. Zeng, E. Peh, and A. T. Hoang, "Sensing-throughput trade-off for cognitive radio networks," IEEE Trans. Wireless Commun., 7(4) , 2008, 1326-1337.
9   I. Mitola, J. and J. Maguire, G. Q., "Cognitive radio: making software radios more personal," IEEE Personal Commun. Mag., 6(4), Aug. 1999,13–18.

10   P. Ravinder Kumar, Archena, Subhash S Kulkarni and Sandeep V M, " Effect of Sensing Time on Performance of OFDM Based Opportunistic Cognitive Radio" International Journal of Emerging Science and Engineering (IJESE), 4(2), December 2015, ISSN: 2319–6378 .

11   LI Jianwu, FENG Zebing, FENG Zhiyong, ZHANG Ping, "A Survey of Security Issues in Cognitive Radio networks", China Communications, March 2015, 132-150.

12   Ian F.Akyildiz, Won-Yeol Lee, Mehmet C.vuran and Shantidev mohanty, "A Survey on Spectrum management in Cognitive Radio Networks", IEEE Communications Magazine, April 2008, 40-48.

Author's Profile

P. Ravinder Kumar, M.Tech (WMC) from Vardhaman College of Engineering, B.Tech (ECE) from Jayaprakash Narayana College of Engineering. Currently he is working as Associate Professor at Jayaprakash Narayan college of Engineering. His areas of interest include computer networks and communications, wireless networks, signal processing.

Dr. Sandeep V.M. completed Ph.D in Faculty of Electrical and Electronics Engineering, Sciences, from Visveswaraiah Technological University, Belgaum, and M.E from Gulbarga University and B.E from Gulbarga University. His research interests are in the areas of Signal and Image Processing, Pattern Recognition, Communication, Network Security, Cloud Computing, Electromagnetics. He is Reviewer for Pattern Recognition Letters (PRL). He acted as Reviewer for many International Conferences. He is member of LMIST – Life Member Instrument Society of India (IISc, Bangalore).

Dr. Subhash S Kulkarni, Professor & Head, Dept of Electronics and Communication Engineering PESIT – Bangalore South Campus, Bengaluru. He received BE from Gulbarga University, M.Tech from IISc Bangalore and Ph.D from IIT Kharagpur, India. He is Fellow of IETE and Fellow of Institution of Engineers. He has published more than 70 papers in reputed international and national journals. He is highly sought after as commendable resource person in Signal and Image Processing. His research interests are mathematical models in signal and image processing and computational architectures based on Vedic math's. He is an inspiration to the researchers in this area and field.

Priyanka Parida[1], Tejaswini P. Deshmukh[2], and
Prashant Deshmukh[3]

Analysis of Cyclotomic Fast Fourier Transform by Gate level Delay Method

Abstract: The critical path delay (CPD) of the circuit structure can be determined by the gate level delay computing method (GLDC).The main aim of GLDC is to determine the delay in hardware implementation of constant matrix multiplication over Galois Field GF (2^m). In this paper, we have applied GLDC method to the Cylotomic Fast Fourier Transform over GF (2^m). For this, we have used the advantage of cyclotomic decomposition for the CFFT. The computing method mainly focuses and is also based on initial delay matrix. The method is adopted as it is suitable for the implementation with computer. Experimental results have been shown for GLDC method over GF (2^m).

Keywords: Gate level delay computing, Constant matrix multiplication, cyclotomic decomposition, critical path delay, Galois field (GF).

1 Introduction

Multiple constant multiplications (MCM) is a cost effective way of executing several constant multiplication with the consistent input data. By using shifts, adders and subtracters, the coefficients are being expressed. MCM is extensively used in digital signal processing (DSP) applications for example linear transformation, image processing etc. Most existing work on MCM has been proposed for minimizing the area in [1].

Previously the work on fast fourier transform (FFT's) is based on Cookey Tookey algorithm and also on prime factor algorithm over a complex field [4]. These algorithm adapted to descrete fourier transform (DFT's) over a finite field were having high multiplicative complexity. Currently CFFT have been proposed and in that it has lower multiplicative complexity [2]. So due to the lower

1 Electronics Engineering, Yeshwantrrao Chavan College, Nagpur,India
parida.priyanka0@gmail.com
2 Electronic Engineering , Yeshwantrao Chavan College, Nagpur,India
tejaswini.deshmukh@gmail.com
3 Electronic Engineering , Yeshwantrao Chavan College, Nagpur,India

multiplicative complexity [2] of the CFFT it has an enormous attraction. DFT have a vast application in error control code and cryptography. In digital application error correcting code has been widely used to recover from random errors. Inverse DFT's are used to regain the transmitted codeword in transform domain decoders.

In this paper, gate level delay computing method has been presented. This GLDC method is being applied on the CFFT matrix over GF (2^m). By this method the critical path delay of constant matrix multiplication (CMM) over GF (2^m) in CFFT is being evaluated. The method proposed in this paper is based on initial delay matrix, where as similar method was proposed in [3], which was based on the restriction graph. The method adopted in this paper because of the implementation of this method is appropriate on the computer.

The paper is organized as follows. In section II, a quick summary of CFFT proposed in [2] is being presented. The GLDC method proposed in [1] is being discussed in section III. The implementation of the GLDC method on CFFT is shown in section IV. Finally, some conclusions are drawn in section V.

2 Cyclotomic Fast Fourier Transform

The method suggested in this paper for the CFFT is the decomposition of an original polynomial [2] into sum of linearized polynomials and then evaluating them at a set of basis point .Many method has been previously proposed for the computation of fourier transform over GF (2^m), but the method proposed in [2] is efficient for narrow FFT length.

The fourier transform of a polynomial proposed in [2] is given as;

$$f(x) = \sum_{i=0}^{n-1} f_i x^i$$

which is of degree deg $f(x)$ = n-1, n| (2^m-1), in the field GF (2^m) are the collection of elements.

$$F_j = f(\alpha^j) = \sum_{i=0}^{n-1} f_i \alpha^{ij}, j \in (0, n-1)$$

where α is an order of element n in the field GF (2^m).

A linearized polynomial over GF (2^m) is a polynomial of the form
$$L(x) = \sum_i l_i x^{2^i}, l_i \in GF(2^m)$$

Linearized polynomial satisfies the equation

$$L(a+b) = L(a) + L(b)$$

This property leads to lemma 1 [5].

The method for the computation of Fourier Transform over GF (2^m) for CFFT is as follows:

Step1: To resolve the cyclotomic cosets for the given Galois Field and then grouping them according to the cyclotomic coset. The coset which has same size will be approached by other computational blocks.

For example, we will consider CFFT of GF (2^3), the cosets taken from [2] are

$$C_0 = \{f0\}$$
$$C_1 = \{f_1, f_2, f_4\}$$
$$C_2 = \{f_3, f_5, f_6\}$$

Step2: In this step the $f(\alpha^i)$ is being developed by the formula from [2],

$$f(\alpha^j) = \sum_{i=0}^{l} L_i(\alpha^{jk_i})$$

where i=0,1,........,n-1, l = (number of cyclotomic cosets)-1, k_i is the first term of cyclotomic coset, n=(2^m-1) and j ranges from 0 to n-1.

From [2],

$$f(\alpha^0) = L_0(\alpha^0) + L_1(\alpha^0) + L_2(\alpha^0)$$

In similar way for GF (2^3), till $f(\alpha^6)$ can be calculated by above formula.

Step3: $L_i(y)$ is developed in this step, where L_i is the cyclic convolution between normal basis and input f_i. And then by the matrix-vector multiplication, CFFT matrix will be obtained. Suppose F is the CFFT matrix then,

$$F = (AQ)*(C*(Pf))$$

Where AQ is the matrix and C*Pf is the vector. The GLDC is performed on the matrix itself i.e. on AQ, so only AQ is essential.

For GF (2^3), the AQ matrix [2] obtained is as

$$\begin{bmatrix} 1\,1\,0\,0\,0\,1\,0\,0\,0 \\ 1\,0\,0\,1\,1\,1\,0\,1\,1 \\ 1\,0\,1\,1\,0\,1\,1\,1\,0 \\ 1\,1\,0\,1\,1\,0\,1\,1\,0 \\ 1\,0\,1\,0\,1\,1\,1\,0\,1 \\ 1\,1\,1\,0\,1\,0\,0\,1\,1 \\ 1\,1\,1\,1\,0\,0\,1\,0\,1 \end{bmatrix} \quad (1)$$

3 Gate Level Delay Computing Method

In this section, gate level delay computing method is being presented. As this method is based on initial delay matrix, similar to this a method has been proposed [3] which is based on restriction graph. The computing process of GLDC method is as follows:

1. Set up an initial delay matrix: First we have to consider a constant matrix M, in which the rows of the matrix will be the output variable and the columns of the matrix will be the input variable. The row of matrix M will be represented as '1' which means the participation of the input variable for the computation of output variable, where as '0' represents there is no participation of the input variable.

Now for the initial delay matrix M_d, the input variables will be represented by '0'. The initial delay matrix is being obtained by the formula as

$$M_d = M - M_1$$

Where, M_1 is the matrix with all the elements are '1'. In initial delay matrix '- ' will be considered as invalid value.

2. Compute delay value: To determine the critical path delay (CPD) of the circuit structure, computation of the delay value is essential. In other words, we can say that the CPD of the circuitis based on the delay value. To compute the delay value the steps involved are as follows:

a)   Appending a row of delay value d_0, d_1 , ...,d_n (n≥2).

b)   Sort the rows in increasing order.

c)   Select the smallest two positive delay value d_i and d_{i+1}.

d)   The new value of d_i and d_{i+1} is being updated as

         $d_i = -1$,

         $d_{i+1} = max(d_i, d_{i+1})+1$;

e)   The steps from b-d are repeated until there is one positive value in that row.

4 Computation of GLDC on CFFT

In this section GLDC method is implemented on CFFT and this is explained by taking an example of GF (2^3).

The CFFT matrix for GF (2^3) on which GLDC will be applied is as follows

$$M = \begin{bmatrix} 1 & 1 & 0 & 0 & 0 & 1 & 0 & 0 & 0 \\ 1 & 0 & 0 & 1 & 1 & 1 & 0 & 1 & 1 \\ 1 & 0 & 1 & 1 & 0 & 1 & 1 & 1 & 0 \\ 1 & 1 & 0 & 1 & 1 & 0 & 1 & 1 & 0 \\ 1 & 0 & 1 & 0 & 1 & 1 & 1 & 0 & 1 \\ 1 & 1 & 1 & 0 & 1 & 0 & 0 & 1 & 1 \\ 1 & 1 & 1 & 1 & 0 & 0 & 1 & 0 & 1 \end{bmatrix}$$

Now GLDC method will be applied to the matrix M.
1) Set up the initial delay matrix
$M_{d1} =$

$$\begin{bmatrix} 1\,1\,0\,0\,0\,1\,0\,0\,0 \\ 1\,0\,0\,1\,1\,1\,0\,1\,1 \\ 1\,0\,1\,1\,0\,1\,1\,1\,0 \\ 1\,1\,0\,1\,1\,0\,1\,1\,0 \\ 1\,0\,1\,0\,1\,1\,1\,0\,1 \\ 1\,1\,1\,0\,1\,0\,0\,1\,1 \\ 1\,1\,1\,1\,0\,0\,1\,0\,1 \end{bmatrix} - \begin{bmatrix} 1\,1\,1\,1\,1\,1\,1\,1\,1 \\ 1\,1\,1\,1\,1\,1\,1\,1\,1 \\ 1\,1\,1\,1\,1\,1\,1\,1\,1 \\ 1\,1\,1\,1\,1\,1\,1\,1\,1 \\ 1\,1\,1\,1\,1\,1\,1\,1\,1 \\ 1\,1\,1\,1\,1\,1\,1\,1\,1 \\ 1\,1\,1\,1\,1\,1\,1\,1\,1 \end{bmatrix}$$

$$M_{d1} = \begin{bmatrix} 0 & 0 & -1 & -1 & -1 & 0 & -1 & -1 & -1 \\ 0 & -1 & -1 & 0 & 0 & 0 & -1 & 0 & 0 \\ 0 & -1 & 0 & 0 & -1 & 0 & 0 & 0 & -1 \\ 0 & 0 & -1 & 0 & 0 & -1 & 0 & 0 & -1 \\ 0 & -1 & 0 & -1 & 0 & 0 & 0 & -1 & 0 \\ 0 & 0 & 0 & -1 & 0 & -1 & -1 & 0 & 0 \\ 0 & 0 & 0 & 0 & -1 & -1 & 0 & -1 & 0 \end{bmatrix} \quad (2)$$

2) Compute the delay value

For obtaining the delay values the evaluation of the steps for GF (2^3) is as follows:

a) Appending a row of delay value

$$0\ 0\ \text{-}1\ \text{-}1\ \text{-}1\ 0\ \text{-}1\ \text{-}1\ \text{-}1$$

b) Sort the row in an increasing order

$$\text{-}1\ \text{-}1\ \text{-}1\ \text{-}1\ \text{-}1\ \text{-}1\ 0\ 0\ 0$$

After computing steps c, d and e, the appended row becomes as

$$\text{-}1\ \text{-}1\ \text{-}1\ \text{-}1\ \text{-}1\ \text{-}1\ \text{-}1\ \text{-}1\ 2$$

Similarly, by computing the delay values for all the rows of matrix M_{d1}, the matrix obtained will be as;

$$\begin{bmatrix} -1 & -1 & -1 & -1 & -1 & -1 & -1 & -1 & 2 \\ -1 & -1 & -1 & -1 & -1 & -1 & -1 & -1 & 3 \\ -1 & -1 & -1 & -1 & -1 & -1 & -1 & -1 & 3 \\ -1 & -1 & -1 & -1 & -1 & -1 & -1 & -1 & 3 \\ -1 & -1 & -1 & -1 & -1 & -1 & -1 & -1 & 3 \\ -1 & -1 & -1 & -1 & -1 & -1 & -1 & -1 & 3 \\ -1 & -1 & -1 & -1 & -1 & -1 & -1 & -1 & 3 \end{bmatrix} \quad (3)$$

5 Result and Conclusion

The critical path delay is determined by GLDC method which is based on initial delay matrix. The results obtained by implementing the GLDC method on CFFT for GF (2^m) i.e. for 2^3 and for 2^4 [5] can be seen from fig.(a) and fig(b). In previous works area has being determined for the CFFT, now by GLDC method critical path delay can also be obtained. By this method, it will be easy to find out the CPD for each output variable. If GLDC method is combined with the previous work proposed for optimizing the area of the circuit than area delay product can be obtain.

The critical path delay of each output variable of the circuit can be decomposed by the matrix (3). One can directly determine the CPD obtained for each output by observing the delay values of the matrix from the last column. Therefore the graphs obtained for GF (2^3) and (2^4) is as follows,

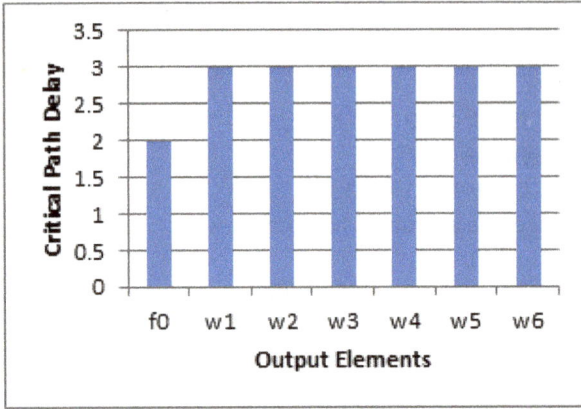

Figure 1. Critical path delay of CFFT for GF (2^3)

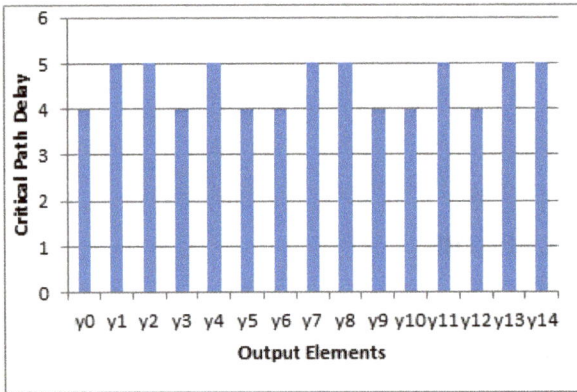

Figure 2. Critical path delay of CFFT for GF (2^4)

Acknowledgment

The authors would like to thank Prof. P.V Trifonov for providing the details about CFFT. They are also very grateful to Prof. Ning Wu for introducing the GLDC method based on initial delay matrix.

References

1   Ning Wu, Xiaoqiang Zhang, Yunfei Ye and Lidong Lan, "Improving Common Subexpression Ellimination Algorithm with a New Gate Level Delay Computing Method", Proceedings of the World Congress on Engineering and Computer Science 2013 Vol II.
2   P.V. Trifonov and S.V.Fedorenko, "A Method for Fast Computation of the fourier transform over a Finite Field", Probl.Inf.Transm.vol.39,no.3,pp.231-238, 2003[Online].Available:http://dcn.infos.ru/~petert/papers/fftEng.pdf
3   N.Chen, and Z.Y.Yan, "High- Performance Design of AES Transformations," IEEE International Symposium on Circuits and System(ISCAS 2009) ,2009,pp.2906-2909.
4   S.V.Fedorenko and P.V.Trifonov, "Finding roots of Polynomial over Finite Field," IEEE Trans. Commun. 2002, vol.50,no.11,pp.1709-1711.
5   Ali Al Ghouwayel, Yves LOUET,Amor NAFKHA and Jacques PALICOT, "On the FPGA Implementation of the Fourier Transform over Finite Field GF(2^m)," SUPELEC-IETR Avenue de le Boulaie CS 4760135576 CESSON-SEVIGNE Cedex, FRANCE-2007.

Liji P I[1] and Bose S[2]

Dynamic Resource Allocation in Next Generation Networks using FARIMA Time Series Model

Abstract: The Next Generation Wireless Networks is a packet based IP network that supports anytime, anywhere service and provides Always Best Connected (ABC) state. Main feature of ubiquitous NGN wireless communication system is seamless mobility. Mobile terminal in the network will be roaming in the vicinity of the heterogeneous wireless network. According to the Received Signal Strength (RSS) the mobile terminal in the network will hand over from one technology to another. As there is frequent handoff from one technology to another, the performance of the mobile terminal will get degraded because enough resource is not available. The stringent Quality of Service (QoS) parameters like delay, delay variance and packet loss will also be affected. It is desired to have a resource allocation scheme which can satisfy the QoS constraints while maximizing the utilization of the network resources with minimizing the packet loss and delay in the network. The mobile terminal in the network carry integrated real time multimedia data (e.g.: Video streaming, video conferencing, IPTV and online gaming) and this type of service should guarantee high Quality of Service (QoS) with minimum delay, packet loss, and jitters. Traffic model can be used as input to analysis resource allocation strategies. Traffic in the network can be modeled in such a way that it should allocate resource efficiently and then reduce end to end delay, packet loss and jitters in the NGN environment to meet the QoS given by the Service Level Agreement (SLA).The high variability data in the network is bursty in nature and conventional traffic modeling like Poisson or Markovian process is inappropriate to model traffic in the networks. The burstiness can be represented as Self similarity with long range dependency. FARIMA a self- similar time series model can represent the higher priority traffic in the network with short range and long range dependency and can predict the future frame from the present and past history of traffic.

1 Research Scholar, Department of Computer Engineering, Anna University, Chennai
lijianil@cet.ac.in
2 Department of Computer Science Engineering. Anna University, Chennai
bosesundan@gmail.com

According to predicted value, resource mainly bandwidth can be allocated on demand.

Keywords: Self-similarity models, FARIMA, QoS, NGN, Heavy tailed distribution

1 Introduction

Next Generation Networks is an IP based infrastructure that supports heterogeneous access technology. The NGN would have a service provider which is equipped with multiple interfaces in the network. The mobile terminal in the network operate in cellular network technology and get handed over to a satellite based network and back to a fixed wireless network, depending upon the network coverage and preference of charging. The service provider in the network is equipped with multiple interfaces (WiMAX, WLAN, GPRS etc) and mobile node in the network has seamless mobility. The mobility of the node in the networks will provide frequent handover from one technology to another and this can degrade performance of the networks due to packet loss, delay and jitters in the packet level. In order to characterize seamless mobility in the network, the resource has to be allocated efficiently .Real time multimedia service in next generation networks should support different traffic characteristics and different Quality of Service (QoS) parameters. For the analysis of traffic the high priority real time video traffic are being considered. This traffic in the network consume huge amount of bandwidth and so raw traffic is first compressed to VBR traffic and then transmitted over the network. Even after applying compression on the raw multimedia traffic [11], these applications involve transmitting huge quantities of compressed VBR traffic with strong correlation and high variability. The aggregated traffic with high variability and strong correlation shows self-similarity property along with long range dependency [1],[2].The Self similar along with LRD traffic in the network can degrade the network performance. So traffic in the network has to be modelled. Conversional traffic models like Poisson or Markovin model [5] are inappropriate for modelling the bursty traffic network and the memoryless model will smooth out the bursty traffic. Self similarity modeling also results in over estimation of performance and insufficient allocation of network resource. Hence a model has to be selected to an appropriate fit in the real time traffic and can improve the performance of the network. The most widely studied second order self-similar processes are the FGN (fractional Gaussian noise) and the FARIMA (Fractional Auto Regressive Moving Average also referred to as ARFIMA) processes. Both models can be used to represent the real time VBR traffic. The real time VBR traffic has both long lag and short lag, but FGN can represent the long term lag in the network [3].The asymptotic self similar model FARIMA which is a stochastic time series model can represent both short term and long term lag along with high variability heavy tailed distribution [3].FARIMA (p,d,q) can model the short range dependency with ARMA(p,q) and long range dependency along with high variability d – 1/α [2], [3].Mobile terminal behaviour in the network can predicted [3] [4],along with mobile node behaviour, the current traffic and past history of data the next N frame in the network can be predicted and the resource can be dynamically allocated .Prediction of traffic reserves bandwidth dynamically and the highly correlated input traffic will get changed to short memory or white noise. This will increase the network utilization and decrease the buffer size. This paper is organized as follows. Section II related works, followed by the mathematical representation of self-similarity, long range dependency and heavy tailed distribution. In Section IV, FARIMA model and modeling of traffic with FARIMA model, traffic prediction then followed by resource allocation Section V finally presents validation results.

2 Related Works

NGN support a wide variety of traffic service, user mobility should guarantee the QoS at anytime anywhere. To guarantee seamless mobility and to maintain the QoS several challenges need to overcome, these include application traffic type, network traffic characterization, network capacity and mobility management. Network traffic characterizations are responsible for maintaining an acceptable quality of service (QoS) level that is deliverable by the network. The congestion control schemes will decide to accept or reject new connections based on their traffic characteristics and available network resources. Performance models require accurate traffic models which can capture the statistical characteristics of actual traffic. If the traffic models do not accurately represent actual traffic, one may overestimate or underestimate network performance [7].

Conventional traffic model like Poisson model will underestimate NGN traffic model. Paxson and Floyd [5] studied how Poisson processes fail as accurate models for WAN packet arrival processes. They found that only user initiated TCP session arrivals such as remote login and file transfer are well modeled as Poisson processes. Real time traffic in the next generation networks is bursty in nature and several studies have already shown that IP traffic may exhibit properties of self-similarity and/or long-range dependence (LRD)

[1],peculiar behaviors that have a significant impact on network performance. Heavy tailed distributions have been widely observed in the high speed network communication networks [3].This has non-negligible impact on the network performance in terms of network throughput, queue stability, and system. Self-similar property on wide range of time scale is different from properties of traditional models based on Poisson, Markov modulated Poisson, and related processes scalability. Garrett and Willinger [10] and Rose [12] showed that models for VBR video traffic using heavy tailed distributions with marginal distribution. The performance of queuing models with self-similar inputs can be dramatically different from the performance predicted by traditional models of tele-traffic based on Markovian processes [3], [8], [7][9].

In [3][17]analysis has been done with storage capacity ,the Gaussian self-similar process shows long range dependency property, impact of Hurst parameter. The fractional Brownian motion has only one parameter, controlling the correlation function, and therefore there is no flexibility in short-range dependence modelling .Major issues in this analysis is that only long range dependency is analysed, but in real scenario the real time traffic exhibit both LRD and SRD [14] [9] showed that FARIMA can model both LRD and SRD.

3 Background of Self-similarity, Long range dependency and Heavy tailed distribution

3.1 Introduction to Self- similarity model

The higher priority VBR traffic in the real time data has significant variance (burstiness) over a wide range of time scale. The burstiness in traffic on many or all timescales can be described statistically by self-similarity. The self similar wide sense stationary process used to capture the fractal behavior of traffic model which is a ubiquitous phenomenon in the networks [1], [2], [3].The aggregated traffic X(t) from multiple sources over wide range of time scales, would maintain its bursty characteristics and this can be analysed as a stochastic time series data in distributional sense. The wide sense stationary stochastic process X(t) has mean μ_x,variance σ^2 and autocorrelation function $\rho_{x(k)}$.The autocorrelation function can be represented as

$$\rho_{x(k)} \approx k^{-\beta} L_1(k), k \to \infty$$

where $0 < \beta < 1$ and L_1 slowly vary to infinite. A continuous time stochastic processes $X(t): 0 < t < \infty$ is self similar if the non overlapping m-aggregated series by summing X over blocks of size m have an infinite dimensional distributions.

$$X(mt) = m^H X(t)$$

where H, Hurst parameter that determine degree of self similarity and value will be ranging $\frac{1}{2} > H > 1$,H→1 the degree of self- similarity increases. If $\rho^m{}_x(k) = \rho_x$, $m \to \infty$, the process is called asymptotically second-order self-similar and for exactly second order self similar processes $\rho^m{}_x(k) = \rho_x$, $k \geq 0$,this implying that the sum of auto correlation diverge. The co-variance function decays hyperbolically Main features of self similarity are 1. Slowly decaying variance 2. the auto correlation is not summable 3.spectral density obey power law in the origin [2].Self-similarity along with long range dependency and the heavy tailed distribution has got significant impact in the queuing analysis of the network.

3.2 Long Range Dependency and Heavy Tailed Distribution

The high speed traffic in the network exhibit a property of correlation over a wide range of time scale and this can be represented as long memory or long range dependency process. Statistical analysis of bursty real time data collected from networks shows the property of self- similarity with long range dependency. Let X_t = 1, 2, ... be a stochastic self similarity process with long range dependency and the autocorrelation function can be represented as $\rho(k) = \frac{Cov(X_t X_{t+k})}{var(X_t)}$.If this function is not summable (e.g., when it decays hyperbolic decay), then it is referred as a long range dependency process with $\sum_{j=0}^{\infty} \rho(k) = \infty$ If autocorrelation function is summable then process is a short range dependent as shown in fig :1.

The main characterize of LRD are: 1. the autocorrelation function not being absolutely summable 2. the spectral density function becoming unbounded as the frequency tends to zero. The aggregated traffic from the multiple sources exhibits a Long Range Dependence and this is a statistical phenomenon observed in time series. Major cause of LRD in network traffic is due to infinite variances and high variability. The high variability non Gaussian bursty traffic can be represented as heavy tailed distributions .The high variability shows non Gaussian bursty traffic possess heavy tailed marginal distributions. The tail of

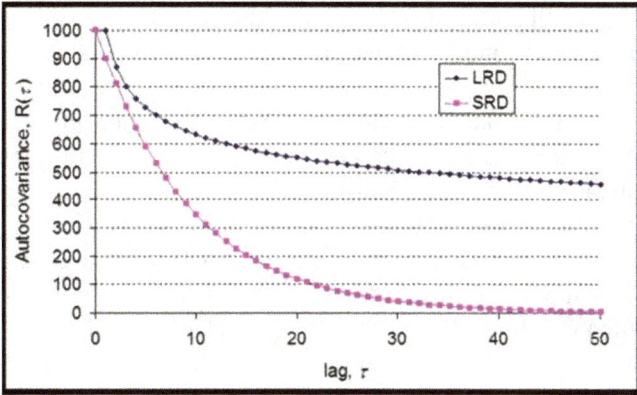

Figure 1. Auto Correlation of LRD and SRD

the distribution decays much more slowly than in the case of an exponential distribution. The Pareto, Weibull and lognormal are some examples of a random variable drawn from a heavy-tail distribution which possess infinite variance or infinite mean depending on the value of the tail parameter α. Heavy tailed distribution shows self- similarity along with long range dependency [3], [4] and the tail of the distribution decays hyperbolically as $P[X > x] \sim x^{-\alpha}$ where X has a distribution with a heavy tail with tail index α and the distribution will be skewed to left as shown in fig:2.

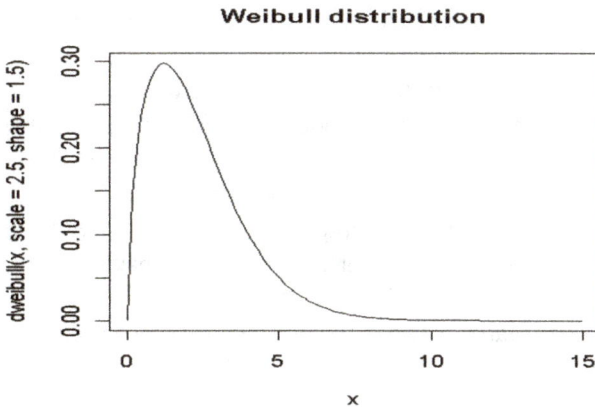

Figure 2. Heavy Tailed Distribution

Garrett and Willinger [10] used a two-hour VBR video, Star Wars, and proposed a hybrid gamma/Pareto model based on the F-ARIMA process. They found that the tail behavior of the marginal distribution can be accurately described using the heavy-tailed Pareto distributions. They also found that the autocorrelation of the VBR video sequence decays more hyperbolically than exponentially and can be modelled using self-similar processes. The main characteristic of a random variable obeying a heavy tailed distribution is that it exhibits extreme variability. The high variability heavy tailed distributed self-similar VBR traffic in the next generation network can be modeled as Fractional Gaussian Noise, FARIMA or M/G/∞ model. The M/G/∞ model can used to model the bursty traffic it is able to model both short range and long range dependence behaviours, inadequacies in modelling the marginal distribution [?],FGN will not able to model the short range lag in the VBR traffic. There for FARIMA is better choose for VBR traffic model [3].FARIMA model can represent both short range dependency and long range along with heavy tailed distribution traffic with parameters (p, q) and d – 1/α [8] [9]. The heavy-tailed behavior of marginal distribution and the sub-exponential decay of autocorrelation function exhibited by video traffic have a significant impact on queuing performance [1], [9], hence real time traffic with high persistence had to be modelled.

4 Traffic Modelling and Prediction using FARIMA Time Series Models

FARIMA an asymptotically second order self similarity is an extension of ARIMA time series model [16], [13].This is a class of long memory that can explicitly account for persistence to incorporate the long term correlation in the data. Compare to other times series model FARIMA models can simulate an autocorrelation with short-range dependencies at small lags as well as long-range dependencies for long lag. The short range dependency defined by (p,q) is an ARMA model and long range dependency represented by d along with high variability heavy tailed distribution parameter α .The traffic is having infinite variance the value of tail index can be calculated as H = d + 1/α and for finite variance H = d +1/2.The aggregated traffic in the network can be considered as stochastic time series processes X_t;t = ..., −1, 0, 1, 2....This series can be described as:

$$\phi(B)\Delta^d X_t = \theta(B)a_t$$

FARIMA(p,d,q) process with where a_t:t =..., −1, 0, 1, 2,.,.,. is a white noise with mean 0 and variance σ^2.Both $\phi(B)$ and θ(B) are polynomials in complex variables with no common zeros and Δ^d fractional differencing operator defined by means of binomial expansion. The degree of differencing d is allowed to take non-integrals values. The high bursty traffic can be modelled by transferring the FARIMA problem to an ARMA problem. Estimate the parameter d [12] using R/S method. The tail index will not have any significance impact, so this can be ignored. For fitting FARIMA time series model following steps has to follows:

Algorithm 1:For fitting FARIMA time series model

1. Pre-processing the measured traffic trace to get zero-mean time series X_t
2. Obtaining an approximate value of d according to the relationship d = H − 0.5 [12]
3. Calculate $Y_t = \Delta^d X_t$
4. Apply the Box-Jenkins algorithm [15]$\phi(B)$ and θ(B) of the ARMA(p,q) model
$$\phi(B)Y_t = \theta(B)a_t$$
5. Model identification: Determining p and q for fitting ARMA models
6. If residual value is a white noise, go to 7 else repeat step5
7. Estimate parameters $\phi(B)$ and θ(B)

From the real time VBR traffic the FARIMA parameter (p, q) and d value had been estimated ARMA(2,1) and d value as 0.36, $\phi_1,$ $\phi_2,$and θ_1 , as shown in table 1.

Table 1. Parameter Estimated for FARIMA

Model	ϕ	θ	d
FARIMA(2,0.36,1)	(0.88,0.01)	0.77	0.36

FARIMA(p,d,q) model to predict the future values of a time series from present and past value. The time series

$$X_t = \sum_{j=0}^{\infty} \Psi_j\, a_{i-j}$$

$$a_t = \sum_{j=0}^{\infty} \pi_j\, X_{i-j}$$

where

$$\sum_{j=0}^{\infty} \Psi_j = \phi^-(B)\theta(B)\Delta^{-d}$$

and

$$\sum_{j=0}^{\infty} \pi_j = \phi(B)\theta^-(B)\Delta^d$$

Linear prediction can be done with FARIMA model. Let $\hat{X}_t(h)$ denote the h-step forecast made at some future time t+h (h is called the lead origin t)

$$X_t = \sum_{j=0}^{\infty} \Psi_j a_{i-j}$$

Linear prediction for FARIMA can be calculated with

$$\hat{X}_t(h) = \sum_{j=0}^{\infty} \pi_j \hat{X}_{i-j}$$

the mean square error of the h-step forecast

$$\sigma^2 = E(X_{i-j} - \hat{X}_{i-j}) = \sigma^2 \sum_{j=0}^{\infty} \Psi_j^2$$

h-step forecast by adding a biase value ξ_μ with $\hat{X}_t$. FARIMA is fitted with (2,0.36,1) and the h-step forecast is calculated as shown in Fig:3.

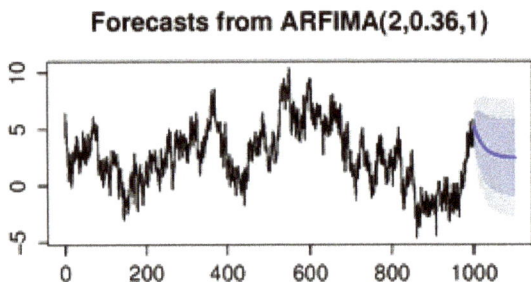

Figure 3. Forecast traffic model with ARFIMA Model

For call admission control will use the upper probability limit to specify the accuracy of traffic prediction, low probability does not contribute any packet loss. Predicting the next value of time series using mean square error by adding ξ_μ

$$\hat{X}_t^\mu = \hat{X}_t + \xi_\mu$$

where

$$P[e_1(h) < \xi_\mu] = \mu, 0.5 > \mu > 1$$

For the traffic model traffic prediction one step unit=0.1second under the upper probability limit 85%.

5 Dynamic Bandwidth Allocation

The IP traffic in the next generation network is bursty in nature. The queuing analysis of long persistence bursty traffic causes large queues, large delays and the network performance will get degraded. This will have a negative impact on the bursty real time VBR video traffic with stringent Quality of Service (QoS) parameters. The active mobile terminal roaming in the integrated heterogeneous network will keep search for high bandwidth. If enough resource is allocated in advance, call will be accepted. Certain admission policy should be ensured for a new connection or handoff connection in order to meet the QoS requirement for the traffic. To meet the QoS constraints in the network, high utilization of resource is needed. If allocation scheme allocates bandwidth equal to the peak rate is not desirable because a significant amount of bandwidth may be wasted due to the bursty nature of the VBR video traffic and mean value allocation may results in traffic congestion in the network. The Call admission control determine whether to accept or reject new call or handoff call and the bandwidth allocation should be neither average nor peak bandwidth allocation. It should have an effect value in between the average and peak value. Effective bandwidth can be considered for call admission control and the bandwidth allocated in the network. The resource in the network can be allocated on demand so as to achieve high utilization with less packet loss, delay and variance. The dynamic bandwidth allocation strategy can be used to allocate bandwidth on demand. In order to reserve network resource in advance, network should use current and past information to predict the future behavior of traffic. Prediction can allocate bandwidth dynamically. If we are able to predict N frame in advance and allocate the resource for the N frame, network performance can be improved. Parsimonious FARIMA can predict future N frame from present and past history of traffic.

5.1 Queue Length Simulation

The actual VBR traffic which is highly correlated and heavy tailed is simulated using a single queue FIFO queue. Queue length simulation for the fitted FARIMA model is given by

$$\log Q = a_1 + a_2\phi_1 + a_2\phi_2 + a_3\theta_1 + a_4d + a_5$$

is a hyperbolic decay function [8] [9].The long range dependence makes the queue longer and the performance of the network degrade. The persistence of long queue (period) give rise to poor quality of service (QoS) parameters. Predicting the future traffic from present and past history the highly correlated input traffic stream is changed from highly correlated input to short memory or white noise. Therefore the buffered value in the queue will be Mean Square Error. If the predicted value is less than the actual value, the difference is buffered. If the predicted value is more, extra bits can be transmitted from the buffer. Since the errors resemble white noise or at most short memory, then smaller buffers, less delays, and higher utilization are expected when compared to traditional fixed rate reservations.

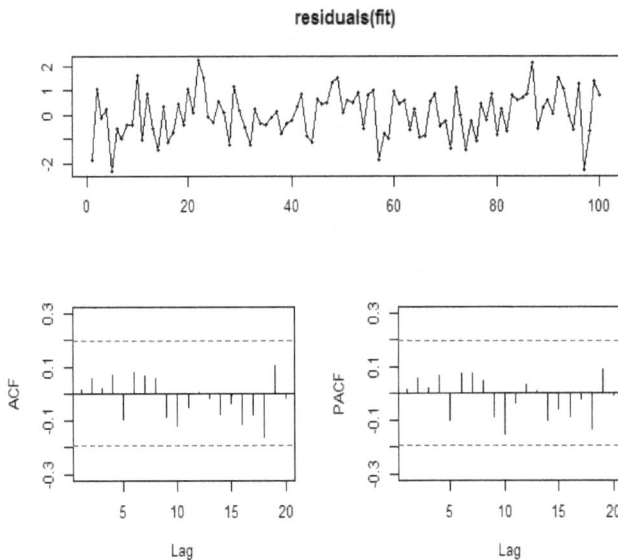

Figure 4. The residue or white noise in fitted model

The residue of fitted model of buffered value shows that this is white noise or short memory and the Hurst value is less than 0.5 as shown in Fig:4.The residuals are uncorrelated and often assumed normally distributed N(0,1).

6 Prediction Based Resource Allocation

Prediction based resource allocation measure the maximum bandwidth. When accepting a new call or handoff call in

link, it will not over subscribe the link bandwidth [2].Therefore, the diverse QoS requirements in each traffic class can be fulfilled. In this algorithm threshold values that are assigned in the bandwidth allocation policy make the admission/rejection decision for an incoming connection request. Denies the sum if the flow request $\rho v < v_s + r_\mu$ and current usage would exceed the target utilization level where ρ link bandwidth and v link utilization target v_s of the transmitted packets over a period S. The predict the rate using FARIMA model over the previous period T Real time VBR video often has stringent QoS requirements on delay and loss, dynamic bandwidth allocation will allocate bandwidth and the buffer size need to be limited.

Algorithm 2 : Prediction Based Resource Allocation

Input:$n_0, n_1, ..., n_{n+k}$,

$n_k, \leftarrow n_{k+1}$,

Measure v_s over a period S

Prediction $\hat{X}_t$ over a period of T

For a = 0 to k-1 do

If traffic trace a is not for the best effort traffic, then

If $\rho v < v_s + r_\mu$, then

Reject request

$n_k, \leftarrow n_{k+1}$,

Endif

Endif

EndFor

Accept request

Real time VBR video often has stringent QoS requirements on delay and loss, dynamic bandwidth allocation will allocate bandwidth and the buffer size needs to be limited.

Conclusion

Real time VBR bursty network traffic is complex in next generation, as it exhibits strong dependence and self-similarity, models of time series such as Poisson and Markovin processes are not appropriate for its modeling. Maintaining high utilization of the bandwidth is the objective for efficient traffic management, which include CAC, policing, scheduling, buffer management, and congestion control etc. The high variable and highly correlated VBR traffic in the network can be modeled with self similarity models like FARIMA or FGN. Mostly VBR traffic in the network exhibits both short range and long range dependency along with heavy tailed distribution can be modeled with FARIMA time series model with parameter (p, d, q). FARIMA time series traffic model can predict traffic and allocate bandwidth dynamically. This model is flexible enough to parsimoniously capture the statistical property of traffic can allocate bandwidth on demand and mathematically analysis the queuing performance. As future work the QoS parameter like packet loss, delay and variances can be analyzed in the network.

References

1   Mark E.Crovella, and Azer Bestavros,"Self-Similarity in World Wide Web Traffic: Evidence and Possible Causes"IEEE/ACM Transactions on Networking, Vol.5, No.6,PP. Dec. 1997.
2   S.Jamin, et al.,"A Measurement-Based Admission Control Algorithm for Integrated Service Packet Networks,"IEEE/ACM Transactions on Networking,Vol.5,no.1,pp.56-70, Feb.1997.
3   Xiaolong Jin,Geyong Min, Speros Velentzas, and Jianmin Jiang "Quality of-Service Analysis of Queuing Systems with Long-Range-Dependent Network Traffic and Variable Service Capacity", IEEE Transactions on Wireless Communication, Vol. 11, No.2 pp.562-570,2012
4   W. Leland, W. E. Taqqu, et al. On the self-similar nature of Ethernet traffic (extended version),"IEEE/ACM Transactions on Networking, Vol.2, No.1, 1994. pp. 1–15,1994.
5   V. Paxson, S. Floyd,"Wide-Area Traffic:The Failure of Poisson Modeling," IEEE/ACM Transactions on Networking, vol. 3, no. 3, pp. 226–244, 1995
6   M. M. Krunz and A. M. Makowski, "Modeling video traffic using M/G/∞ input processes: A compromise between Markovian and LRD models," IEEE Journal on Selected Areas in Communications, vol. 16, no. 5, pp. 733–748, June 1998.
7   Abdelnaser Adas "Traffic Models in Broadband Networks" IEEE Communications Magazine pp 82-89 July 1997
8   A.Adas."Supporting real time VBR video using dynamic reservation based on linear prediction.", Proc. IEEE INFOCOM'96, pp. 1476-1483, 1996
9   A.Adas and A.Mukherjee,"On Resource Management and QOS Guarantees for Long-Range Dependent Traffic" Proceedings of INFOCOM95,pp. 779-787, Boston 1995
10 M. W. Garrett and W. Willinger, "Analysis, Modeling and Generation of self-similar VBR video traffic", in Proceedings *SIGCOMM 94, pp. 269-280, 1994.*
11 K.Park, G. Kim, and M. Crovella,"On the relationship between file sizes, transport protocols, and self-similar network traffic,"in Proc. IEEE International Conference on Network Protocols , pp. 171-180, October 1996.

12 Richard G.Clegg"A Practical Guide To Measuring The Hurst Parameter "Dept. Of Mathematics, University of York, YO10 5DD

13 Yantai Shu,Zhigang Jin, Lianfang Zhang,Lei Wang"Traffic Prediction Using FARIMA Models" IEEE ICC'99, Vancouver, Canada, June 6-10, 1999, S22-6

14 William Su, Mario Gerla,"Bandwidth Allocation Strategies For Wireless ATM Networks Using Predictive Reservation" Proc. IEEE Global Telecommunications Conference, vol.4 pp. 2245 – 2250, 1998.

15 Yantai Shu,Huifang Feng,Hua Wang and Maode Ma "FARIMA Model Based Admission Control to Support QoS Service in the Networks with WiFi Access" 2nd International Conference on Mobile Technology, Applications and Systems, pp. - 6 ,2005

16 Box, G.E.P.,Jenkins,G.M.and Reinsel, G.C.,". Time series analysis: Forecasting and control,"Pearson Education,Delhi."(1994)

17 Park K., Willinger W., "Self-similar Traffic and Performance Evaluation, John Wiley & Sons,Inc."IEEE(2000).

Ms Shanti Swamy[1], Dr.S.M.Asutkar[2] and Dr.G.M.Asutkar[3]

Classification of Mimetite Spectral Signatures using Orthogonal Subspace Projection with Complex Wavelet Filter Bank based Dimensionality Reduction

Abstract: The high dimensional hyperspectral images contains a lot of redundant Information and the signal information is usually concentrated on lower dimensional subspaces .After the dimension reduction based on complex wavelet filter bank approach, the mimetite material spectral signatures are classified using the orthogonal subspace projection method. The correlation of first twenty spectral signatures of sample is found out and with the fixed threshold, the redundant information is ignored .With this combination of dimension reduction and classification, the results would be quite better. The undesired spectral signatures are efficiently rectified with the help of OSP. The three different materials are present in an HIS,mimetite,scorodite and coloamanite materials .First the mixed pixels are studied for dimension reduction and then orthogonal subspace projection is applied on the reduced one.

Keywords: dimension reduction OSP classifier, matlab implementation

1 Introduction

Classification of a hyperspectral image data amounts to identifying which pixels contain various spectrally distinct materials that have been specified by the user.For example, the roofs of some buidings and the roads can be made by the same material (alsphalt). Therefore, contextual information, geometrical features for example, is necessary for classification task.Images are classified as supervised and unsupervised classification. The supervised classification is based on reference samples to identify unknown pixels in terms of classes.The unsupervised classification is based on identifying the groups.

1 Asso Prof,KITS,Ramtek
2 Asso Prof,MIET,Gondia
3 Principal,PIET,Nagpur

2 Related work

Classification approaches such as minimum distance to the mean (MDM) and Gaussian maximum likelihood (GML) [14] can be applied to this data as well as correlation/matched filter-type approaches such as spectral signature matching [16] and spectral angle mapper [17]. However,these techniques have difficulty accounting for mixed pixels (pixels that contain multiple spectral classes). A possible solution is to use the concept of orthogonal subspace projection (OSP) [15], [16]. This can be done by finding the desired spectral signatures from undesired one.

3 Motivation

Remote sensing technology is concerned with the determination of characteristics of physical objects through the analysis of measurement taken at a distance from these objects. One important problem in remote sensing is the characterization and classification of (spectral) measurements taken from various situations on the earth surface. As a result, hyperspectral data permit the expansion of detection and classification activities to targets previously unresolved in multispectral images.

4 Problem domain

With the increased use of hyperspectral image data comes the need for more efficient tools that can dimensionally reduce, classify, and reveal important information content of the data. As previously mentioned, classification of a hyperspectral image data amounts to identifying which pixels contain various spectrally distinct materials that have been specified by the user.[17]

The idea of the orthogonal subspace projection classifier is to eliminate all unwanted or undesired spectral signatures (background) within a pixel, then use a matched filter to extract the desired spectral signature (end member) present in that pixel. The methods already existing would be studied to understand the concept of dimensionality with classification and matching of selected features of hyperspectral images. Some of the algorithms are considered here.

5 Problem definition and statement

Hyperspectral imagery provides richer information about materials than multispectral imagery. The new larger data volumes from hyperspectral sensors present a challenge for traditional processing techniques. Principal component analysis (PCA) has been the technique of choice for dimension reduction. Spectral data reduction and estimation of matched hyperspectral imagery using complex wavelet filter bank with perfect reconstruction can be considered for better results.

Performance of wavelet reduction can be better for larger dimensions. This property is due to the very nature of wavelet compression, where significant features of the signal might be lost when the signal is under sampled.

The principle of the proposed method is to apply a signal matched complex wavelet filter bank concept to hyperspectral data in the spectral domain and at each pixel location. This does not only reduce the data volume, but it also can preserve the characteristics of the spectral signatures. The advantage of complex wavelet filter bank would be considered in terms of directionality and shift invariant nature which would give better results in terms of proper reduction and matching after the classification done.

6 Innovative content

6.1 Orthogonal subspace projection

1) Available spectral signatures from ISRO ,RRSSC,Nagpur.
2) dimension reduction with the help of wavelet complex filter bank.
3) mimetite,scorodite and colomanite material samples are considered.
4) Other two materials , scorodite and colomanite are undesired materials.
5) Matlab code is written for the classifier operator based on orthogonal subspace projection[1]
6) The mathematical equations are as under:-
7) An operator P which eliminates the effects of U, the undesired signatures. an operator that projects r onto a subspace that is orthogonal to the columns of U.(undesired)

8)   3 vectors or classes, each 3 elements or bands long. The vectors are in re-
      flectance units and can be seen below. 8)twenty samples of three different
      materials are taken for the dimension reduction.
9)   The idea is to nullify the undesired signatures with the help of orthogonal
      subspace operator P .

6.2 Mathematical equations

The projection operator becomes

$$\mathbf{P} = (\mathbf{I} - \mathbf{U}\mathbf{U}^{\dagger})$$

[1]

where I is the identity matrix.
With P, it does a good job at minimizing the effects of U.

PU=null   [2]

Where U∓ is the pseudo inverse of U,denoted by

[3]

$$\mathbf{U}^{\dagger} = (\mathbf{U}^{T}\mathbf{U})^{-1}\mathbf{U}^{T}$$

PU=[0 0]   [4]

6.3 About algorithm

1)   Dimension reduction is done with the help of wavelet filter bank. It is tested
      for twenty samples and based on correlation algorithm, reduced to only
      three samples.
2)   The desired material is retained with the help of OSP and above classifica-
      tion steps.
3)   The OSP P is found out using the above equation(1) with the help of U∓.
4)   Finally ,PU is found out which is to be null matrix.
5)   In this HSI,spectral signatures are considered for mimetite scorodite and co-
      lomanite and following are the results and conclusions.

Results and conclusion

$$U = \begin{bmatrix} 7.3400 & 73.5900 & . \\ 7.5000 & 73.6500 & . \\ 7.4300 & 73.8300 & . \end{bmatrix} \quad U^{\mp} = \begin{bmatrix} -6.1398 & 6.7606 & 0.6243 & . \\ 0.6230 & -0.6765 & 0.0674 & . \\ . & & . & . \end{bmatrix}$$

$$UU^{T} = \begin{bmatrix} 0.7823 & -0.1627 & 0.3793 \\ -0.1627 & 0.8784 & 0.2834 \\ 0.3793 & 0.2834 & 0.3393 \end{bmatrix} P = \begin{bmatrix} 0.2177 & 0.1627 & -0.3793 \\ 0.1627 & 0.1216 & -0.2834 \\ -0.3793 & -0.2834 & 0.6607 \end{bmatrix}$$

PU=1.0e-010 *

$$\begin{bmatrix} -0.0725 & -0.7411 \\ -0.0747 & -0.7178 \\ -0.0730 & -0.7267 \end{bmatrix}$$

PU approximately zero which reduces the effect of undesired signatures.The desired signature is thus classified.

Justification of results

The following results indicate that the undesired signatures or materials are removed from the same HSI as per the classifier .The classification can be extended to the high dimension data. As per the algorithm, the matrix PU is approximately null.

References

1   "shantiswamy,SMAsutkar,GMAsutkar", Dimension reduction techniques in hyperspectral imagery: a signal processing perspective at International conference Chennai ,15-17 Dec, 2014
2   "shantiswamy,SMAsutkar,GMAsutkar",Complex wavelet filter bank algorithm for Hyperspectral mimetite spectral signature –correlation compared at JIT,Nagpur,International conference May 19-20,2015.
3   R. H. Yuhas, A. F. H. Goetz, and J. W. Boardman, "Discriminationn among semi-arid landscape endmembers using the spectral angle mapper (SAM) algorithm," *Summaries 3rd Annu. JPL Airborne Geosci. Workshop,* June 1992, R. O. Green Ed., Publ. 92-14, vol. 1, Jet Propulsion Laboratory, Pasadena, CA1992, pp. 147-149.
4   M. O. Smith, P. E. Johnson, and J. B. Adams, "Quantitative determination of mineral types and abundances from reflectance spectra using principal components analysis," in *Proc. 15th Lunar and Planetary Sei. Conf., Part 2, Geophys. Res.,* vol. 90, suppl.,pp. C797-C804, Feb. 15, 1985.
5   R. Nishii, S. Kusanobu and S. Tanaka, "Enhancement of Low Spatial Resolution Image Based on High ResolutionBands,"*IEEE Trans. on Geoscience and Remote Sensing,*
6   K. Rasche, R. Geist, and J. Westall, "Detail preserving reproduction of color images for monochromats and dichromats," IEEE Comput. Graph. Appl., vol. 25, no. 3, pp.22–30, May-June 2005.

7   M. Ichikawa, K. Tanaka, S. Kondo, K. Hiroshima, K. Ichikawa , S. Tanabe, and K. Fukami, "Preliminary study on color modification for still images to realize barrier free color vision," in Proc. 2004 IEEE Int. Conf. Syst., Man, Cybern.,vol. 1, Oct. 2004, vol.34, no.5, pp.1151-1158, 1996.

8   S. Haykin, Ed., *Advances in Spectrum Analysis and Array Processing, Volume II.*

9   *Adaptive Filter Theory*, 2nd ed. Englewood Cliffs, NJ: Prentice- Hall, 1991.

10   O. L. Frost III, "An algorithm for linearly constrainedadaptive array processing," *Proc. IEEE* 60, 926–935 ,1972

11   B. D. Van Veen and K. M. Buckley, "Beamforming: aversatile approach to spatial filtering," *IEEE ASSP Mag.* , 4–24 Apr. 1998.

12   C.-I. Chang and H. Ren, "Linearly constrained minimum variancebeam forming for target detection and classification in hyperspectral imagery," in *IEEE Int. Geoscience and Remote Sensing Symp. '99*,pp. 1241–1243, Hamburg, Germany ,1999

13   H. Ren, "A comparative study of mixed pixel classification versus pure pixel classification for multi/hyperspectral imagery," M S Thesis,Department of Computer Science and Electrical Engineering, Universityof Maryland Baltimore County, Baltimore ,May 1998

14   S. Haykin, *Adaptive Filter Theory*, 3rd. ed., Prentice-Hall, Englewood Cliffs, NJ ,1996.

15   A tuturial on Hyperspectral Image Classification Using Orthogonal Subspace Projections: Image Simulation and Noise Analysis Emmett Ientilucci, Center for Imaging Science ,April 23, 2001

16   Tu, T.-M, Chen, C.-H, Chang, C.-I, "A posteriori least squares orthogonal subspace projection approach to desired signature extraction and detection", IEEE Transactions on geoscience and remote sensing, vol. 35, pp. 127-139, January 1997.

17   Harsanyi, J.C, Chang, C.-I, "Hyperspectral image classification and dimensionality reduction: An orthogonal subspace projection approach", IEEE Transactions on geoscience and remote sensing, vol. 32, pp. 779-785, July 1994.

Sharmila Kumari M[1], Swathi Salian[2] and Sunil Kumar B. L[3]

An Illumination Invariant Face Recognition Approach based on Fourier Spectrum

Abstract: This paper describes the representation of human faces using the amplitude information of the 2-D Fourier spectrum in order to overcome the problem of illumination variation. Illumination variation is one of the major challenges in face recognition system. In this paper, we propose the technique which extracts amplitude information of the 2-D Fast Fourier Transform (FFT) of the facial image and then applies Principal Component Analysis (PCA) on it. The results of FFT-PCA are compared with the existing PCA, Discrete Cosine Transform (DCT) and FFT based face recognition techniques. Through extensive experimentation, we conclude that, with the reduced dimension of the face images, the combined FFT-PCA approach has improved performance. The experiments are carried out on the standard face datasets such as ORL and UMIST and comparative analysis is presented along with conventional PCA and frequency domain based approaches.

Keywords: Pattern Recognition, Image Processing, Fourier transform, Principal Component Analysis, Fourier coefficients, Eigen face, Face recognition

1 Introduction

Biometric based pattern recognition systems are widely used in security area. These systems consider the physiological/ behavioral characteristics of the human such as face, fingerprint, iris, hand geometry, retina etc for identification purpose. Among the various biometrics, the face based recognition systems has been successfully accepted in various applications such as person authentication, access control to buildings etc. In such systems, an individual is recognized by matching the templates of all the users in the database [1]. However

1 Department of Computer Science and Engineering, P A College of Engineering, Nadupadavu, Mangalore, sharmilabp@gmail.com
2 Department of Information Science and Engineering, Srinivas institute of technology, Valachil, Mangalore, salian.swathi@gmail.com
3 Department of Information Science and Engineering, Srinivas institute of technology, Valachil, Mangalore, blsuny@gmail.com

there are several challenging issues to address in face recognition. An efficient system must deal with changes in expression, pose, illumination etc.

The face recognition system involves several steps: i) face image acquisition: uses a face dataset consisting of face images of various persons with varying pose, illumination etc. ii) preprocessing: involves resizing, converting into grey scale form etc. iii) feature selection: significant features of the face image are extracted either in spatial domain itself or in the transformed domain as a part of face representation. iv) Template matching: extracted features of the test image are compared with all the images of the face dataset using different distance measures [2].

It is noted from the literature that the Eigenspace based approaches is the most successful option available for face representation. In [3], Kirby and Sirovich first showed that PCA can be effectively used to represent images of human faces by means of reduced dimensionality using Eigenspace based approach. In [4], it is shown that PCA is also an efficient face recognition technique, where it used Euclidean distance as a similarity measure. Over the years, there are many improvements brought in the conventional PCA algorithm in order to improve its performance. In modular PCA [5], the face images are divided into number of blocks or sub-images and the PCA is applied onto these blocks individually, thus handling the large variations in facial features. In two dimensional PCA [6], the recognition rate is improved by eliminating the need of converting 2D face image matrices into 1D vector. The 2D PCA uses more coefficients for feature representation. In order to overcome this, bidirectional PCA [7] and $2D^2$ PCA [8] were proposed.

There are some face recognition techniques which works in transformed domain of the face images rather than directly processing the gray scale images (spatial domain). In DCT based face recognition approach [9], the 2D DCT function is applied onto the face images and then the most significant DCT coefficients are extracted as a part of feature representation. The robustness of the technique is improved by employing certain normalization methods which best handles the problem of illumination variations [11]. In [12], it is proved that the higher reductions in the dimension of the face images are achieved along with improvements in the recognition rate by combining PCA and DCT together. The Fast Fourier transform (FFT) [13] on the other hand, transforms the signal or image to the frequency domain which is the best way to represent an image. It is a faster form of discrete Fourier transform (DFT), developed by Cooley and Tukey around 1965. The Fourier representation of a signal involves complex numbers, i.e., the amplitude and phase parts. Based on the phase information of the Fourier spectrum a new concept for face representation is used, to overcome the effect of illumination [14]. The phase of the 2D FFT of face images preserves the location of edge information. In [15], [16] effort is made to combine the amplitude information of one image with the phase of another to produce a mixed image and thus improving the performance.

Our work focus on preprocessing of face images by using FFT, which also improves the performance of feature extraction of PCA technique. In this approach, named FFT-PCA, we extract the amplitude information of 2D Fourier spectrum of the face image and then PCA is applied on this extracted information (FFT coefficients). The results obtained from this technique will be compared with the other face recognition approaches such as PCA, DCT and FFT based techniques. The experiments are conducted with varying dimension of features on a variety of face databases such as AT&T, UMIST, which include pose, illumination and occlusion problems. The effect of varying training samples is addressed in our study. In the rest of the paper, we discuss various face recognition approaches, present our approach, experimental results are given and analyze experimental results with other face recognition techniques.

2 Face recognition using PCA and DCT based techniques

The PCA method of Turk and Pentland [4] has been popularly used for feature extraction and face recognition purposes. It treat the face images as 2-D data, and classifies the face images by projecting them to the eigenspace which is composed of eigenvectors obtained by the variance of the face images. This variance is obtained by getting the eigenvectors of the covariance matrix of all the images. In the standard eigenface procedure suggested by Turk and Pentland, Euclidean distance is used for the classification of test images.

Suppose $p_1, p_2, p_3,....,p_M$ be M training samples each of size m x n. Each 2-D image matrix p_i (i=1,2,...,M) is represented as1-D image vector of size N x 1 (N= m*n). The images are then mean centered by subtracting the average of all the images from each image vector.

$$\bar{p}_i = p_i - avg \,, \text{ where avg}= \frac{1}{M}\sum_{i=1}^{M} p_i \tag{1}$$

These vectors are then concatenated side by side to form data matrix $\bar{X}$ of size N x M (M is the number of images).

The data matrix and its transpose are multiplied to obtain the covariance matrix C as:

$$C= \bar{X}\,\bar{X}^{\,T} \tag{2}$$

The Eigen vectors of the covariance matrix are computed and k Eigen vectors corresponding to the k largest Eigen values are ordered from high to low to form the eigenspace V. Each of the centered training image vectors ($\bar{p}_i$) is then projected into the Eigen space V.

$$\tilde{p}_i = V^T \bar{p}_i \qquad (3)$$

In order to recognize a test image, it is mean centered by subtracting the average image from the test image vector. The mean centered test image vector is then projected into the obtained Eigen space.

$$\bar{t}_i = t_i - m \text{ , where } m = \frac{1}{M}\sum_{i=1}^{M} p_i \qquad (4)$$

And

$$\tilde{t}_i = V^T \bar{t}_i \qquad (5)$$

The comparison is done between the projected test image and each of the training images using similarity measure. The training image found to be closest to the test image is used to classify the test image. The most well known similarity measure is Euclidean distance.

The second technique considered in the comparative study is DCT. It is an invertible linear transform that can express a finite sequence of data points in terms of a sum of cosine functions oscillating at different frequencies. The original face image is converted to frequency domain by applying 2-D DCT. The original image can be restored back from the DCT coefficients by using invert 2-D DCT [9, 10].

In DCT based face recognition, important frequency components of the training image, obtained by DCT function is extracted. The feature vectors for the training images contain this extracted, more relevant information. In the classification stage, the feature vectors obtained for the training images and test image are compared using a similarity measure, mostly Euclidean distance. A training image with minimum distance is used to classify the test image.

Given an r by c image $f(r, c)$, 2-D r by c DCT is defined as follows:

$$F(u, v) = \alpha(u)\alpha(v) \sum_{i=0}^{r-1} \sum_{j=0}^{c-1} f(i,j) \cos\left[\frac{u(2i+1)\pi}{2r}\right] \cos\left[\frac{v(2j+1)\pi}{2c}\right] \qquad (6)$$

Where, $\alpha(u) = \sqrt{\frac{1}{r}}$ for u=0,

$\quad = \sqrt{\frac{2}{r}}$ for u=1, 2, 3...r-1

And $\alpha(v) \quad = \sqrt{\frac{1}{c}}$ for v=0,

$$= \sqrt{\frac{2}{c}} \text{ for } v=1, 2\ldots c\text{-}1$$

3 FFT and FFT-PCA based approach

The Fourier transform based face recognition model discussed above is another approach considered in this paper for comparative analysis. Because we are dealing with images, it is appropriate to mention the 2D DFT function which is given in (7). In this technique, the 2D DFT function is applied to spatial domain image to obtain the frequency domain coefficients. These coefficients which comprise the information such as amplitude and phase angle [15], are used for representing the face images.

Let the dimensions of an image be m and n. There are two frequencies u and v corresponding to the two coordinates x and y. If f(x, y) is the grey value at location (x, y) then the two dimensional Discrete Fourier Transform values f'(u,v) are given by

$$f'(u, v) = \sum_{x=0}^{m-1} \sum_{y=0}^{n-1} f(x, y) e^{2\pi j \left(\frac{xu}{m} + \frac{yv}{n} \right)} \quad j = \sqrt{-1} \tag{7}$$

$$= f'_{real}(u, v) + f'_{imag}(u, v)$$

$$= |f'(u, v)| e^{\theta(u,v)}$$

$$\text{Where } |f'(u, v)| = \sqrt{\left(f'_{real}(u, v) \right)^2 + (f'_{imag}(u, v))^2} \tag{8}$$

and

$$\theta(u, v) = \arctan\left[\frac{f'_{imag}(u,v)}{f'_{real}(u,v)}\right]$$

are the amplitude and phase of Fourier transform respectively.

In the proposed method, we first apply FFT on the training faces, which contain the amplitude and phase information as mentioned above. The main purpose of our work is to address the problem of changes in illumination in face recognition. Here we made an attempt to use only amplitude information of FT coefficients of the faces. Next step is to apply PCA on this amplitude information of face images, as a part of feature extraction, in order to improve the recognition rate. By using PCA, we are able to obtain features of faces of reduced size. Our work combines the FFT and PCA approaches, hence able to extract the illumination invariant features of the faces. In the similar manner, the illumination invariant features of test image is extracted and compared with all other face images of the dataset. The Euclidean distance is used as similarity measure. The experiments are carried out on the different approaches discussed in the paper using the standard datasets ORL and UMIST. The results obtained using

FFT-PCA technique is analyzed by comparing with PCA, DCT and FFT based approaches.

4 Experimental results

In this section, we present experimental results obtained by the FFT-PCA. The experiments are performed on standard ORL and UMIST face databases. We also evaluate the performance by conducting the same set of experiments on the other face recognition systems mentioned, i.e. PCA, DCT and FFT based approaches and provide a comparative analysis. In all our experiments, we used the Euclidian distance as the similarity measure for classification.

4.1 Experiments on ORL database

The ORL face database consists of gray-scale images of 40 individuals each with 10 samples. They represent some variation in facial expressions, facial details, scale and also limited rotation. All images are cropped to size of 112 x 92 pixels. Fig. 1 shows the subset of one such subject of the ORL database.

Figure 1. Ten images of one person in ORL

The experiments are carried out by varying the number of training samples and testing samples under each subject. We have chosen five different samples of each person for training such as (1) alternate samples (1, 3, 5, 7, 9 and 2, 4, 6, 8, 10), (2) last 5 samples, (3) initial 5 samples and remaining as testing. Similarly, experiments are conducted with seven random samples (i.e. total 280 training samples) and three (i.e. 120 training samples) random samples of each individual for training and the remaining samples for testing. In all the cases, the recognition accuracy is measured for differing dimensions such as 5,10,15,20 and 25. The results obtained due to the FFT-PCA are shown in Fig. 2.

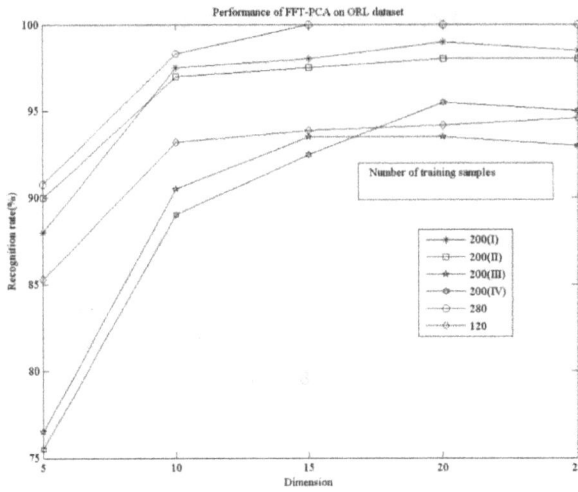

Figure 2. Recognition accuracy of FFT-PCA for ORL dataset

4.2 Experimentation on UMIST face dataset

The UMIST face dataset consists of 564 images of 20 people with large pose variations. In our tests, we considered a partial set of face images consisting of 15 images each of 20 different individuals from the UMIST face database. The Fig. 3 shows 15 such samples of a single person in UMIST database.

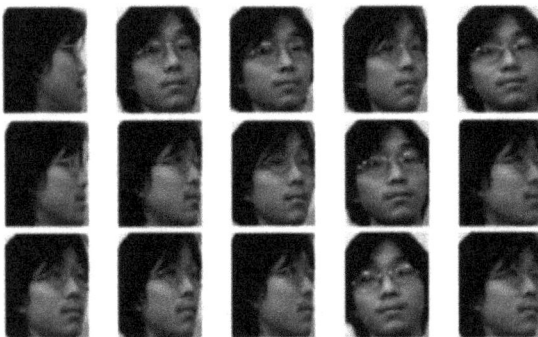

Figure 3. Samples of a person in UMIST

As in the case of ORL, We have conducted the experiments by considering alternate samples (even and odd), continuous samples (last 8 and first 7 samples) and random samples (11 and 4 samples of each person) for training, separately and remaining samples for testing. In all above experiments, we recorded the recognition rate for FFT-PCA for different dimensions. The experimental results are shown in Fig.4.

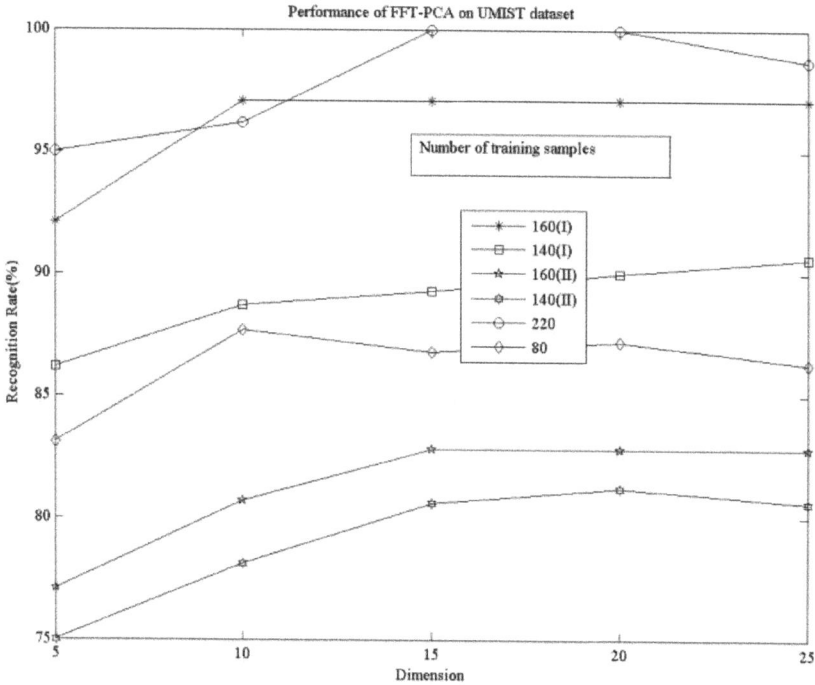

Performance of FFT-PCA on UMIST dataset

Number of training samples

* ── 160(I)
* ── 140(I)
* ── 160(II)
* ── 140(II)
* ── 220
* ── 80

Figure 4. Recognition accuracy of FFT-PCA for UMIST dataset

For the purpose of providing a comparative study, we have performed the same set of experiments mentioned above on PCA and DCT models, using different face datasets such as ORL, UMIST datasets. The recognition accuracy of all experiments for these models is recorded.

The recognition performances of PCA and DCT techniques with varying dimensions of feature vectors for ORL dataset are plotted in Fig. 5, Fig. 6 respectively and for UMIST dataset are shown in Fig. 7, Fig. 8 respectively.

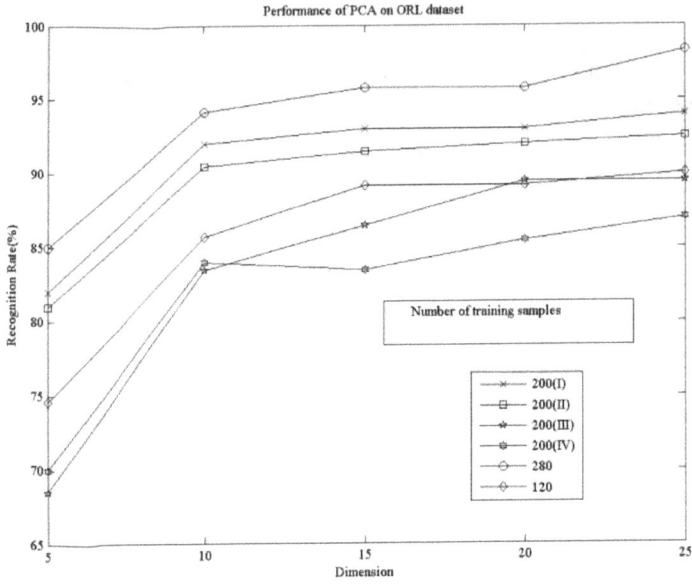

Figure 5. Recognition accuracy of PCA for ORL dataset

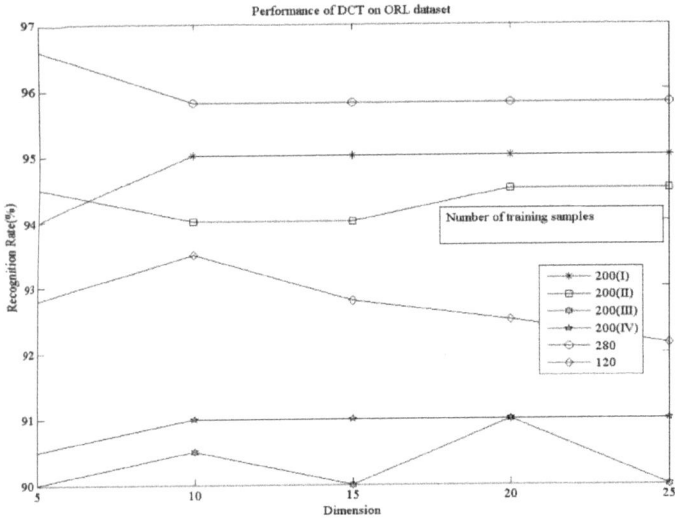

Figure 6. Recognition accuracy of DCT for ORL dataset

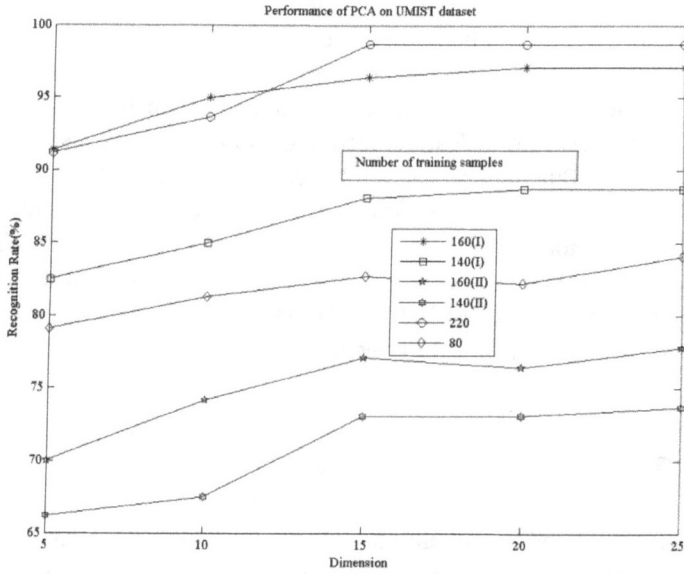

Figure 7. Recognition accuracy of PCA for UMIST dataset

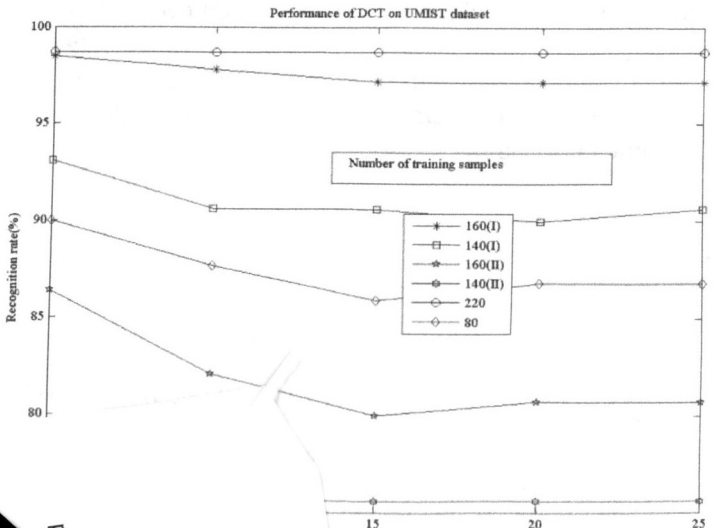

6. J. Yang, D. Zhang, A.F. Frangi, J.Y. Yang, "Two-dimensional PCA: a new approach to appearance based face representation and recognition", IEEE Trans. Pattern Anal. Mach. Intell, 26 (1), 2004, pp. 131–137.
7. W Zuo, D Zhang, K Wang, "Bidirectional PCA with assembled matrix distance metric for image recognition", IEEE Transaction on Systems, Man, And Cybernetics-Part B, 36(4), 2006, pp. 862–872.
8. Daoqiang Zhang, Zhi-Hua Zhou, "(2D² PCA): Two-directional two-dimensional PCA for efficient face representation and recognition", Neurocomputing, 69(1), 2005, pp. 224–231.
9. Anand, Najan and Mrs. A. C. Phadke, "DCT Based Face Recognition", International Journal of Engineering Innovation & Research, 1(5), 2012, pp. 415-418.
10. Shermina, J, "Illumination invariant face recognition using discrete cosine transform and principal component analysis.", Emerging Trends in Electrical and Computer Technology (ICETECT), 2011 International Conference on. IEEE, 2011, pp. 826-830.
11. Ziad m. Hafed and Martin D. Levine, "Face recognition using the discrete cosine transform", International Journal of Computer Vision 43(3), 2001, pp. 167–188.
12. Dandpat, Swarup Ku, and Sukadev Meher. "New Technique for DCT-PCA Based Face Recognition." (2011).
13. Samra, Ahmed Shabaan, Salah Gad El Taweel Gad Allah, and Rehab Mahmoud Ibrahim, "Face recognition using wavelet transform, fast fourier transform and discrete cosine transform.", Circuits and Systems, 2003 IEEE 46th Midwest Symposium on. Vol. 1. IEEE, 2003, pp. 272-275.
14. Sao, Anil Kumar, and B. Yegnanarayana. "On the use of phase of the Fourier transform for face recognition under variations in illumination." Signal, image and video processing, 4(3), 2010, pp. 353-358.
15. Dehai Zhang, Da Ding, Jin Li, Qing Liu, "A Novel Way to Improve Facial Expression Recognition by Applying Fast Fourier Transform", Proceedings of the International MultiConference of Engineers and Computer Scientists, Vol 1, 2014.
16. Dehai, Zhang, et al. "A PCA-based Face Recognition Method by Applying Fast Fourier Transform in Pre-processing.", 3rd International Conference on Multimedia Technology (ICMT-13), Atlantis Press, 2013.

Arlene Davidson R[1] and S. Ushakumari[2]
Optimal Load Frequency Controller for a Deregulated Reheat Thermal Power System

Abstract: Load Frequency Control is a very important issue in the operation and control of a power system. Major changes have been introduced into the structure of the electric power industry with deregulation. Hence LFC in deregulated power systems assumes importance. This paper deals with the modelling, design and simulation of optimal integral controller for load frequency control in an interconnected two area deregulated power system consisting of reheat thermal power plants. Genetic algorithm optimization technique is used to tune integral gain constants of integral controller subject to minimization of a suitable quadratic index. Performance assessment of the system is done using Integral Absolute Error (IAE) and Integral Square Error (ISE) performance indices.

Keywords: Deregulation, Load frequency control, Genetic algorithm, Bilateral contract, Performance index.

1 Introduction

The electric power industry has been witnessing a tremendous change in the way of operation ever since mid-eighties globally. The vertically integrated structure which was prevalent in electric power industry and performed the functions of generation, transmission and distribution was mostly owned by federal or state governments. With deregulation of electric power industry, this structure is transformed to a horizontally integrated one in which the functions of generation, transmission and distribution are peformed by distinct entities named as Generation Companies (GENCOs), Transmission Companies (TRANSCOs) and Distribution Companies (DISCOs). The motive behind the transformation is the need for inducing better efficiency in production and delivery of

1 Research Scholar, Dept. of Electrical Engineering, College of Engineering Trivandrum, Kerala, India
arlenerosaline1@gmail.com
2 Professor, Dept. of Electrical Engineering, College of Engineering Trivandrum, Kerala, India
ushalal2002@gmail.com

electrical energy, improving service standards and bringing about competition into the industry with the introduction of market structure.

Load Frequency Control (LFC) is one of the significant problems in electric power system design and operation. Any deviation in frequency with sudden load perturbation can directly influence power system operation and system reliability. A large frequency deviation can cause an unstable condition for a power system. The primary objectives of load frequency control include maintaining system frequency at nominal value and minimizing unscheduled tie-line power flows between neighbouring control areas [1]. With restructuring of electric power industry, attention has been given to LFC of deregulated power systems. The operational structures resulting from deregulation is explained in [2]. A ramp following controller for a deregulated system is given in [3]. Donde et al. [4] have taken into account the effect of bilateral contracts in modelling the system and simulation done considering bilateral contracts and contract violation. In [5], LFC synthesis problem is formulated as a mixed H_2/H_∞ static output feedback control problem to obtain a desired PI controller. A decentralized Radial Basis Function Neural Network (RBFNN) controller for LFC in a deregulated power system is proposed in [6]. Genetic algorithm is used for optimization of integral gains and bias factors in AGC for a three area power system after deregulation in [7]. Genetic Algorithm optimization techniques are employed for tuning PID controller gains in [8] for a four area deregulated power system. Optimal output feedback control and reduced order observer are made use of in [9] for a simplified model of the deregulated system. Two degree of freedom Internal Model Control (IMC) method is used to tune decentralized PID type load frequency controllers in [10] for a deregulated environment. A load following controller in deregulated scenario has been designed in [11]. Structured singular value method is used for robustness analysis in [12]. Fractional Order PID controller is applied to AGC of multi area thermal system with reheat turbines under deregulated environment in [13]. Optimal output feedback controller is used in [14] for LFC in deregulated environment for multi-source combination of hydro, reheat thermal and gas generating units in each control area. [15] discusses the use of Thyristor Controlled Series Compensator in improving the dynamic response of deregulated power systems. Optimal load frequency controller for a two area deregulated non-reheat thermal power system is detailed by the authors in [16]. This paper deals with the modelling, design and simulation of integral controller for a two area deregulated power system. The power system considered consists of two GENCOs and two DISCOs in each control area. All the GENCOs are of reheat thermal type. A suitable quadratic performance index is minimized using Genetic Algorithm (GA) optimization so that minimum overshoot and settling time is ensured. Finally, performance of the system is assessed using IAE and ISE performance indices.

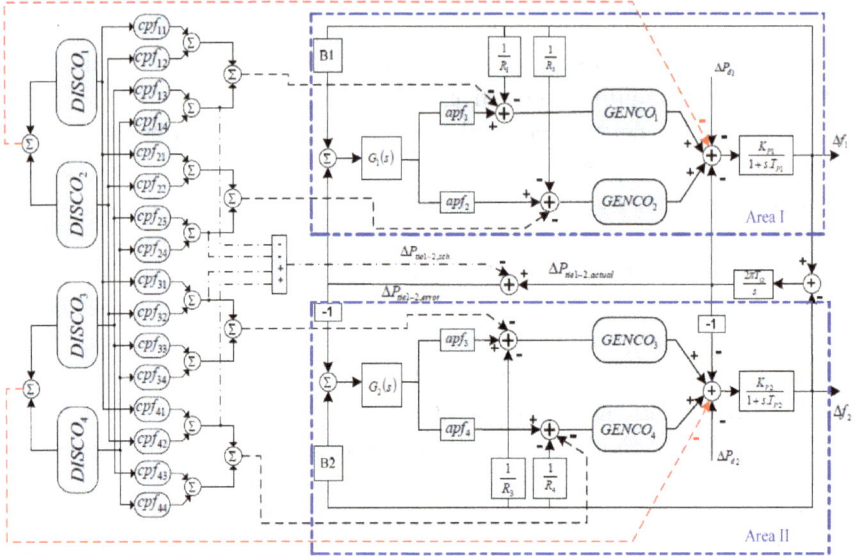

Fig 1. Schematic block diagram of a two area deregulated power system

2 LFC in Deregulated Power System

Generation Companies are responsible for operating and maintaining generating plants and in most cases are the owners of the plant. Transmission Companies own and maintain the transmission lines but the operation is done by the Independent Service Operator (ISO). Distribution Companies maintain the distribution network and provide facility for electricity delivery and retailers may be present who sell electricity directly to consumers. The ISO is entrusted with the responsibility of ensuring the reliability and security of the whole system. There are three basic ways in which system operation can take place in a deregulated power system. (1) Poolco which is a governmental or quasi-governmental agency which buys power for all customers on the basis of lowest cost bids (2) Bilateral Exchange in which multi-sellers and multi-buyers make a deal to exchange power under mutually agreed conditions (3) Power exchange which acts as a forum to match electrical energy supply and demand based on bid prices.

The time horizon of this market based system ranges from half an hour to a week or longer. The usual pattern found in deregulated markets is a day ahead market system to facilitate energy trading one day ahead of each operating day.

Fig.1. shows the deregulated system considered for LFC. There are two GENCOs (GENCOs 1 and 2 in Area 1 and GENCOs 3 and 4 in Area 2) and two DIS-COs (DISCOs 1 and 2 in Area 1 and DISCOs 3 and 4 in Area 2) in the deregulated power system considered. All GENCOs comprise reheat thermal systems. In order to meet the Poolco-based and bilateral transactions, a Disco Participation Matrix (DPM) is used [3]. Each element of DPM is called contract participation factor cpf_{kl} which expresses the contract of power between kth GENCO and lth DISCO and is computed as the fraction of the total load power contracted by DISCO 'l' from GENCO 'k'. For a two area system with two GENCOs (GENCO 1, GENCO 2) and two DISCOs (DISCO 1, DISCO 2) in area 1 and two GENCOs (GEN-CO 3, GENCO 4) and two DISCOs (DISCO 3, DISCO 4) in area 2, the DPM is given by

$$
\begin{bmatrix}
cpf_{11} & cpf_{12} & cpf_{13} & cpf_{14} \\
cpf_{21} & cpf_{22} & cpf_{23} & cpf_{24} \\
cpf_{31} & cpf_{32} & cpf_{33} & cpf_{34} \\
cpf_{41} & cpf_{42} & cpf_{43} & cpf_{44}
\end{bmatrix}
\tag{1}
$$

Sum of all entries in a column of DPM is unity.

ie., $\sum_{k=1}^{Ng} cpf_{kl} = 1$; for $l = 1,2,.....N_d$ $\tag{2}$

The generation of each GENCO must track the contracted demands of DISCOs in steady state. The expression for power contracted by DISCO 'l' from GENCO 'k' is given by

$$\Delta P_{gc,k} = \sum_{l=1}^{Nd} cpf_{kl}\Delta P_{L,l}; \text{ for } k = 1,2,.....Ng \tag{3}$$

where $\Delta P_{gc,k}$ is the contracted power of kth GENCO and $\Delta P_{L,l}$ is the total load demand of lth DISCO.

The scheduled steady state power flow on the tie-line, $\Delta Ptie_{1-2,sch}$ is expressed as the difference of total power exported from GENCOs in control area 1 to DISCOs in control area 2 and total power imported by DISCOs in control area 1 from GENCOs in control area 2 and is given below.

$$\Delta Ptie_{1-2,sch} = \sum_{k=1}^{2}\sum_{l=3}^{4} cpf_{kl} \Delta P_{L,l} - \sum_{k=3}^{4}\sum_{l=1}^{2} cpf_{kl} \Delta P_{L,l} \tag{4}$$

The tie-line power error, $\Delta Ptie_{1-2}$ is defined as

$$\Delta Ptie_{1-2,error} = \Delta Ptie_{1-2,actual} - \Delta Ptie_{1-2,sch} \tag{5}$$

At steady state, tie-line power error, $\Delta Ptie_{1-2,error}$ vanishes as the actual tie-line power flow reaches the scheduled power flow. This error signal is used to generate the respective Area Control Error (ACE) signal as in the conventional power system.

$$ACE_1 = B_1\Delta f_1 + \Delta Ptie_{1-2,error} \tag{6}$$
$$ACE_2 = B_2\Delta f_2 + a_{12}\Delta Ptie_{1-2,error} \tag{7}$$

Where $a_{12} = -\frac{Pr1}{Pr2}$ where P_{r1}, P_{r2} are the rated area capacities of area 1 and area 2 respectively.

The total load of the kth control area ($\Delta Pd, k$) is expressed as the sum of the contracted and uncontracted load demand of the DISCOs of the kth control area.

$$\Delta P_{dk} = \sum_{k=1}^{Nd} \Delta P_{L,k} + \Delta P_{UL,k} \tag{8}$$

where $\Delta P_{L,k}$ is the contracted load demand of the kth DISCO and $\Delta P_{UL,k}$ represents the uncontracted load demands of DISCOs in k^{th} area.

2.1 Design of controller

Integral controller for each area is designed as per the equation

$$u_i(t) = - K_i \int ACE_i dt \tag{9}$$

where i is the area and the integral gains K_1, K_2 are tuned optimally to obtain the area frequencies and tie-line power exchange with minimum overshoot and minimum settling time.

A quadratic performance index was selected as

$$J = \int_0^t (\Delta f_1^2 + \Delta f_1^2 + \Delta P_{tie1-2,error}^2) dt \tag{10}$$

which is minimized for 10% load demand on each DISCO to get optimum values of K_1 and K_2 using genetic algorithm [17].

3 Simulation Results and Discussion

For the two-area deregulated reheat thermal power system having two GENCOs and two DISCOs in each area whose parameters are mentioned in Appendix A, three contract cases were considered and time domain simulations done using Matlab.

3.1 Case 1 – Unilateral Contract (Poolco based transaction)

In unilateral contract, DISCOs in an area have contract of power with GENCOs of the same area only. Let DISCO 1 and DISCO 2 have power contract with GENCO 1 and GENCO 2 in area 1 as per the DPM given.

$$DPM = \begin{bmatrix} 0.6 & 0.4 & 0 & 0 \\ 0.4 & 0.6 & 0 & 0 \\ 0 & 0 & 0 & 0 \\ 0 & 0 & 0 & 0 \end{bmatrix} \qquad (11)$$

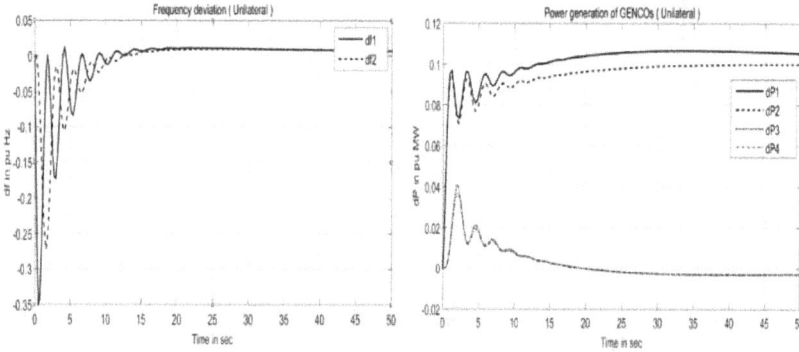

Fig 2. (a) Frequency deviation plot (Unilateral case); Fig. 2. (b) Change in power generation of GENCOs (Unilateral case)

The corresponding frequency deviation plots for the two areas are given in Fig. 2. (a). The computed values of power output of GENCO 1 and GENCO2 are 0.1 pu MW each while that of GENCO 3 and GENCO 4 are zero as per (3). The power outputs of the different GENCOs are plotted in Fig. 2. (b).

3.2 Case 2 – Bilateral Contract

In bilateral contract, a DISCO can have contract of power with GENCOs in any area. The bilateral transactions between the GENCOs and DISCOs are simulated based on the DPM given as follows.

$$DPM = \begin{bmatrix} 0.5 & 0.25 & 0 & 0.3 \\ 0.2 & 0.25 & 0 & 0 \\ 0 & 0.25 & 1 & 0.7 \\ 0.3 & 0.25 & 0 & 0 \end{bmatrix} \qquad (12)$$

The frequency deviation plots are shown in Fig. 3. (a) for the two areas. As per (3), power output of GENCO 1 at steady state is 0.105 pu MW while that of GEN-CO 2 is 0.045 pu MW, GENCO 3 is 0.195 pu MW, GENCO 4 is 0.055 pu MW. The power outputs of the different GENCOs are plotted in Fig. 3. (b).

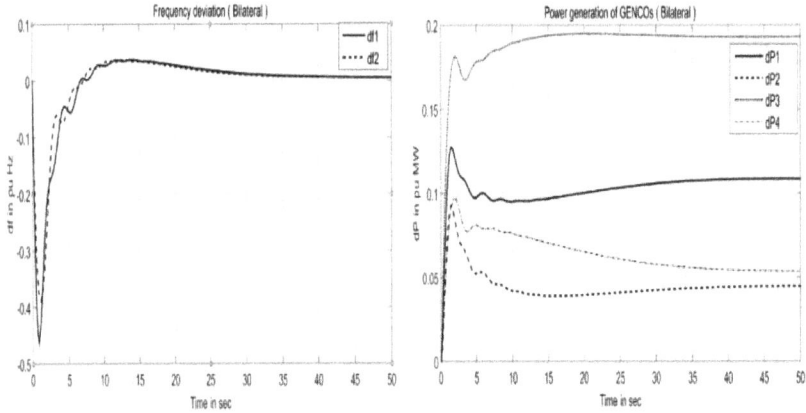

Fig. 3. (a) Frequency Deviation plot (Bilateral case) ; Fig. 3. (b) Change in power generation of GENCOs (Bilateral case)

3.3 Case 3 – Contract violation

In some situations, a DISCO may violate a contract by demanding excess power. This uncontracted power must be supplied by the GENCOs in the same control area. This kind of demand is reflected as a local load of the control area. Consider that DISCO 1 demands 0.1 pu MW excess power. DPM is taken the same as in Case 2. The uncontracted load of DISCO 1 is reflected in the generations of GENCOs 1 and 2 in proportion to their ACE Participation factors (*apf*s). The related frequency deviation response graphs are shown in fig. 4.(a). The power output of GENCOs are plotted in Fig. 4. (b). It can be seen that the sum of the generated power of the GENCOs adds up to the total load demand of the DISCOs.

The optimized integral constants for the three contract cases of operation and the respective cost functions are found in Table 1.

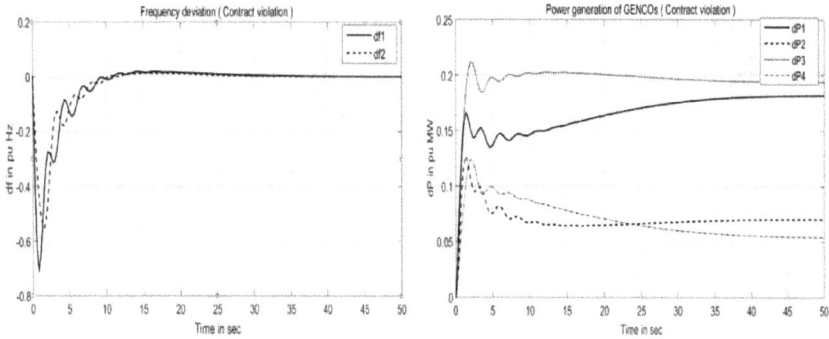

Fig. 4. (a) Frequency Deviation plot (Contract violation case) ; Fig. 4. (b) Change in power generation of GENCOs (Contract violation)

Table 1. Optimized gain parameters for three contract cases of operation

Unilateral contract		Bilateral contract		Contract violation	
J=0.00007417		J=0.0048119		J=0.0048119	
K$_1$	K$_2$	K$_1$	K$_2$	K$_1$	K$_2$
0.0135	1		1	0.0802	0.1312	0.1961

3.4 Computation of Performance Indices

Two performance indices are computed for LFC problem in this deregulated system. They are,

1   IAE (Integral Absolute Error), $J = \int_0^\infty |e(t)| dt$

2   ISE (Integral Square Error), $\;\; J = \int_0^\infty |e(t)|^2 dt$

     It is noteworthy that each performance index specifies different aspects of system response. Large errors contribute more in ISE than IAE. TABLE 2 gives the computed values of performance indices for three contract cases of operation of a deregulated power system. It is desirable that the performance indices computed with the controller in the system give minimum values so that good performance is ensured. In this problem, the error values taken correspond to Area Control Error (ACE) values for the respective areas.

     From TABLE 2, it is observed that for Area 1, IAE shows minimum for bilateral case and ISE shows a much lesser value for all three cases with the minimum value corresponding to bilateral case. Hence this controller tuned for bilateral case is satisfactory for Area 1. For Area 2, IAE shows minimum value

Table 2. Computed values of performance indices for three contract cases

		Unilateral	Bilateral	Contract violation
Performance Index		GA optimized Integral controller	GA optimized Integral controller	GA optimized Integral controller
Area 1	IAE	1.264	0.5207	0.9041
	ISE	0.0901	0.0496	0.1528
Area 2	IAE	0.2947	1.299	1.376
	ISE	0.0143	0.1037	0.1265

under unilateral case and ISE gives minimum value under unilateral case followed by bilateral case and contract violation case. Since all three values are small for ISE, it can be concluded that for Area 2, this controller is suitable.

Conclusions

In this work, load frequency control is applied to a two area reheat thermal power system under deregulated environment. The integral gain constants have been tuned using genetic algorithm optimization procedure considering a quadratic performance index. Extensive analysis is done for studying load frequency control in deregulated environment considering unilateral, bilateral and contract violation cases. It is seen that in all the cases, the frequency deviation becomes zero at steady state conditions. Also, the simulated values of generated power output of GENCOs agree with the calculated values, at steady state.

Computation of performance indices shows that the controller is best considering ISE index. For area 1, the controller gives the best results under bilateral conditions considering IAE and ISE indices and for Area 2, the controller gives the best results under unilateral conditions considering both IAE and ISE indices. For Area 2, the ISE indices are better compared with IAE indices. Considering both the indices, it can be concluded that genetic algorithm optimized integral controller is suitable for load frequency control of two area reheat thermal deregulated power system. Further work in this system could be done by introducing uncertainties in system parameters and a robust controller could be designed so as to accommodate the uncertainties in the system.

Appendix A

(i) System Data

$K_{p1} = K_{p2} = 127.5$ Hz/pu MW

$T_{p1} = 25$ s ; $T_{p2} = 31.25$ s

$R_1 = 3$ Hz/pu MW ; $R_2 = 3.125$ Hz/pu MW ;

$R_3 = 3.125$ Hz/pu MW ; $R_4 = 3.375$ Hz/pu MW

$B_1 = 0.532$; $B_2 = 0.495$

$K_{r1} = K_{r2} = K_{r3} = K_{r4} = 0.5$

$T_{r1} = T_{r2} = T_{r3} = T_{r4} = 10$ s

$T_{g1} = 0.075$ s ; $T_{g2} = 0.1$ s ; $T_{g3} = 0.075$ s ;

$T_{g4} = 0.0875$ s

$T_{t1} = 0.4$ s ; $T_{t2} = 0.375$ s ; $T_{t3} = 0.375$ s ;

$T_{t4} = 0.4$ s

$T_{12} = 0.07$

$apf_1 = 0.75$; $apf_2 = 0.25$; $apf_3 = apf_4 = 0.5$

(ii) Genetic Algorithm Parameters

Population size = 100

Crossover = 0.8

No. of Generations = 200

Elite count = 2

Mutation Rate = 0.2

Penalty factor = 100

References

1 P. Kundur, "Power System Stability and Control," McGraw Hill Inc, 1994 Edition.
2 Richard D Christie and Anjan Bose, "Load Frequency control Issues in Power System Operations after Deregulation," IEEE Transactions on Power systems, vol. 11, no. 3, August 1996.
3 Bjorn H Bakken and Ove S Grande, "Automatic Generation Control in a Deregulated Power System," IEEE Tr. on Power Systems, vol. 13, no. 4, November 1998.
4 Vaibhav Donde, M A Pai and Ian A Hiskens, "Simulation and Optimization in AGC Systems after deregulation," IEEE Transactions on Power Systems, vol. 16, no. 3, August 2001.
5 Bevrani H, Mitani Y and Tsuji K, "Bilateral based robust load frequency control," Energy Conversion and Management 2004; 57:2297-2312.
6 Shayeghi H, Shayanfar H A and O P Malik, "Robust decentralized neural networks based LFC in a deregulated power system," Electric Power systems Research 77 (2007) 241-251.

7   Demiroren A and Zeynelgil H L, "GA application to optimization of AGC in three-area power system after deregulation," Electrical Power and Energy Systems, 29 (2007) 230-240.

8   Praghnesh Bhatt, Ranjit Roy and S P Ghoshal, "Optimized multi area AGC simulation in restructured Power systems," Electrical Power and Energy Systems 32 (2010) 311-322.

9   Elyas Rakhshani and Javad Sadeh, "Practical viewpoints on load frequency control problem in a deregulated power system," Energy Conversion and Management, 51 (2010) 1148-1156.

10 Wen Tan and Hongxia Zhang, Mei Yu, "Decentralized load frequency control in deregulated environments," Electrical Power and Energy Systems 41 (2012) 16-26.

11 Rajesh Joseph Abraham, D Das and Amit Patra, "Load following in a bilateral market with local Controllers," Electrical Power and Energy   Systems 33 (2011) 1648 – 1657.

12 Wen Tan and Hong Zhou, "Robust analysis of decentralised load frequency control for multi-area power systems," Electrical Power and Energy Systems 43 (2012) 996-1005.

13 Sanjoy Debbarma, Lalit Chandra Saikia and Nidul Sinha, "AGC of a multi-area thermal system under deregulated environment using a non-integer controller," Electric Power Systems Research 95 (2013) 175-183.

14 K P Singh Parmar, S Majhi and D P Kothari, "LFC of an interconnected power system with multi-source power generation in deregulated power environment," Electrical Power and Energy Systems 57 (2014) 277-286.

15 M Deepak and Rajesh Joseph Abraham, "Load following in a deregulated power system with Thyristor Controlled Series Compensator," Electrical Power and Energy Systems 65 (2015) 136–145.

16 Arlene Davidson R, S. Ushakumari, "Optimal Load Frequency Controller for a Deregulated Non-Reheat Thermal Power System," Proceedings of the IEEE International Conference on Control Communication and Computing ICCC 2015.

17 Mo Jamshidi, Renato A Krohling, Leandro dos Santos Coelho and Peter J Fleming, "Robust Control Systems with Genetic Algorithms," Taylor and Francis CRC Press, 2003 Edition.

Chandana B R[1] and A M Khan[2]

Design and Implementation of a Heuristic Approximation Algorithm for Multicast Routing in Optical Networks

Abstract: Efficient and reliable network transport technologies are the key features in optical network. With the introduction of new multicast based applications and high bandwidth intensive applications, there has been a possibility of tuning the optical properties of the signals so that it becomes more easy to use heuristic algorithm to address the problem of physical impairment constrained routing. Several methods have been developed that do not provide a unified solution for routing and wavelength assignment with physical impairment constraint awareness. Here we present a heuristic algorithm with new power allocation scheme for reliable multicasting in optical networks.

Keywords: multicasting, optical networks, routing, wavelength assignment

1 Introduction

In the past few years traffic data carried by network has increased significantly due to rapid growth in number of internet users and multicast based network applications. The network providers find it difficult to satisfy the traffic demands by merely increasing the network capacity. The type of demand varies from unicast to multicast that includes applications like digital media broadcasting, digital media distribution and streaming.

Optical networks employing wavelength division multiplexing (WDM) offer the promise of meeting the high bandwidth requirements of emerging communication applications, by dividing the huge transmission bandwidth of an optical fiber (50 terabits per second) into multiple communication channels with bandwidths (10 gigabits per second) compatible with the electronic processing speeds of the end users.

1 Research Scholar, Dept. of Electronics, Mangalore University
chandanaramya1@gmail.com
2 Professor and chairman, Dept. of Electronics, Mangalore University
asifabc@yahoo.com

There has been great interest in WDM networks consisting of wavelength routing nodes interconnected by optical fibers. Such networks carry data between access stations in the optical domain without any intermediate optical to/from electronic conversion. To be able to send data from one access node to another, one needs to establish a connection in the optical layer similar to the one in a circuit-switched network. This can be realized by determining a path in the network between the two nodes and allocating a free wavelength on all of the links on the path. Such an all-optical path is commonly referred to as a *light path* and may span multiple fiber links without any intermediate electronic processing, while using one WDM channel per link. The entire bandwidth on the light path is reserved for this connection until it is terminated, at which time the associated wavelengths become available on all the links along the route [3].

In the absence of wavelength conversion, it is required that the light path occupy the same wavelength on all fiber links it uses. This requirement is referred to as the *wavelength continuity constraint*. However, this may result in the inefficient utilization of WDM channels. Alternatively, the routing nodes may have limited or full conversion capability, whereby it is possible to convert an input wavelength to a subset of the available wavelengths in the network [1].

The paper is organized as follows: In the section II multicasting in optical networks has been highlighted. In the section III all the previous problem investigation has been done. In the section IV problem formalization has been discussed. Section V describes the relation between maximum allowable distance and signal power. ILP formulation and branching limitations details are found in Section VI and section VII. Results and conclusions are presented in section VIII.

2 Multicasting in Optical Networks

Multicasting is defined as the ability to transmit information from a source to multiple receivers. It is also referred as point-to multipoint, multipoint, multi destination or one-to-many communication. The set of nodes to which the multicast packet is directed is known as the *multicast group*, g, with a cardinality given by $|g|$ that ranges from 1 to N-1, where N is the total number of nodes in the network.

Multipoint communication could be achieved using unicast or broadcast service. Employing unicast service involves multiple transmissions of the same packet to all members of g which results in wasting the network resources. Using broadcast service is another extreme. Although the source node transmits a single copy of the packet, this broadcasting operation will flood the network and may also cause wastage in the entire bandwidth, especially if the duration of the multicast session is long and the group membership size is relatively small with respect to the total network size. Between these two extremes, multicasting techniques should be developed to satisfy near optimal usage of the various network resources, such as sources, link bandwidth and receivers.

The multicast IP protocols had been a good area of development in the last few years where many efficient algorithms were developed to be reliable protocols. From this, and taking into consideration the need of compatibility between IP and non IP networks, the best mechanism to deploy multicasting principles in optical domains is to adapt the same protocols on optical networks. Since optical cross connects and ordinary routers have a lot of differences in their packet processing, some modifications must be done to deploy the same protocols. Also, the quick evolution of "all optical switches" [4], along with their efficient performance in doing switching at the physical layer, demonstrates that their use will reduce the time loss in the processing of packets from the optical layer to the network layer. Multicasting over optical networks is simply a layer 1 switching and splitting mechanism that need to be developed into algorithms and standardized into protocols [6].

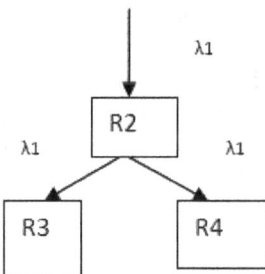

Figure 1. Multicast tree

3 Related Work

Peter Soproni et.al investigated and evaluated the performance of multicast routing in grooming capable multi-layer optical wavelength division multiplexing (WDM) networks. New wavelength – graph models were proposed for network equipments capable of optical layer branching of light-paths. The author has evaluated the cost-effectiveness of electronic and optical-layer multicast and unicast as well. The high scalability of multicast is shown. All routing and technical constraints are formulated in ILP (integer linear programming) formulation and verified in network simulator.

Asuman E Ozdaglar et.al formulated several novel optimization problem that offer the promise of several radical improvements in optical networks. It addressed highly efficient linear programming methods and yield optimal RWA policies. An efficient algorithmic approach has been developed for optimal routing and wavelength assignment for optical networks that can be extended to networks with sparse wavelength conversion.

Vasilis A et.al proposed a wavelength routing algorithm that accounts for optical signal to noise ratio degradation and other non-linear effects two establishment algorithm were developed which can be either shortest path or shortest widest path.

Pinar Kirci et.al investigated on multicast routing protocols. A new multicasting protocol structure was introduced .it was also proved that best performance is gained with multicasting on JET protocol with least packet drop rates.

Peter Soproni et.al investigated the problem of survivable all optical routing in WDM networks with physical impairments. It was assumed that each demand requires full capacity of a wavelength. It also evaluated the propositions with regard to shared and dedicated protection scenarios.

Antonakopoulas et al showed a logarithmic approximation for cost minimization with a goal to aggregate low bandwidth traffic streams to effectively use high bandwidth media. The work also discovered the special case of half-wavelength demands having rich combinatorial propagation closely related to graph problem.

4 Problem Formalization

A two-layer network is assumed where the upper, electronic layer is time switching capable, while the lower, optical layer is a wavelength switching capable one. The two layers are assumed to be interconnected, i.e. the control plane has information on both layers and both layers take part in accommodating demands. Wavelength Graph (*WG*) model is used for routing in these networks as shown in Figure 2.

Network topology, physical constraints and the number of wavelengths are assumed to be given as well as the description of demands, capacity of wavelength channels and lengths of the links. Static traffic consisting   of multicast demands is assumed. The objective is to reach all destination nodes for all demands, while observing all routing and technical constraints and minimizing power usage.

A sub-graph of a versatile physical device is depicted in Fig. 2. The equipment is a combination of an OXC with WL-conversion and an OADM: it can originate and terminate traffic demands, as well as perform space-switching. WL-conversion is possible only through the electronic layer. This is illustrated by an electronic node in the sub-graph, while other (pair of) logical nodes correspond to interfaces. Fig. 2 assumes two fibers connected to the device, and two WLs per fiber, which results in two input and two output interfaces – because all edges are directed. This complex node is used in the simulations.

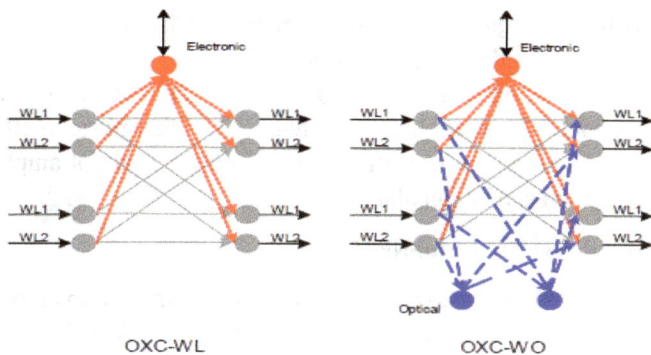

Figure 2. Sub-graph of OXC-WL device in wavelength graph model

The OXC-WO (OXC with WL-conversion and Optical splitting) devices should be considered as an extension of the OXC-WL type. It introduces a new functionality: Grooming-Enhanced Multicast in optical networks with optical splitting of light-paths. The branching function is represented by dashed edges in the sub-graph. By adding these specific logical edges into the sub-graph, we can accurately determine the cost of optical branching. The in-degree of splitting nodes is always less than or equal to one, while the out-degree is more than or equal to two. Branching in this device is, however, possible in the electronic layer as well. To accomplish electronic branching, the demand must be first routed up to the electronic layer. The returned (branched) light paths should not necessarily use the same WLs. Thus WL-conversion can also be performed in addition to branching and 3R processing of the signal [7].

5 New Power Allocation Scheme

PICR injects different channel powers into the same optical fiber according to the distance the ligthpath has to take. Most of the new problems caused by this new power allocation scheme are mentioned and solved in [4].To investigate the relation between signal power and maximum allowed distance we consider a noise limited system where other physical effects can be taken into account as power-penalties. It is possible to prove by analytical calculations that there is a linear relationship between the channel power and the maximum allowed distance of a light path [5]: $L = Lc * P_{mw}$, where P_{mw} is the input power in mw, L is the maximum allowable distance, and Lc is the linear factor between them. The effect of a node to the signal quality is equal to a given length of fiber. Thus, it is modeled with an additional edge length: LPhyNode.

To investigate the relation between the signal power and the maximum allowed distance, we consider a noise limited system where other physical effects can be taken into account as power-penalty: Considering a chain of amplifier the OSNR of the end point can be calculated as follows:

$$OSNR = 58 + Pin - \Gamma (db) - NF - 10 \log N - M \text{---------EQ} \tag{1}$$

where noise figure (NF) is the same for every amplifier and the span loss (Γ (dB)) is the same for every span. Pin is the input power in dBm, M is the margin for other physical effects, and N is the number of spans.

$$Qdb = OSNR (db) + 10 \log(B_0 / B_e) \text{-----EQ} \tag{2}$$

where Bo is the optical bandwidth and Be is the electronic bandwidth of the receiver. The logarithmic Q dB and the linear Q have the following relation:

$$Q = 20 \log Q \text{ dB } \text{----------------------- EQ} \tag{3}$$

Substituting equation (1) and (2) into (3) we obtain the linear relation between the maximum allowable distance and the signal power.

It has been proved analytically that there is linear relationship between channel power and maximum allowable distance of a light path.

L= Lc * Pmw; where PmW is the input power in mW, L is the maximum allowable distance, and Lc is the linear factor between them.

Lc= 1/10 [20log q-10log (BO/Be)-58+Γdb+NF+M/10]

6 ILP Formulation

We use the following ILP formulations to route multiple multicast trees in the network.

$$\sum_{JI}^{OR} Z - \sum_{IK}^{OR} Z = \begin{cases} -1 & \text{if } i = S^r \\ 0 & \text{if } i \varepsilon \ \{S^r, t^{or}\} \text{ --- EQ} \\ +1 & \text{if } i = t^{or} \end{cases}$$ (4)

$Z_{ij}{}^{or}$ ε {0, 1} : indicates whether the sub demand 0 of multicast tree uses edge (i,j) or not .x

$X_{ij}{}^r$ ε {0,1} : indicates whether r uses edges(i,j)

y_{ij} ε {0,1} : indicates if edge (i,j) is used by any of the demands.

V_i^+ : denotes the set of nodes

V_i^- : set of nodes reachable from i by one directed edge.

A, V, V$_e$, 0, R denote set of edges, nodes , electronic nodes, sub-demands and the set of multicast trees respectively.

The source of tree is denoted by S^r while targets are denoted by t^o . 0 is the corresponding sub-demand.

$Z_{ij} <= X^r_{ij}$ v (i,j) ε A ------------------EQ   (5)

X_{ij} ε {0, 1} v (i,j) ε A , V_r ε R -------EQ   (6)

Y_{ij} ε {0, 1} v (i,j) ε A ---------------EQ   (7)

Objective function: Minimize$\sum_{v(i,j) \varepsilon A} C_{ij} Y_{ij}$ - ------------------- EQ   (8)

EQ (5) indicates flow conservation of each sub-demand of the tree in all nodes. EQ (6) indicates that edge must be allocated if any sub-demand of tree wants it. The objective function expresses that the total cost of allocated edges should be minimized.

7 Branching Limitations

Most of the switching devices are not able to perform branching of the signal at the optical or electronic layer due to technical constraints. The electronic layer branching can be constrained using the following inequality:

$$\sum_{ij}^{r} x \leq \propto ; \forall_i \in V_E , V_r \varepsilon R \quad \text{-------------------EQ} \tag{9}$$

The branching limit in node i is denoted by content $\propto i$.

Multicast tree size limitations: The optimal solutions of multicast tree produces long paths between source and some of the targets. It also implies higher delay which is unacceptable of the applications as it harms the QOS (Quality of service).

The length of path can be limited by the following formula: depth-limit constraint;
$$\sum_{ij}^{or} Z \leq \beta^{or} ;$$

The number of links contained by the tree can be limited by the following formula:

$$\sum_{ij}^{r} x \leq \mu^r ; \quad V_r \varepsilon R ; \text{---------------- EQ} \tag{10}$$

Where μ^r : constant value related to the tree size limit.

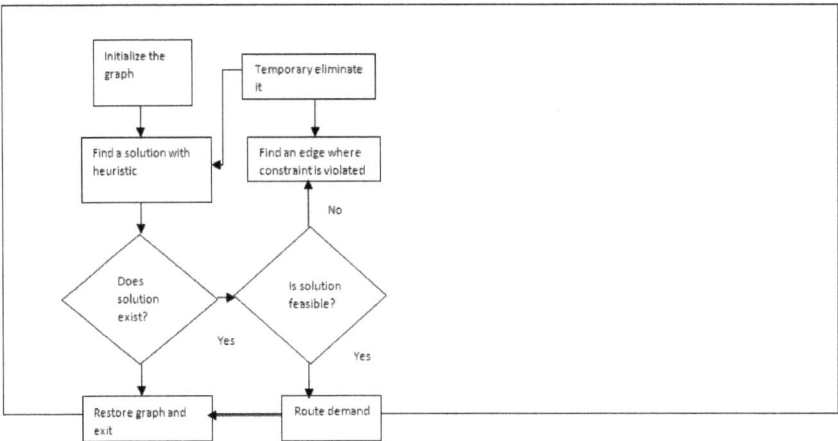

Figure 3. Application of heuristics

There are various heuristic methods such as dijkstras algorithm, minimum cost spanning trees such as Prims Algorithm, Kruskals Algorithm and an optimal Randomized Algorithm. These methods can be applied to the multicast routing problem with an additional search loop.

The main steps to route a demand are as follows:

1) Try to find a path from source to destination with a given method.
2) If solution doesn't exist, then exit.
3) Otherwise check the constraint on every affected edge.
4) If all the constraints are held, route according to their path.
5) If power constraints are violated find an edge with most power and temporarily remove it.

Then search for another path in modified graph. Repeat until path is found or exit the loop.

8 Result

During the simulations multicast scenarios were generated with 1 to 5 multicast demands in each, with 1 to 25 target nodes. Nodes were generated with uniform distribution. The original network size was decreased to its 12, 5% in multiple steps. Three methods were compared: the new heuristic method, non-PICR aware and the Accumulated shortest path (*ASP*) equivalent of the demand, i.e. a dedicated demand is used for every destination. An additional check is done against these results whether the given path is feasible for *PICR* or not. If it is not feasible it will be marked as unrouted.

Fig 5 shows that the new method is able to reach nearly all the destinations in all the simulations whereas in non-PICR method number of destination increases with reduction in scaling ratio and ASP method fails to reach a larger group of destination for a high scaling ratio value.

Fig 6 shows that while the cost-effectiveness and resource usage of multicast routing was done with optical routing and without; it was found that optical routing performs better than electronic layer branching. Optical branching outperforms the electronic branching only if the no of nodes participating in the trees is low compared to number of nodes in the network. As the number of nodes increases in the tree; benefits of optical branching disappears.

The difference is the most significant in case of *ASP* if the network is large shows the overall power that should be reserved according to the used method.

In case of non *PICR* aware algorithms it also contains the power of infeasible paths. *ASP* is the most power wasteful.

Conclusion

The proposed work shows that optical branching outperforms the electronic branching. The new heuristic algorithm for multicast routing has been proved to be able to reach all the targets in all network scaling ratio while observing technical constraints. It evaluated the cost and found out that cost could be saved in networks that are incorporated with optical branching capability.

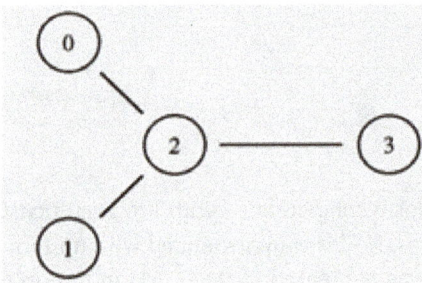

Figure 4. Simulation result of multicast tree design

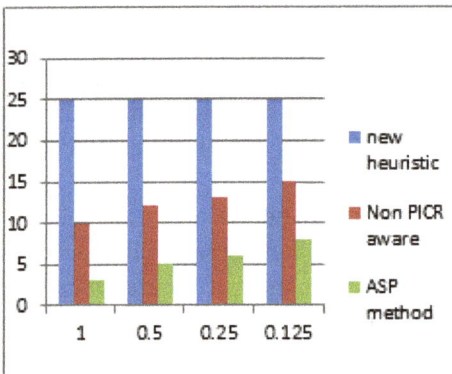

Figure 5. Reachable destination versus scaling ratio

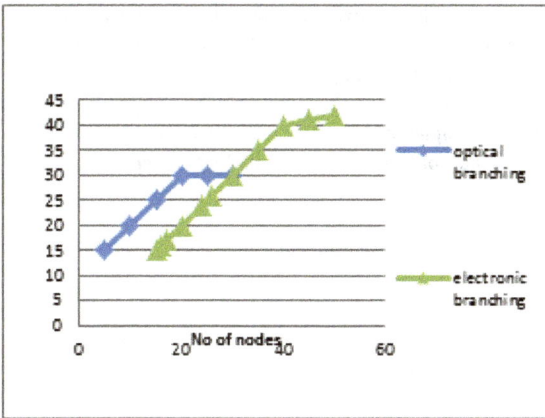

Figure 6. Comparison between optical and electronic branching as a function of resource versus the number of target nodes

References

1   Bharath T Dsohi, Subramanyam, Yafee Wang "Optical Network design and restoration", *Bell labs technical journal*, Jan-March – 1999.
2   Neeraj Mohan, Seema "A comprehensive study of protection and restoration in optical networks," *International journal of research in engineering and technology*, vol2, no .1 , 2013.
3   C. Ou, J. Zhang, and H. Zhang, "New and improved approaches for shared-path protection in WDM mesh networks," *J. Lightw.Technol.*, vol. 22, no. 5, pp. 1223–1232, May 2004.
4   Zhe Liang, Wanpracaha Art "Redundant multicast routing in multilayer networks with shared risk resource groups: Complexity, models and algorithms", *Science Direct, Computer and operation research*, 2010.
5   Yi Zhu and Jason P. Jue, "Reliable and collective communication in weighted SRLGs in optical networks," *IEEE/ACM transactions on networking*, vol 20, no 3 , June 2012.
6   Peter Saproni, Tibor Cinkler, "Preplanned restoration of multicast demands in optical networks", Computer networks, 2012,
7   Peter Saproni, Tibor Cinkler, "Methods of physical impairment constrained routing with selected protection in all optical networks." *Telecommunication system* ,177-188, 2014
8   Antonakopouolas,"Approximation algorithm for grooming in optical network design", *INFOCOM, IEEE*, 2009.
9   Chandana.B.R, A.M.Khan,   Dept of electronics, Mangalore university, "A review of basic optical networking approaches on functionalities of various devices and technologies.", 2013, *Conf. Proceedings,SIT,Tumkur.*
10 Chandana.B.R, A.M.Khan, Dept of electronics, Mangalore university "Restoration methods in WDM optical networks.-A review."*Elseveir Publications,* Conf. Proceedings, 2013.

11 Chandana.B.R, A.M.Khan, Dept of electronics, Mangalore university "A Comparative analysis of reliable multicast protocols in an optical network", First International online conference (IOCRSEM), *Conf proceedings,* August 20, 2014.

12 Chandana.B.R, A.M.Khan, Dept of electronics, Mangalore university "A Comparative analysis of logical topology design and development of heuristic algorithm for reliable multicasting in an optical network", *National conference at Manipal, Conf Proceedings,* 2014.

Sneha Sharma[1] and P Beaulah Soundarabai[2]

Infrastructure Management Services Toolkit

Abstract: The object of designing and creating the Infrastructure Management Service Toolkit is to facilitate the L1 Analysts to complete the day-to-day tasks in a faster and much efficient manner. The tool provides the analyst to access the access the remote folders resources located in a Windows File server without having to initiate the logon procedure to a server. It allows modifying, assigning and revoking the security permissions stored in the Access Control List (ACL) of folder structure with ease of a Graphical User Interface command buttons. It allows the analysts to access the Active Directory (AD) database through Windows PowerShell, the necessary AD objects can be viewed and their memberships can be modified based on the customer requirement. It is pertinent to note here that analyst will not be provided access to Active Directory due to complexity of database maintenance and is duly managed by System Administrators.

Keywords: SMB 2.0, LDAP, ACL, Windows PowerShell, .Net Assembly forms, Active Directory, File Servers

1 Introduction

Infrastructure is the backbone of the enterprise IT solutions and operations. The IT infrastructure lays the core platform for all the enterprise users by providing constant network connectivity, software and hardware resources that enables smooth functioning of the day to day business. Infrastructure Operations adhere and function in tandem with the process guidelines and standards provided by the IT Infrastructure Library (ITIL). The ITIL is a standard which talks about best practices for IT Service Management and highlights on aligning IT Services with the business needs. Processes, procedures, checklists and tasks are not organization-specific, but organizations apply this methodology for integrating the organization's strategy, thereby delivering value and maintaining minimum or controlled level of competition. An organization through ITL can establish a

---
1 Student –MS (CSA), Department of Computer Science, Christ University, Bangalore
coolsneha1208@gmail.com
2 Associate Professor, Department of Computer Science, Christ University, Bangalore
beaulah.s@christuniversity.in

baseline, through which it can plan, implement and measure the productivity. And thus demonstrates compliance and is a measure for improvement. The ITIL best practices are currently detailed within five core publications as shown in the below diagram. The analysts who are working on various L1 tasks which involve creating users , groups and assigning permission's to appropriate network directories. The tool also fetches the network location of the directories like the server on which it is hosted and its IP configuration. All the aforesaid activities involve repetitive manual work which cost valuable resource time and effort that could be utilized more effectively for better productivity.The tool facilitates all the analysts to perform the below folder and Active Directory operations.

Folder Operations
- Create new folder
- Assign permissions to the new folder
- Retrieve single /multiple folder permissions.
- Retrieve the current folder structure in the network directory.
- Retrieve the drive and shares of servers.

AD Operations
- Add members to group
- Group location in Active Directory
- Status of the AD user
- Retrieve members of group.
- Retrieve AD groups of the user.

It also facilitates creating reference documents for the referenced tickets and server details like location and Operating System.

2 Literature Study

Managing file server security permission is a tedious task in an organization. Remotely viewing and applying permissions is not incorporated as a feature in file server. The specific folder needs to be accessed directly from the server to view and change the permissions. There is sufficient amount of manual work involved when setting up new fileserver and also in making the shared folder available to respective users. Below are the available tools to manger NTFS permissions and Active directly objects like users, groups, computers, etc. One such power in-house tool of Microsoft is ADMT(Active directly migration tool) which helps in fully migrating the AD object , FSMO tools and restructure domain hierarchy if necessary .Please find the refernces of each tool in the end of the document.

2.1 Vyapin

Vyapin NTFS Security Auditing Solution provides a single comprehensive solution for an Enterprise's reporting NTFS permissions. It can handle the below given challenges with ease.

Server and Workstations Audit across network is performed to track who has access to what and what are the actions they can perform on these files and folders and Sheer Volume of data in the form of Access Control List is the big challenge here. Accounts such as users and groups have Access Control Entities and they run in hundreds of thousands of entries across files and folders and these accounts are granted permissions. The volume of entries discourages manual analysis of NTFS permissions on files and folders. Hence all these challenges are dealt by the Vyapin tool.

2.2 AD Plus Manger

NTFS folder permissions report for Active Directory is generated by the Manager Plus. The list of users and groups that have access to the folder for a specific share path is listed in the NTFS permissions report. And the access control entities that are associated to it indicate the level of access that the user/group has on the folder and also indicates the inheritable permissions (permissions that are carried from the parent folder to the child folders) if there are any. This report of File and Directory Management can be used as the NTFS permissions analyzer for folders that will ensure that the folders have been given right permissions. Hence, administrators can monitor the NTFS permissions on the folders and easily secure the folders.

2.3 Hyena

Hyena has inbuilt tools for directory searching and filtering, managing object properties, customization queries, security auditing, advances attribute management and other AD mixed and native mode functionality. Hyena has an Explorer-style interface for all the operations including right mouse click, pop-up menus for objects, management of users, groups (which can be local or global in nature), shares, domains, computers, events, services, files, printers and print jobs, disk space, messaging, exporting, job scheduling, processes and printing are also supported.

Active Directory Migration Tool

Active Directory Migration Tool is used to migrate objects in the AD forests. The tool incorporates wizards that automate migration tasks like migrating users, groups, computers and trusts and security translation. Any migration process involves change of domain structure known as "Domain Restructuring" and involves either consolidating domains or addition of domains and this can be done in a forest or between forests. The ADMT tasks is performed using ADMT console, a command line or through scripts. When the ADMT is run through the command line it is good practice to use an option file to specify the command-line options, this will increase the overall efficiency. Each migration activity has a section that illustrates options that are specific to the task. When ADMT is run via command-line the section name will correspond to the task name.

3 Functional specification

The user logs in into the server or his desktop using the AD authentication credentials though which he connects to the Windows Domain Network .Please note once the authentication is successful a single sign-on secure token is created and assigned to 'username ' and that is used to authenticate the analyst while launching/accessing any application in the Windows Domain Network.

3.1 Single Folder Access

The security descriptor of any file or folder can be obtained using the Get-Acl cmdlet. It contains the access control lists (ACLs) that specifies the permissions for users and groups to access any resource. Hence the he existing security permissions of the destination network folder in any Windows server across the network can be retrieved.

3.2 Multiple Folder Access

The tools allows to fetch the security permission details of multiple folders via a single user GUI console without having to login to each server individually. This is useful when the analyst receives a request to grant access to parent folder which contains a multiple no of subfolders having separate security permissions. The tool will gather the security groups from all its subfolder and will give two output files, one showing the entire subfolder path hierarchy and other with security permissions for each subfolder.

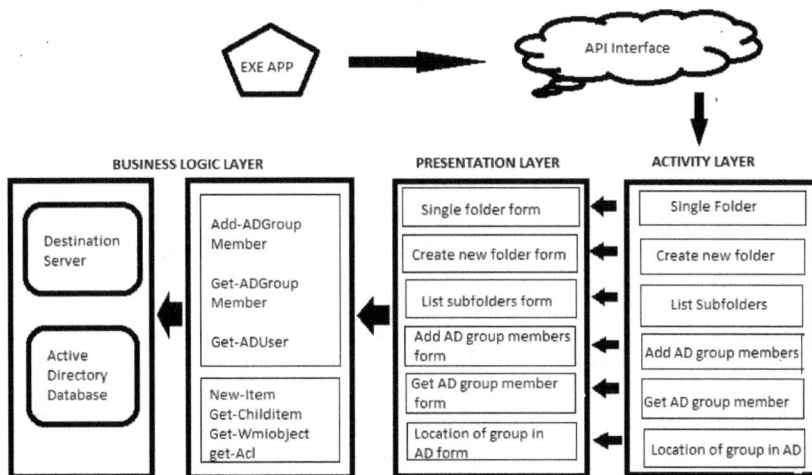

Fig 1. Tool Architecture Diagram

3.3 New Folder Creation

The tool creates a new folder and assign read/write security permissions to that folder by creating appropriate read and write groups in active directory. It creates the groups with proper naming convention and applies that newly created groups to the destination network folder. It also generates a text file with all information of the new folder that was created which can be used for updating the requester.

3.4 Creation of Hidden Folders

The tool also allows the creation hidden folder shares where the shared folder will not be available for all users. When the share name of the folder is accompanied by a '$" symbol it make the share invisible to all users connecting to the server. On creation of the hidden shared only the respective user will be added to the ACL thus providing highly restricted user access. It also generates a text file with all information of the new hidden share that was created and can be used for updating the requester.

3.5 List the Subfolders

This options provides us with the list of all the subfolders in a given network share path. It generates the output text file with the list of all the subfolders under the parent directory.

3.6 Modify/Revoke

The tool also allows us to modify and revoke access of an existing network folder, if folder is already available in the network we add can read/write permissions based on the requirement.

3.7 Adding users to AD group

With the help of the tool we can add members to a security group in the Active Directory (AD). The group name and the list of user ID's are given as input and the output text file will display the list of users who are added successfully and it also shows the list of failed users along with reason of failure such as 'user ID is already exists'.

3.8 Location of Group Object

Identifies the location of the group in the AD hierarchical database which is the Distinguished Name (DN) of the AD object, DN allows us to maintain the integrity of the created group the AD structure The input is given as the group Name and the will output test file gives the location of the group in AD, which helps to check access based on OUs.

3.9 Groups Owner and membership

It displays the status of the AD user object if it's active or disabled. It retrieves membership information of an AD group and shows the current users and groups that are members of the group. It also gathers and outputs the group membership of any given AD user object.The toolkit in Figure 2 is launches after the Windows Authentication and calls the .NET assembly forms which form the home screen of the application. The home screen provides the analyst with

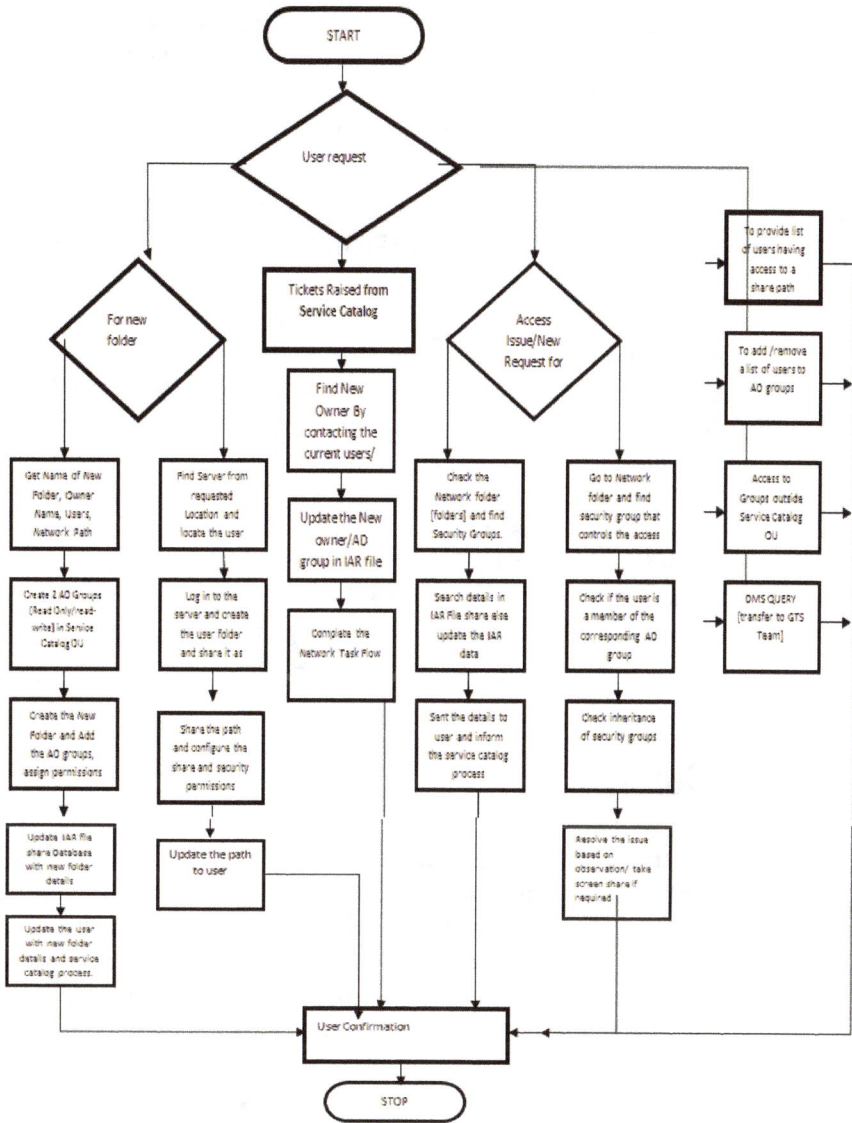

Fig 2. Block diagram of Toolkit

Folder and AD operations and by clicking each button the analyst invokes a separate subroutine. The Folder permission related operations rely on the network communication between the analyst machine and the Destination file server which users SMB protocol. When the analysts invoke the subroutine related to AD operations the tool communicates directly with the Active Directory and performs the requested operation, Lightweight Directory Access Protocol (LDAP) is used for the communication with Active Directory.

4 Design and Implementation

Integrating Windows PowerShell with the .NET assembly forms which are the building blocks of the Graphical User Interface. Adding the necessary buttons to the GUI and separate modules of code are invoked by the analyst when the analyst clicks each button. The SMB protocol and ACL cmdlets are used to make the toolkit communicate with the remote server which can be located across the globe. The Lightweight Directory Access Protocol is used by the cmdlets when making AD related operations like adding user to a group or creation of a group object. The New Folder option creates new folder in a file server across the network via SMB and also it interacts with the Active Directory to create the necessary groups ,in addition to that the newly created groups are added to the destination folders ACL and there by securing the folder immediately.

Conclusion

Thus the Infrastructure Management Services Toolkit is a PowerShell tool which eases performing the monotonous and tiring tasks like Active Directory and folder permission management in just few clicks. As the analyst is able to perform his daily tasks through a console of the tool, which in turn connects with various other repositories like Active directory, Database like Service catalog for example. Hence IMS toolkit will act as the right-hand for all the administrative tasks.

Advantages and Limitations

The tool has been designed to provide the easy access to the Active Directory and to create, modify and revoke the permissions of the folder residing in a Windows server. The daily tasks performed by the analyst involve time taking operations that increase the duration spent for solving simple requests , often these are repetitive and might involve the analyst will have to login to separate servers for a single request and in turn it increases the time spent on a single request. The IMS tool kit offers a GUI interface where the user inputs the necessary details which retrieves the output within matter of minutes. The tool allows the analysts to view the current object information in the Active Directory and also allows the modification of user and groups properties which are only available to AD System Administrators. The tools is platform specific and can run only of Windows based Domains, user analyst should use a healthy PC or Server to login to the fast domain network The tool relies on the LAN network speed and the connectivity of the server to provide faster outputs. The network ports used for SMB and LDAP need to opened at the firewall to allow the communication.

Future Enhancements

In the future the tool can be enhanced further to include more functionalities and thus making it more useful for the user. The proposed enhancements are listed below:
- Retrieving all the permission detail for all the network folders and create repository file. This can be used to re-apply the permissions in case of file server migrations
- Retrieving installed server hardware and software information
- Include an extensive ADSI edit-like capability to manage any Active Directory attribute for one or more directory objects.
- Define our own attribute sets for existing or newly created objects and to be used and visually see what the values are before and after making any changes
- Managing the new fine-grained password policies and password settings objects in Windows 2008/2012 Server
- Managing AD objects across multiple Windows domains ,also able to integrate AD objects between Windows 2008 and Windows 2012

References

1   http://www.vyapin.com/whitepapers/file-server-security-audit
2   http://www.systemtools.com/hyena/

3   https://download.manageengine.com/products/ad-manager/manage-user-life-cycle-in-active-directory.pdf
4   https://technet.microsoft.com/en-us/library/cc974332(v=ws.10).aspx
5   http://manageengine.com.mx
6   www.mediafi.org
7   www.dbl.co.uk
8   www.linksoft.com.tw
9   www.systemtools.net
10 www.softwareparadise.co.uk
11 http://techlauve.com

Divyansh Goel[1], Agam Agarwal[2] and Rohit Rastogi[3]

A Novel Approach for Residential Society Maintenance Problem for Better Human Life

Abstract: This research paper presents a digital solution to all the generic problems faced by people living in Indian co-operative societies. An optimal way to find a solution to all the possible problems with ourselves using our smartphones and tablets. This paper proposes an efficient method which sorts out the communication gap between different service provider and you such as Electrician, Plumber, sanitary, etc. Moreover, it deals to resolve an issue of communicating with the person whose car is creating a trouble for your vehicle to park. This application ensures that no vehicle can make any sort of trouble for you in your co-operative society. Overall, this paper is a smart solution for all the mysterious issues of the residents of society for their more luxurious and comfortable lifestyle in their society.

Keywords: Society Maintenance System (SMS), ParkAlert, Parking Problems, Co-Operative Societies, Module.

1 Introduction

Advancement in Technology has a significant impact on the human life. Rapid and wide technological applications have made the human life luxurious and easy going as never before. With the growing population and rapidly increasing in the number of nuclear families, people are rushing to purchase Flats in Co-operative Societies. The high pace at which population is increasing is also making the buying power of people manifold.

Residing in Flats in Co-operative Societies has become the sudden craze in Residential Property. Thus, current scenario surely tells about the boom of Co-Operative Society Culture. Today, people wants to live a healthy and comforta-

---

1 ABES Engineering College/C.S.E Department, Ghaziabad (UP), INDIA
E-mail: divugoel@gmail.com
2 ABES Engineering College/C.S.E Department, Ghaziabad (UP), INDIA
E-mail: agamaggarwal11@gmail.com
3 ABES Engineering College/C.S.E Department, Ghaziabad (UP), INDIA
E-mail: rohit.rastogi@abes.ac.in

ble life. Facilities including power backup, water backup, and vehicle parking has become one of the major issues in many top cities and the problem is expected to increase more in the future that's why to get rid of these problems people are migrating to these Co-operative Societies because they provide 24X7 power and water backups and also ensuring the parking space for the resident's vehicle but are not ensuring its proper implementation. Facilities such as Plumber, Electrician and Sanitary are also available but to avail their services Residents have to communicate the issue to Society Maintenance Department which increases the complexity of the whole procedure. This paper aims to ensure Smart solutions to the above problems in the Co-Operative Societies and to make all these facilities available for residents in a very efficient and friendly manner so that residents can avail and grab these benefits without any headaches and frustration.

2 Literature Survey

Some of the existing approaches related to this work have been there which focuses on reserving your parking area likewise

[1] Lalitha Ayer; Manali Tare; Renu Yadav; Hetal Amrutia in their paper "Android Application for Vehicle Parking System: "Park Me" presents a mobile application to examine the number of free slots for parking in an area. In their application driver can pre book a parking slot and administrator allocates the slots to the users in the queue.[2] Soumya Banerjee; Hameed Al-Qaheri proposed a paper "An Intelligent Hybrid scheme for optimizing parking space: A tabu metaphor and rough set based approach." In which they solve the parking area problem using a search technique i.e. tabu search assisted by rough set. [3] Akmarina Izza; Mokeri stated a paper "Smart Parking via RFID Tag" in which they used the RFID tag to alert the driver from wrong parking and vacant parking space along with the indication of the vehicle location. [4] Ramneet Kaur and Balwinder Singh presented "Design and Implementation of Car parking system on FPGA" which includes two basic modules modelled in HDL one to identify the visitor and another for checking the slot status. These modules are implemented on FPGA. [5] Kun-Chan Lan; Wen-Yuah Shih in their paper "An intelligent driver location system for smart parking" Presents a phone based system which will track the driver's path to know when the parking slot can get free. [6] Kuo-Pao Yang; Ghassan Alkadi; Bishwas Gautam; Arjun Sharma; Darshan Amatya; Sylvia Charchut; Matthew Jones presents "Park-A-Lot : an automated parking management system" which can send the

status of the parking space to web interface after all processing on the Arduino board. The tier system reads the value of ultrasonic sensor echo distances, finds the average and then decides whether parking slot is free or taken. [7] VAHAN is an e-service in which one has to send a text SMS with vehicle registration number to some unique number to get some details about that vehicle. However, giving vehicle owners personal details to someone via SMS jeopardize with the privacy of peoples.

3 Our Approach

The smart and user friendly solution proposed in this paper overcomes the limitations discussed in this section. Our application is basically divided into different modules:
- ParkAlert
- Digital Registration of Visitors
- Society Maintenance Services

3.1 ParkAlert

ParkAlert is the Smart Solution to directly communicate and alerting the problem creating user whose vehicle is wrongly parked in the society. This approach ensures a quick and fast solution with 100 % privacy of the details of every User.

3.2 Visitor's Registration Digitally

We have made the Digital Registration of the Visitor's coming to the society and linked that data with the data of the ParkAlert module so the Visitor's Vehicle can also be searched in ParkAlert which in result ensures that Visitor's Vehicle cannot create any problem for the Resident and removal of Paper related formalities thus making the Record Maintenance of Visitor in Societies easier.

3.3 Society Maintenance Services (SMS)

SMS provides a direct one to one communication with the residents and the service providers of that society which reduces the lengthy and irritating method for availing the same. In this module, the user can see all the service providers that are allotted in the society for a particular task and can also directly communicate through the message providing the information about the problem they are facing. A big advantage of this approach is that the user can also see the Pending requests of different service providers so providing an option to choose the one having less number of requests for that day. After the completion of the problem user will update that in their profile so that pending requests can be minimized for others and will also benefit the service providers to get more and more requests. *"Smartness is required to make the World Smarter"*

4 Implementation

4.1 Resident's Login

4.1.1 Registration / Login

Resident needs to be authenticated by the Society to complete the registration process and to avoid the registration of fake entries. All these details will be stored in the database of the Society from which details will be fetched when User login.

Figure 1. Login and Register interface

4.1.2 Resident's Account

Resident's profile contains all the facilities that will ease the life of the residents including ParkAlert, Society Maintenance Services (Electrician, Plumber, Sanitary, etc.). Resident can easily switch to the desired module with a single click **(Figure 2).**

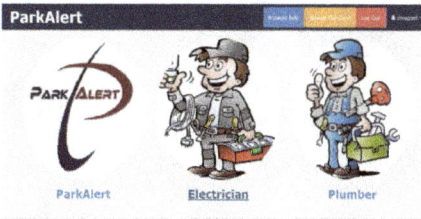

Figure 2. Residents Account Interface

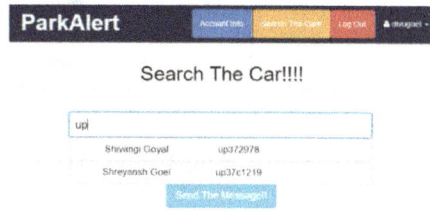

Figure 3. ParkAlert Interface where a resident can Search the car number

4.1.3 ParkAlert

In this module, there is a Live Search on which the Resident who is facing the problem will input the problem creating Vehicle's number and the results will be shown related to that using AJAX. After inputting the Vehicle's Number User will click the button below (Send the Alert) and after clicking that an automated message gets delivered to the Mobile Number registered with that Vehicle's Number alerting to move their Vehicle as soon as possible **(Figure 3).**

4.1.4 Society Maintenance Services

This module contains the list of services that a particular society wants to facilitate their Residents with like Electrician and Plumber.

A. Electrician / Plumber: - This contains details of service providers that are selected and registered by the society. The details will be Name, Mobile Number, Occupation and a Button. On clicking the button an automated message gets delivered to the Selected Service Provider. The Profile also contains the pending requests, so Residents can choose from different options and after the problem is resolved the User will update the Request Status to complete, resulting in decreasing the Pending Requests.

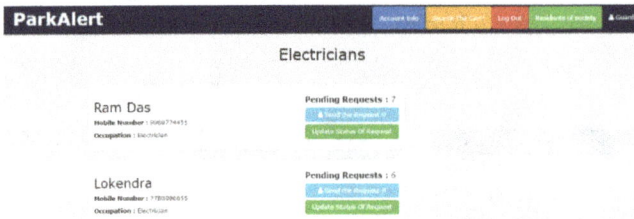

Figure 4. Interface showing different service providers details to visitors

4.2 Society Login

4.2.1 Visitor's Registration / View Visitors Registered

This is a special permission page that will be provided only to the society head who can register the visitors online and can view all the visitor details that are registered **(Figure 5)**.

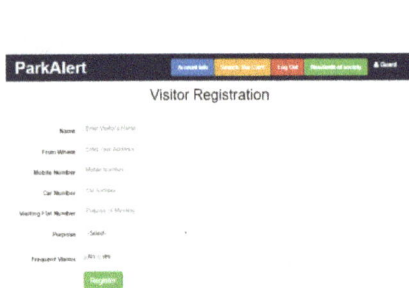

Figure 5. Visitor Registration Interface   **Figure 6.** Service Providers registration by Society Head

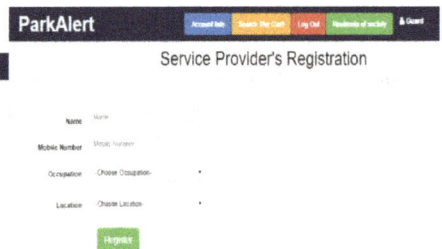

4.2.2 Service Provider Registration

This module can only be accessed by the society head who can add their fixed Service Providers whom they pay the monthly salary and provide free services to the Residents **(Figure 6)**.

5   Results And Analysis

After the implementation of the application in Pratap Vihar Society, Vijay Nagar, Ghaziabad, we come to the analysis of 3 months, which shows the benefit and usefulness of the application which reduces the efforts and stress of the Residents and made their life more comfortable and easy going.

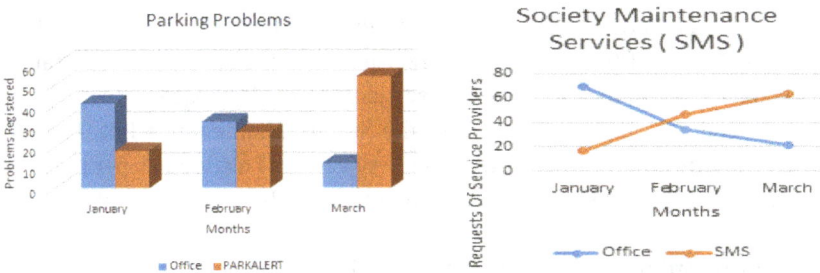

Figure 7. Analysis and Results for 3 months in Pratap Vihar Society

5.1 Benefits Of Our Approach

– Hassle free and user-friendly interface of the application.
– One to one communication with the desired person, no third party involvement.
– Automated alerts delivery to the User whose Vehicle is creating the problem.
– Promote Digitalization as data is now stored directly in the database, not on papers.

5.2 Limitations

- Parking problems coming outside the society will not be resolved by this application as the database is limited to a particular society.
- The Desired User cannot be reached if the user's Mobile Number is Switched Off, or Out of Coverage area.
- Not everywhere the Computer is available for the Guards for the Registration of the Visitors.

If the driver is not carrying the registered mobile number, then he/she can't be alerted for their wrong parking.

6 Future Scope

All the Parking Related Problems will be easily solved within a country if all the Residents of most of the societies gets registered in this application so that we will get nearly get all the Car Numbers registered in the Database.

Different Modules such as Payment of the Society Maintenance Bill, Delivery from the Nearby Shops, Daily Newspaper Services can be added as per the need and demand of the society for the more luxuries and better lifestyle of the Residents.

Voting regarding any Topic and issues in the welfare and Choosing the President of Society can also be done digitally within the Residents Profile so that Resident can Vote directly from anywhere and anytime.

We are also planning to add an automated calling feature in our ParkAlert module for more surety of alerting the person.

7 Acknowledgement

With due respect, we are very much Grateful to all the members that are involved in this formation of the Paper. We are thankful to the Head of the Department, Prof. (Dr.) Shailesh Tiwari, HOD-CSE for his support at every single step and also to Educational ABESEC Infrastructure.

Conclusion

We are trying to make the facilities provided in Indian Apartments more comfortable for the residents to use them in more efficient manner. An application that sorts out all the basic problem in using these facilities of the apartment in a digital and friendly for residents. ParkAlert module ensures that no car can create any problem for you in your society. So this paper deals in providing more luxurious and hassle free lifestyle of increasing residents in Indian cooperative society.

References

1 Lalitha Iyer; Manali Tare; Renu Yadav; Hetal Amrutia ,(2014) "Android Application For Vehicle Parking System: "Park Me" ".International Journal of Innovations and Advancements in Computer Science ,ISSN 2347-8616 Volume 3,Issue 3.
2 Soumya Banerjee; Hameed Al-Qaheri ;(2011)" An Intelligent Hybrid scheme for optimizing parking space: A tabu metaphor and rough set based approach.". Egyptian Informatics Journal, volume 12, 9-17.
3 Akmarina Izza;Mokeri; (2013) "Smart Parking via RFID Tag".Project Report.UTeM,Melaka,Malaysia.
4 Ramneet Kaur; Balwinder Singh;(2013) "Design and Implementation of Car parking system on FPGA".International Journal of VLSI design and Communication System .Volume 4, Number 3.
5 Kun-Chan Lan; Wen-Yuah Shih;(2014) "An intelligent driver location system for smart parking". Expert System with Application Volume 41, Issue 5, Pages 2443-2456.
6 Kuo-Pao Yang;Ghassan Alkadi; Bishwas Gautam;Arjun Sharma;Darshan Amatya;Sylvia Charchut;Matthew Jones;(2013) "Park-A-Lot : an automated parking management system". Computer Science and Information Technology, 1,276-279.
7 S.Banerjee;P.Choudekar;M.K.Muju;,(2011) " Real time car parking system using Image Processing ".Proceedings of the 3rd International Conference on Electronics Computer Technology,Vol.2, 99-103.
8 https://vahan.nic.in/nrservices/

H. Kavitha[1], Montu Singh[2], Samrat Kumar Rai[3] and
Shakeelpatel Biradar[4]

Smart Suspect Vehicle Surveillance System

Abstract: The proposed system uses image pre-processing and character recognition approaches to identify vehicles by automatically reading their license plates. Smart Suspect Vehicle Surveillance (SVS) system consists of three main modules: Vehicle License Plate (VLP) detection, plate number segmentation and plate number recognition. In VLP detection module, an efficient boundary line-based method combining the Hough transforms and Contour algorithm is used. This method optimizes speed and accuracy in processing images taken from different angles. Then horizontal and vertical projection is used to separate plate numbers in VLP segmentation module. Finally, each plate number will be recognized by Optical Character Recognition (OCR) module.

Database is maintained globally allowing the authorization only to the law enforcement officers. Only the authorized users can update the database locally, which later updates automatically in its cluster. Vehicle registration number obtained after processing is searched in the database, which consist of prime suspect vehicle registration numbers. If the vehicle number is found in database then system alerts with the description of the crime. SVS system also consists of a latest news section, where the news is broadcasted.

Keywords: VLP detection, Segmentation, License plate Recognition, Hough Transform, Contour, Optical Character Recognition.

1 Introduction

License plate recognition (LPR) systems have received a lot of attention from the research community [1]. LPR stands for a fast enough operation to not miss a single object of interest that moves through the scene [2, 3, 4]. With the rapid

1 Asst. Prof, Siddaganga Institute of Technology, Tumakuru, India
Email: kavithahalappa@gmail.com
2 Student, Siddaganga Institute of Technology, Tumakuru, India
Email: montu.1si12is018@gmail.com
3 Student, Siddaganga Institute of Technology, Tumakuru, India
Email: samrat.1si12is049@gmail.com
4 Student, Siddaganga Institute of Technology, Tumakuru, India
Email: shakeelpatel.1si12is053@gmail.com

growth in the number of vehicles, there is a need to improve the existing systems for identification of vehicles. A fully automated system is in demand in order to reduce the human effort. License Plate Recognition is a combination of image pre-processing, character segmentation and recognition technologies used to identify vehicles by their license plates [1]. Since, only the license plate information is used for identification, this technology requires no additional hardware to be installed on vehicles. LPR technology is constantly gaining popularity, especially in security and traffic control systems [5]. License Plate Recognition Systems are utilized frequently for access control in buildings and parking areas, law enforcement, stolen car detection, traffic control, automatic toll collection and marketing research. A tedious job for the Police authorities is to track and find lost vehicles or vehicles involved in crimes manually. Smart Suspect-Vehicle Surveillance System can be used to perform these tasks faster and more efficiently. Blocking of roads looking for suspected vehicles can also be avoided without disturbing the general public and also reducing the dependency on labor [6]. VLP recognition technology is constantly gaining popularity, especially in security and traffic control systems. License Plate Recognition Systems are utilized frequently for access control in buildings and parking areas, law enforcement, stolen car detection, traffic control, automatic toll collection and marketing research.

2 Proposed system

The proposed System is shown in Figure 1. Images Captured either by real time video or the static images are given as input to the system. Processing of the image is done to identify the number plate of the vehicle. Using OCR technique characters are found in the recognized license plate. Number plate so obtained is re-verified by checking the sequence of alphanumeric characters license number so obtained is queried into suspected database for any criminal records. If the records are found the detailed crime is displayed and officer takes a necessary action. If no records were found the entire process is auto repeated for next image or video. Authorized persons are provided with the authorization ID, with the help of which officers can update the crime status.

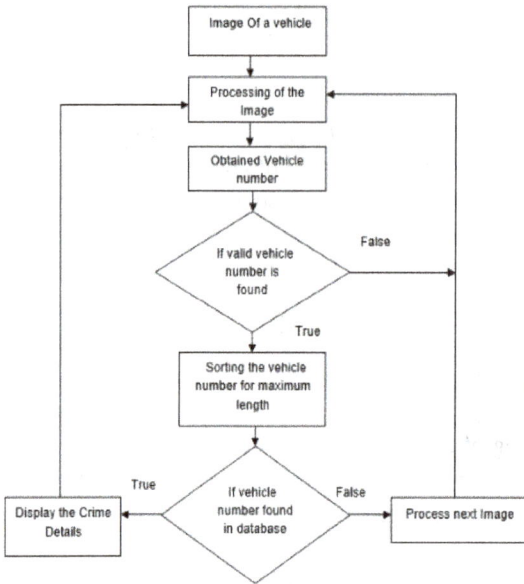

Figure 1. Architectural of proposed system

2.1 Database

The database consist of 300 four wheeler and 50 two wheeler vehicle images as shown in Figure 2, which mainly consist of detailed records of crime uploaded by authorized persons. Database concentrates on the vehicle related crimes involved, where vehicle number plays a vital role in solving the crime.

Figure 2. Input Images

2.2 VLP Modules

SVS system consists of four modules: Pre-processing, Vehicle License plate (VLP) detection, character segmentation, and optical character recognition (OCR), as shown in Figure 3. The last three modules deal with three main problems of a VLP recognition domain and are discussed in the following subsection.

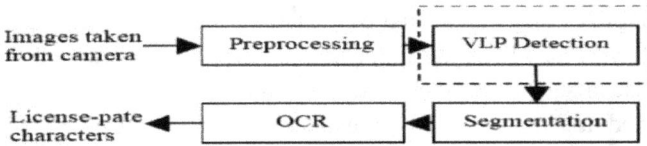

Figure 3. Main modules in SVS system

2.2.1 Pre-processing

Images taken from camera are processed by the pre-processing module. The purpose of this module is to enrich the edge features [7]. Because our detection method is based on the boundary features, it will improve the success rate of the VLP detection module. The algorithms sequentially used in this module are greying, normalizing and histogram equalization [8]. After having obtained a greyscale image, Sobel filter is used to extract the edge of an image and then the image is threshold to binary form by local adaptive threshold algorithm to optimize its speed and make it suitable to real time applications [1]. The resulted images are used as inputs for the VLP detection module.

2.2.2 VLP Detection Algorithm

In boundary-based approach, the most important step is to detect boundary lines. One of most efficient algorithms is Hough transform that is applied to the binary image to extract lines from images. However, the drawback of this approach is that the execution time of the Hough transform requires too much computation when being applied to a binary image with great number of pixels. The algorithm used in this system is the combination of the Hough Transform and Contour algorithm [5] which produces higher accuracy and faster speed so that it can be applied to real time systems.

The approach is as follows: for the extracted edge, the contour algorithm is used to detect closed boundaries in image. These contour lines are transformed to Hough coordinate to find two interacted parallel lines (one of 2parallel lines holds back the other 2-parallel lines and establishes a parallelogram-form object) that are considered as a plate-candidate. Since, there are quite few (black) pixels in the contour lines, the transformation of these points to Hough coordinate required much less computation [7]. Hence, the speed of the algorithm is improved significantly without the loss of accuracy.

However, some plates may be covered by glasses or decorated with headlights. These objects may also have the shape of two interacted 2 parallel lines and therefore, are also falsely detected as plate-candidates. To reject such incorrect candidates, a module is implemented for evaluating whether a candidate is a plate or not.

2.2.3 Plate-candidates Verification

From the two horizontal lines of a candidate, its inclination with respect to horizontal coordinates is calculated. Then rotate transformation is applied to adjust it to straight angle. After this process, these straight binary plate candidate regions are passed for evaluating. The evaluating plate-candidates algorithm is based on evaluating the ratios between the height and the width of the candidates.

In this stage, the ratios of width to height are checked and those satisfying pre-defined constraint are selected [1]. Since there are two main types of Indian plates: 1-row and 2-row, there are two adequate constraints for two types. $2.5 < W/H < 3.5$ with 1-row plate-candidates and $1.0 < W/H < 2.5$ with 2-row plate-candidates. Those candidates which satisfy one of the two above constraints are selected and passed to the next evaluation.

2.2.4 Segmentation

To correctly recognize characters, binary plate image is segmented to set of images which only contain one character. These character images will be passed to the OCR module for recognition. The common algorithm for this task is applying projections. However, in some cases, it does not work correctly. So a different approach in segmentation by adding some enhancements to this method is described. A horizontal projection is used to detect and segment rows. Because, binary plate images adjust their inclined angles to zero, the result of row segmentation is nearly perfect. The positions with minimum values of horizontal projection are the start or the end of a row in plate.

Different form row segmentation, character segmentation is more difficult due to many reasons such as stuck characters, screws, and mud covered plates. These noises cause the character segmentation algorithm using vertical projection to have some mistakes. In some worst case of bad quality plate images, a character can be segmented into two pieces. Several constraints of ratio of the height to the width of a character are applied.

In this approach the minimum values is searched in the vertical projection and only the minimum positions which give cut pieces satisfing all predefined constraints are considered as the points for character segmentation. By this enhancement, better results are achieved in this task. After this step, a list of character candidates is listed. Not all of the candidates are actually images of characters. For example in Figure 4 (a), there are 9 candidates but only 7 of them are character images. The first and third candidates are images of 2 screws in the plate.

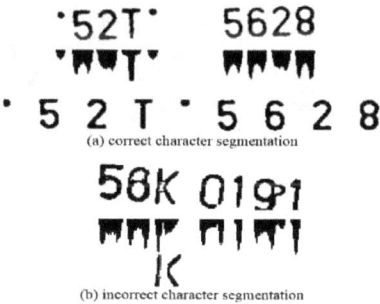

Figure 4. Character segmentation by vertical projection

2.2.5 Training Model

In the recognition module, there is a need to classify a character image into one of 36 classes (26 alphabet letters: A, B, C... and 10 numeric characters: 0, 1, 2...). To train this model, training sets are used which are images of alphabets and digits [9]. In the last step, some specific rules of Indian VLPs are used to improve accuracy. The third character in plate must be a letter, the fourth is sometimes a letter but usually a number and the other positions are surely numbers.

3 Experimental Results

The experiment are done by using images of car taken from different angle which makes the number plate tilted, some part of the number plate is having less brightness and plates having higher brightness. The average time noted to be as 0.3 seconds per image of car for detecting license plate number. The proposed system shown in Figure 5 works well with these type of number plate and hence can be used for real time application. The system works well for small vehicles like motor bikes and scooty, these vehicles have smaller number plate in comparison with cars.

Figure 5. Results of proposed system

The system is also subjected to real time video. To experiment this module USB camera of 1.1 mega pixel is used. This USB camera was mounted on an electronic bot. This electronic bot is a simulation of real world traffic patrolling vehicle. The number plate present in the real time video is also detected even if the number plate or the camera is in movement. This camera can be mounted on the police vehicle and can scan number plate on the street. The details of the result is given in Table 1.

Table 1: SVS system analysis results

Total number of Images tested	Number of plates accurately identified	Number of plates partially or falsely identified	Accuracy in %
4-wheeler 300	283	17	94.3
2-wheeler 50	46	4	92

Conclusion

The system performed well on various types of VLP images, even on scratched and number plates of different sizes. In addition, it deals with the cases of multiple plates in the same image or different types of vehicles such as motorbike plates, car plates or truck plates. However, it still has few errors when dealing with bad quality plates. LPR system has been developed and tested. Spectral Analysis approach and Connected Component Analysis approach are generally used for license plate extraction. It is found that when these techniques are used individually, they failed. Fusion of both Spectral Analysis and Connected Component Analysis gives better results. It is found that the techniques used in this system gives better recognition accuracy compare to other feature extraction methods.

References

1   Lekhana G.C and R.Srikantaswamy, Real Time License Plate Recognition System, IEEE ISSN No: 2250-3536 Volume 2, Issue 4, July 2012.
2   B. Honglisng and L.Changping, "A hybrid license plate extraction method based on edge statistics and morphology," in Proc, ICPR, 2004
3   C. Arth, F. Limberger, and H. Bischof, "Real time License plate Recognition on an embedded DSP-platform," in Proc. IEEE Conf. CVPR, pp. 17-22, Jun, 2007
4   S. Z. Wang and H. J.Lee, "A cascade framework for a real time Statistical plate recognition system," IEEE Trans. Inf. Forensics Security, vol. 2, Jun,2007
5   Tran Duc Duan, Tran Le Hong Du, Tran Vinh Phuoc and Nguyen Viet Hoang, "Building an Automatic Vehicle License-Plate Recognition System", Intl. Conf. in Computer Science RIV-F' 05, February 21-24, 2005.

6   D. A. Duc, T. L. Du, T. D. Duan, "Optimizing Speed for Adaptive Local Thresholding by using Dynamic Programming", The 7th International Conference on Electronics, Information, and Communications, pp. 438-441, Vol 1, 2004.

7   Jun-Wei Hsieh, Shih-Hao Yu, Yung-Sheng Chen, "Morphologybased License Plate Detection from Complex Scenes", 16th International Conference on Pattern Recognition (ICPR'02) Vol 3, 2002.

8   Kwang In Kim, Keechul Jung and Jin Hyung Kim, "Color Texturebased Object Detection: an Application to License Plate Localization", Lecture Notes in Computer Science, pp. 293-309, Springer, 2002.

9   Kamat, Varsha, and Ganesan, "An Efficient Implementation of the Hough Transform for Detecting Vehicle License Plates Using DSP'S", Proceedings of Real-Time Technology and Applications, pp. 58-59, 1995.

10 http://opensource.org

11 http://opencv.org/documentation.html

12 https://en.wikipedia.org/wiki/Automatic_number_plate_recognition

13 http://www.emgu.com/wiki

Takahito Kimura[1] and Shin-Ya Nishizaki[2]

Formal Performance Analysis of Web Servers using an SMT Solver and a Web Framework

Abstract: When developing software on a web server, one should pay attention to its correctness but also its performance. Traditionally, the performance of the server software is analyzed through testing an implemented system in a real situation. In this paper, we propose a new method of analyzing web server software using an SMT solver. Using our method, one can estimate the computational cost of packet processing based on the real measured values. In contrast to traditional testing, our method covers a wider range of inputs and situations.

Keywords: SMT Solver, Web Framework, Performance Analysis.

1 Introduction

1.1 Web Framework

A web server is type of software that provides an HTML document and other kinds of data objects such as image files, as a response to a request from a web browser on a client computer obeying the Hypertext Transfer Protocol (HTTP).

A web application framework (or simply, a web framework) is a software framework for the development of web applications on web servers. A web framework provides activities common among web servers. For example, many web frameworks offer the following functions.

- HTTP message analysis: parsing HTTP messages based on the message format.
- HTTP request queuing: if a new HTTP request is received while processing a previous request, the new request is added to a queue of HTTP requests.

1 Tokyo Institute of Technology, Department of Computer Science, Tokyo, JAPAN
E-mail: takahito.kimura@lambda.cs.titech.ac.jp
2 Tokyo Institute of Technology, Department of Computer Science, Tokyo, JAPAN
E-mail: nisizaki@cs.titech.ac.jp

- User session control: a web framework maintains and destroys user sessions, based on the user session's information in the header part of an HTTP message.
- Database administration and maintenance: a web framework gives a unified API to a relational database management system on the web server.
- Template engine: a web framework combines several templates to generate HTML documents using other data on the web server.

Using the facilities mentioned above, a programmer develops a new web server avoiding redundancy of coding and focusing on the essential part of a web application on the web server.

There are several implementations of web frameworks at the moment: Bottle [8], Flask [9], Django [10,11], and Ruby on Rails [12], which are written in scripting program languages such as Python and Ruby. In the following table, we compare web frameworks from the viewpoint of the number of lines of program codes.

Table 1: Comparison among Web Frameworks

	Lines of Code	Implementation Language	Version No.
Bottle	3620	Python	0.12
Flask	10192	Python	0.10.1
Django	64417	Python	1.7.3
Ruby on Rails	194749	Ruby	4.2

In the second column in this table, we present the total number of lines of the program code for each web framework (except lines of testing code). As can be seen in this table, the web framework is very lightweight.

1.2 SMT Solver

SAT[13] is an abbreviation of the propositional satisfiability problem, that is, the decision problem on whether there exists a valuation of propositional variables which makes true a given propositional formula (or a Boolean expression). The SAT problem is NP-complete, which is known as the Cook's theorem: one can find a solution in exponential time and check a solution in polynomial time, but it is unknown whether one can find the solution in polynomial time or not. From the theoretical viewpoint, no algorithm is known that solves an SAT problem efficiently. However, several high-performance algorithms have been found for solving propositional formulas originating from practical problems in a realistic time. There are many implementations of the SAT solver at the moment, the most famous one being MiniSAT [14].

The satisfiability modulo theories (SMT) problem is a decision problem for a quantifier-free logical formula in the first-order predicate logic with decidable theories, such as linear arithmetic, bit-vectors, arrays, etc. Various implementations of the SMT solver have been released: yices2 [3], Z3 [2], CVC4[15], etc. The SMT solvers are more useful than the SAT solvers since one can describe a domain problem more directly using a theory related to the problem domain.

1.3 Research purpose

In this paper, we propose a new formal approach to performance analysis of web servers which incorporates the analysis algorithm into a web framework. The algorithm for performance analysis is implemented using an SMT solver. The analysis algorithm estimates the performance of a web server based on concrete values obtained by execution of actual implementations of web servers.

2 Formal Performance Analysis of Web servers

In this section, we show a new method of formal performance using a web framework in which an SMT solver is embedded.

2.1 Overview of our system

In this research, we apply our method to the web framework Bottle[8] and the SMT solver yices2[3].

The overview of the process in our formal performance analysis is as follows.
1. One makes a program of a web server using the web framework Bottle in which a constraint-extracting code is embedded.
2. The program of the web server is executed on the web framework. Then, the web framework generates a logical formula based on performance data

which were derived from the experimental executions, which describes the performance requirement for the web server.

3. The SMT solver checks whether the logical formula is satisfied or not. If the logical formula includes free variables that do not correspond to any values in the program, the SMT solver tries to find the maximum values of satisfiability solution.

In Figure 1, we show a diagram of our performance analysis and the web framework.

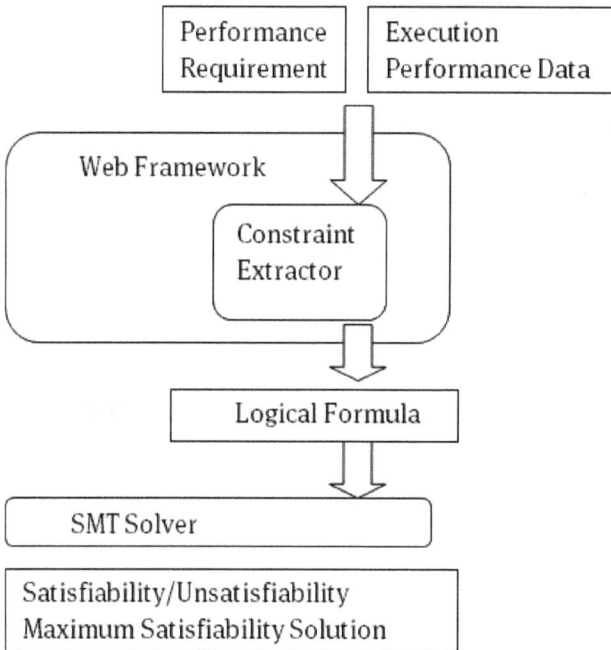

Figure 1. Note how the caption is left aligned

2.2 Execution Performance Data

The execution performance data is obtained through actual execution of sample test programs. The performance data shows the execution time for each function provided by the web framework for its user programmer.

2.3 Input Requirement and Extracted Logical Formula

An input requirement is written in the format of a logical formula for the SMT solver yices2. The following is a sample of an input requirement.

  (define n :: int) (assert (> n 1))
  (define m :: int) (assert (> m 1))
  (define f1::real (+ (* n 'send_video') (* m 'send_video')))
  (define f2::real (* n 'convert_video'))
  (define spec::real (+ f1 f2))
  (assert (< spec 100.000))

The literal 'send_video' corresponds to a function in the web framework. The function is intended to be the execution time for transmission of a video file. The literal 'convert_video' corresponds to a video-encoding function for adjustment of video format. The variable n means the number of videos transmitting with video-encoding, and the variable m means the number of videos transmitting without video-encoding. The requirement means that the total execution time should be less than 100.00. The web framework with the constraint extractor outputs the following logical formula for the SMT solver yices2.

  (define n :: int) (assert (> n 1))
  (define m :: int) (assert (> m 1))
  (define f1::real (+ (* n 1.00488185883) (* m 1.00488185883)))
  (define f2::real (* n 3.00375199318))
  (define spec::real (+ f1 f2))
  (assert (< spec 100.000))

The constraint extractor replaces 'send_video' and 'convert_video' with the concrete values 1. 00488185883 and 3.00375199318, respectively. These values are obtained through the execution of sample programs.

This logical formula is given to the SMT solver yices and returns the following result.

  sat
  (= m 3)
  (= n 24)

This output means that the logical formula mentioned above is satisfied assuming that m is 3 and n is 24. Actually, these solution provides maximum values, which are found by repetitive application of the SMT solver to the logical formula, increasing the solution candidate's values.

2.4 Evaluation of our method

In this section, we show an example of our performance analysis. We made a web server which delivers video-streaming on the web framework, whose code is shown in the Appendix. The following is the performance requirement for the web server.

> (define hd :: real 'convert_video_to_hd ')
> (define fhd :: real 'convert_video_to_fhd ')
> (define sd :: real 'convert_video_to_sd ')
> (define spec :: real (/ (+ (* sd 1) (* hd 10) (* fhd 5))
>   (+ 1 10 5)))
> (assert (< spec 0.5))

The literals 'convert_video_to_fhd', 'convert_video_hd', and 'convert_video_sd' mean necessary times for video conversion to FHD (Full High-Definition), HD (High-Definition) and SD (Standard Definition) format, respectively. In this example, we assume that the ratio of conversion requests for HD, FHD, and SD is 1:10:5. The requirement is that the total amount of time should be less than 5.0.

The result of the performance analysis is in Figure 2, which shows that the requirement is not satisfied.

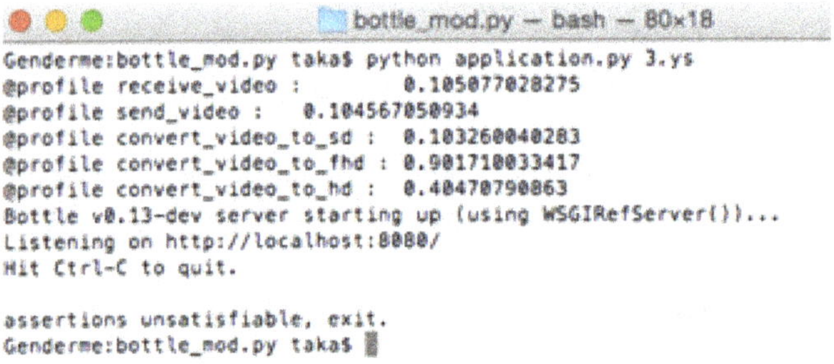

```
● ● ●                    bottle_mod.py — bash — 80×18
Genderme:bottle_mod.py takaS python application.py 3.ys
@profile receive_video :       0.105077028275
@profile send_video :    0.104567050934
@profile convert_video_to_sd :  0.103260040283
@profile convert_video_to_fhd :  0.901710033417
@profile convert_video_to_hd  :  0.40470790863
Bottle v0.13-dev server starting up (using WSGIRefServer())...
Listening on http://localhost:8080/
Hit Ctrl-C to quit.

assertions unsatisfiable, exit.
Genderme:bottle_mod.py takaS ▌
```

Figure 2. Result of Performance Analysis

The following is another requirements.

> (define m :: int)
> (define n :: int)
> (assert (> m 1))
> (assert (> n 1))

```
(define receive :: real 'receive_video ')
(define send :: real 'send_video ')
(define convert :: real 'convert_video_to_sd ')
(define spec :: real (+ (* receive n) (* convert 1) (* send m)))
(assert (< spec 10.0))
```

The literal 'receive_video', 'send_video', and 'convert_video' mean the execution time for a function of receiving a video file, for a function sending a video file, and for a function converting a video file, respectively. The requirement means the total execution time should be less than 10.0. The result of our performance analysis is in Figure 3. We find that the requirement is satisfied with the maximum solution m=92 and n=1.

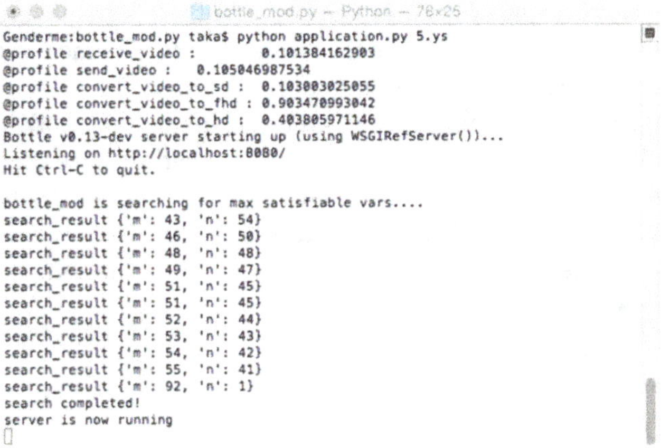

```
Genderme:bottle_mod.py taka$ python application.py 5.ys
@profile receive_video :      0.101384162903
@profile send_video :   0.105046987534
@profile convert_video_to_sd :  0.103003025055
@profile convert_video_to_fhd : 0.903470993042
@profile convert_video_to_hd :  0.403805971146
Bottle v0.13-dev server starting up (using WSGIRefServer())...
Listening on http://localhost:8080/
Hit Ctrl-C to quit.

bottle_mod is searching for max satisfiable vars....
search_result {'m': 43, 'n': 54}
search_result {'m': 46, 'n': 50}
search_result {'m': 48, 'n': 48}
search_result {'m': 49, 'n': 47}
search_result {'m': 51, 'n': 45}
search_result {'m': 51, 'n': 45}
search_result {'m': 52, 'n': 44}
search_result {'m': 53, 'n': 43}
search_result {'m': 54, 'n': 42}
search_result {'m': 55, 'n': 41}
search_result {'m': 92, 'n': 1}
search completed!
server is now running
```

Figure 3. Result of Second Performance Analysis

3 Related Works

Balsamo et al. [1] provided a review of research in the field of model-based performance analysis integrated into the software development process. In this research, we use the SMT solver yices 2 [3], but there are various other solvers such as Z3 [2]. We relate the web framework to the SMT solver and enable it to analyze the performance of web servers. There is another possibility: we could analyze the performance of a web server modeled in the model-checker JavaPathFinder[4], whose model is a JVM bytecode program. However, models described in JVM bytecode derive computational explosion of the enormous number of states.

Iyengar et al. [5] studied web server performance, and they found that web servers should reject enough requests so that the average load on the system is 95 % or less of the maximum capacity in order to prevent latencies from becoming too large. Katz et al. [6] proposed a web server design in which the number of servers can be increased dynamically in order to deal with unpredictable network traffic.

Conclusion and Future Work

We proposed a testing method [7] using model-checking where a model is analyzed through exhaustive checking of symbolic execution paths. It is an interesting and promising research theme to integrate the symbolic testing method into performance analysis with the web framework. In our previous works [16] and [17], we studied formalization of communication protocols in the framework of the process calculus. We will study integration of the formal performance analysis in this paper and the formalization of the communication protocols. In the work [18], we proposed a software design which is resistant to denial-of-service attacks. It would also be interesting to extend our work based on the software design.

Acknowledgement

This work was supported by Grants-in-Aid for Scientific Research (C) (24500009).

Appendix. Sample Program of Video-Streaming Server

```
#!/ usr/bin/env python
# -*- coding:utf-8 -*-
import time
import sys
```

```
from bottle_mod import Bottle
app = Bottle ()
app.yices_input_file = sys.argv [1]
@app.route('/ upload /')
def listen_upload ():
  receive_upload ()
  convert_video ()
  send_video ()
@app.profile ({' receive_video ': [[1], {}]})
def receive_upload(mb =1.0):
time.sleep(mb * 0.1)
@app.profile ({'send_video ': [[1], {}]})
def send_video(mb =1.0):
  time.sleep(mb * 0.1)
@app.profile ({' convert_video_to_sd ': [[], {'resolution ': 'sd '}],
  "convert_video_to_hd ": [[], {'resolution ': 'hd '}],
  "convert_video_to_fhd ": [[], {'resolution ': 'fhd '}]})
def convert_video(mb=1, resolution='sd '):
  if resolution == 'sd ':
    time.sleep(mb * 0.1)
  elif resolution == 'hd ':
    time.sleep(mb * 0.4)
  elif resolution == 'fhd ':
    time.sleep(mb * 0.9)
  else:
    time.sleep(mb)
app.run(host='localhost ', port =8080)
```

References

1   S. Balsamo, A. Di marco, P. Inverardi, and M. Simeoni, "Model-Based Performance Predic-
    tion in Software Development: A Survey," *IEEE Transactions on Software Engineering*, vol.
    30, no. 5, pp. 295-310, May 2004.

2   L. de Moura and N. Bjorner, "Z3: An Efficient SMT Solver," *Tools and Algorithm for the
    Construction and Analysis of Systems*, Lecture Notes in Computer Science, vol. 4963, pp.
    337-340, Springer, 2008.

3   B. Dutertre and L. de Moura, "The Yices SMT solver," tool paper at
    http://yices.csl.sri.com/papers/tool-paper.pdf

4   K. Havelund and T. Pressburger, "Model Checking JAVA programs using Java PathFinder," *International Journal on Software Tools for Technology Transfer,* vol. 2, issue 4, pp. 366-381, March 2000, Springer-Verlag.

5   Iyengar, E. MacNair, and T. Nguyen, "An Analysis of Web Server Performance," Proceedings of IEEE GLOBECOM, pp. 1943-1947, Nov. 1997.

6   E. D. Katz, M. Butler, and R. McGrath, "A scalable HTTP server: the NCSA prototype", Computer Networks and ISDN Systems, vol. 27, issue 2, pp. 155-164, Elsevier Science Publishers, Nov. 1994.

7   H. Kumamoto, T. Mizuno, K. Narita, and S. Nishizaki, "Applying Model Checking to Destructive Testing and Analysis of Software System," *Journal of Software,* vol. 8, no. 5, May 2013. doi:10.4304/jsw.8.5.1254-1261

8   Bottle: Python Web Framework, http://www.bottlepy.org/

9   Flask homepage, http://flask.pocoo.org

10   Django homepage, https://www.djangoproject.com

11   The Django Book, http://www.djangobook.com/

12   Ruby On Rails project homepage, http://rubyonrails.org/

13   M. Huth and M. Ryan, LOGIC IN COMPUTER SCIENCE Modelling and Reasoning about Systems, Cambridge University Press, 2004.

14   N. Eén, Niklas Sörensson, An Extensible SAT-solver, Theory and Applications of Satisfiability Testing, Lecture Notes in Computer Science, vol. 2919, pp. 512-518, Springer-Verlag, 2004.

15   CVC4 homepage, http://cvc4.cs.nyu.edu/

16   S. Nishizaki and Ritsuya Ikeda, Formal Method of Time for Analyzing Denial-of-Service Attacks, International Journal of Advancements in Computing Technology, vol. 5, issue 7, 2013, doi: doi:10.4156/ijact.vol5.issue7.71

17   Ritsuya Ikeda, Shin-ya Nishizaki, "Model Checking of Broadcast Communication via Process Calculus", AISS, Vol. 4, No. 17, pp. 373 ~ 379, 2012, doi: doi:10.4156/AISS.vol4.issue17.43

18   Takayuki Sasajima, Shin-ya Nishizaki, "Blog-based Distributed Computation", IJACT, Vol. 4, No. 15, pp. 354 ~ 361, 2012,doi:10.4156/ijact.vol5.issue7.71

Jisha P Abraham[1] and Dr.Sheena Mathew [2]

Modified GCC Compiler Pass for Thread-Level Speculation by Modifying the Window Size using Openmp

Abstract: Now a days the processors are come in the multi-core form such as dual-core and quad-core processors. To get the maximum utility of the these cores, parallel programs are written. Currently existing applications are sequential in nature and when run on multiple cores, it utilize only one core. To increase the performance of software program , parallelization is an important technique to make efficient use of all the cores. Manual parallelization requires huge effort in terms of time and money and hence there is a need for automatic code parallelization. In this paper we introduce Automated Code Parallelizer using OpenMP, which automates the insertion of compiler directives to facilitate parallel processing on shared memory machines with multiple cores. It converts an input sequential program into a multi- threaded program for multicore shared memory architectures. In this work we focuses on loops and speculatively parallelizes the different iterations of a loop while taking care of data dependency between the different iterations. While executing the various iteration the window size is varied and find the optimal window size.

Keywords: parallel programming, Open mp, window size, Thread-Level Speculation , multi threading, speculative execution

1 Introduction

The trend towards the shifting of multicore processors have greatly affected the software industry. Software should be written in such a way that it take advantage of multiple cores. Even the working platform also support multithreading concepts. Hences programmers have to rethink the way they design and write software applications. Here comes the importance of

1 M A College of Engineering/Department of Computer Science, Kothamangalam, India
E-mail:jishaanil@gmail.com
2 Cochin University of Science and Technology/Department of Computer Science , Kochi, India
E-mail:sheenamathew@cusat.ac.in

parallelism while developing new applications for performance gains on multicores. To wire parallel programs manually is difficult, cost and time consuming. Hence there is a need for tools to convert sequential codes to parallel codes. the various parallelization tools [1] or frameworks take into consideration the hardware architecture,memory architecture, data and control dependencies in the software. Parallelizability of an application is based on the inter-dependency of the tasks within that application. Inorder to identifying this detailed code analysis is required.

OpenMP, MPI, and CUDA are various options that can be used to write parallel code. These tools will specify "how- to"parallelize. An expert in this domain is able to choose the correct tool for parallelizing an application. Manual identification of parallel tasks from sequential code is time consuming and error prone.The proposed Automated Code Parallelizer is an automated parallelizing tool which will handle the problem of porting legacy applications to multi core hardware.

OpenMP[2] is successful in the case of shared memory, which allows the parallelization of user defined code region. But it does not ensure the correct execution of the code according to sequential semantics, making the programmer responsible for such tasks. Possible dependence violations that may occur between iterations during execution need to be addressed by the programmers. Automated Code Parallelizer using OpenMP automate the insertion of compiler directives for parallel processing on shared memory machines.

Automated parallelization offered by compilers only extracts parallelism from loops when there is no risk of a dependence violation at runtime. Only a small fraction of loops falls into this category, leaving many potentially parallel loops unexploited. When the compiler cannot ensure that the loop can be safely run in parallel ,Thread-Level Speculation (TLS)[3] techniques allow the extraction parallelism from fragments of code that cannot be analyzed at compile time. The variables present in the code should be classified "private" variables , and "read-only shared" variables, that are only read and not written in any iteration. If all variables inside a loop are either private or read-only shared, then the loop can be safely parallelized. If a there is a variable found that does not fit in these two categories, then the loop is not parallelizable at compile time. But the loop can be executed in parallel by software-based speculative parallelization. In this method, the loops are executed as if the iterations are independent even if there are dependencies. The TLS runtime library used is based on the same design principles as the speculative parallelization library developed by Cintra and Llanos[4].

1.1 Existing System

One of the techniques currently in use for parallel programming is OpenMP. The appropriate insertion of OpenMP features into a sequential program will allow many, applications to benefit from shared-memory parallel architectures— often with minimal modification to the code. We can exploit considerable amount of parallelism in many applications. But OpenMP fails to parallelize execution when there are data dependencies within the loops. Such loops require speculative execution.

Current speculative techniques are experimental and require manual intervention of expert programmers. These programmers firstly need to extract certain information about the source code that they want to parallelize. Programmers have to manually extract that information, such as variable usages within each loop, or I/O functions that complicate or even preclude the parallelization. It should determine whether it is worth parallelizing a loop or if the thread-management overheads would be larger than the benefit of parallelizing. After the extraction it is added to all the functions and structures[5,6] needed to handle the speculative execution.

1.2 Proposed System

Thread-Level Speculation tries to extract parallelism of loops that cannot be considered fully parallel at compile time. Other loops can be considered as well, but as long as their number of iterations cannot be so easily predicted, the applicability of TLS solutions is limited by scheduling problems.The main problem we face in software-based TLS is how to reduce the time needed to get the most up-to-date value when reading speculative data, and to search for a possible dependence violation when a thread writes on a speculative variable.

The proposed system is a new solution to traverse speculative data in a software-based TLS library. In the current TLS the window size is set to the number of threads plus one . In our work the size of window are varied and we find that optimal result is aaailable when the window size is set to double the number of threads.

1.3 Implementation

All the flags are compleatly removed during the test to get the error free code optimization, hence we are able to make the correct analysis of the change in performance. And as specified earlier the window size will determine the number of slots avilable for the threads. We changed the window size and identified that the optimal result is avilable when the window size is set as doule the size of theards. In the result we tested it with two conditions were the window is equal to the thread plus 1[3]. We varied the window size (WSIZE) as double the number of threads using the function. Which resulted in a perceivable performance improvement. Default valuses where also assigned to threads and block size(BLK) in case they are not specified.Each test program has two varients, multithreaded version and normal version(ie, single threaded).

1.4 Result Analysis

Experiments were performed to study the performance improvement gained by the system on using speculative parallelization. Test programs were executed on various environments and the execution time was analyzed. The experiments were ran on a PC with Processor   : 4x Intel(R) Core(TM) i5-2450M CPU @ 2.50GHz, Memory   : 4 GB OF DDThe experiment was conducted on single core, dual core and quad core environments with one, two, four and eight threads per execution. Table 4.1 to Table 4.2 shows the execution time for different programs under various environments. The graphs in Figure 1 - Figure 5 shows the results of the experiments for different number of cores and threads used for execution.The same set of legend is used in all the graphs.

Table 1. Execution time of pro grammes with window size (n+1)

Execution time with actual window(n+1)

	Program 1 (20k)			Program 2 (22k)			Program 3 (24k)			Program 4 (25k)			Program 5 (30k)		
	1	2	4	1	2	4	1	2	4	1	2	4	1	2	4
1	5.22	5.22	5.24	6.5	6.5	6.5	5.07	5.07	1	2	4	10.53	15.18	15.17	15.14
2	5.79	5.76	5.71	5.95	6	5.86	6	5.94	6	6.6	6.39	6.3	14.11	14.03	13.83
4	10.54	10.23	3.11	10.87	10.47	3.18	10.66	10.74	3.08	10.93	11.04	4.16	22.25	22.11	7.69
8	13.82	13.67	5.91	13.86	13.74	5.98	13.75	13.62	6.16	14.56	14.3	7.26	23	22.91	12.95

Table 2. Execution time of programmes with window size (n+1)

Execution time with window(2n)

Program 1 (20k)			Program 2 (22k)			Program 3 (24k)			Program 4 (25k)			Program 5 (30k)		
1	2	4	1	2	4	1	2	4	1	2	4	1	2	4
1														
5.21	5.22	5.35	7	6.52	6.56	5.24	5.1	5.13	10.8	10.55	10.69	16	15.25	15.31
2														
8.56	5.86	5.84	9.77	6.4	5.89	8.17	6	5.95	13.71	6.26	6.12	19.6	13.96	13.78
4														
9.31	7.44	3.06	10.69	7.8	3.14	9.28	7.59	3.14	13	7.87	3.88	18.75	13.96	7.67
8														
12	7.1	4.95	12.51	7.29	5.19	11.62	7.26	5.16	14.63	7.32	5.45	18.78	13.79	8.05

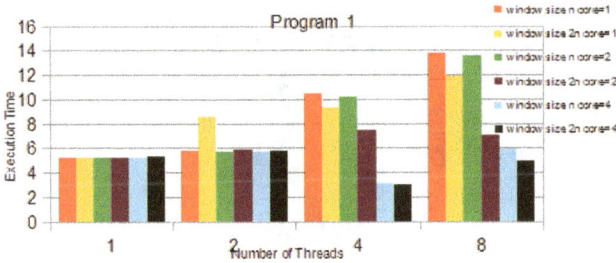

Figure 1. Execution time of program1 with window size (n+1)and 2n

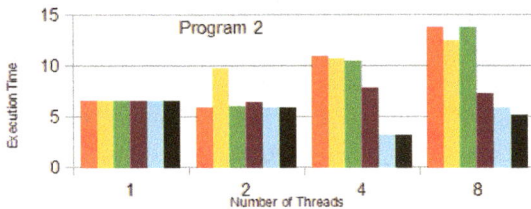

Figure 2. Execution time of program2 with window size (n+1)and 2n

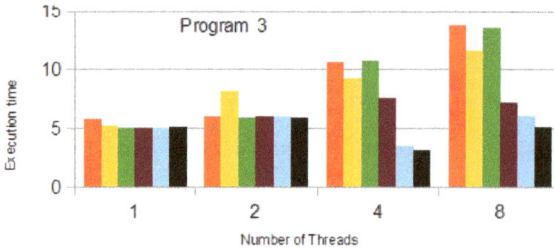

Figure 3. Execution time of program3 with window size (n+1)and 2n

It is clear from the graphs that speculative multi-threaded execution significantly improves the performance of the system. It can also be seen that the performance of multi-core systems are at their best when the number of threads is equal to the number of processing cores. In dual core and quad core systems, two and four threaded executions respectively takes the least time for completion.

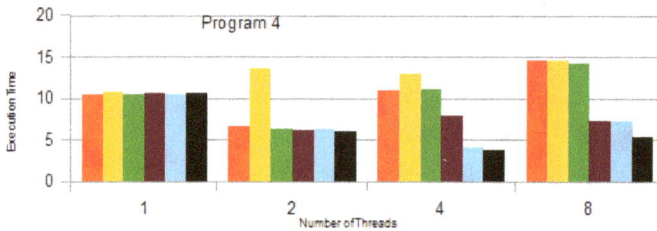

Figure 4. Execution time of program4 with window size (n+1)and 2n

This is the reason why two-threaded executions are sometimes slower in than single threaded execution in dual core systems, as seen in Figure 1 and Figure 5. It can also be seen from the same graph that as the number of threads increases, the performance improvement due to speculative parallelization dominates over the delays caused by speculative overhead and thus completes the execution faster.

In the current multi threaded model, the window size is limited to one more than the number of threads. It was found that a higher performance is achieved if the window size is doubled as is evident in Figure 1 to Figure 5. The upgraded values were reached through trial and error and extensive experimentation.The

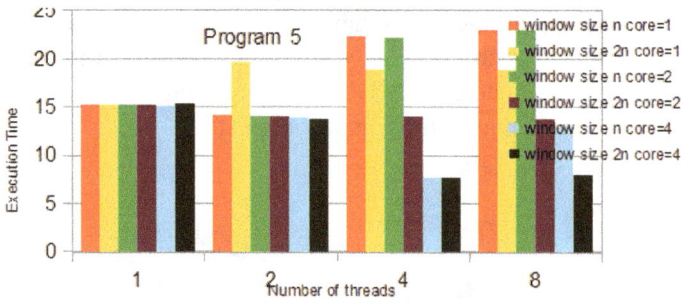

Figure 5. Execution time of program5 with window size (n+1)and 2n

trial process was however limited by the inconsistencies in the designated fields of operation beyond the allowed level of size encapsulation. Further increase in the window size would lead to increased speculative overheads which would in turn reduce the performance.It can be seen from Table 4.1 that increase in the number of threads doesn't reflect much on the performance.It is clear from Figure 1 to Figure 5 that the execution speed increases in a non linear fashion as the program size increases. This is due to the increase in the number of loops with the increase in the file size. It is also observed that the execution speed is the best when the number of cores is equal to the number of threads.The performance improves as the number of threads increases, until the number of threads equals the number of cores. If the number of threads is further increased, the performance gradually decreases. This is because when the number of threads is made higher than the number of processing cores, then multiple threads are sharing the same core. As a result, there occurs some context switching between these threads which costs time.

Conclusion

GCC Compiler Pass for Thread-Level Speculation Using OpenMP is a compile-time system that automatically adds the code needed to handle the speculatively parallel execution of a loop, and uses a new OpenMP clause to find those variables that may lead to a dependence violation. For the generation of the parallel code the programmer have add only one line , instead of the significant amount of lines required by the manual parallelization, which depends on the number of accesses to speculative variables. The system performance will also depend on the relationship between the number of threads and window size. The experiments shows that when the window size is doubled the overall performance of the system is get increased . We get the maximum performance when the window size is set as double the number of threads.

References

1   P. Randive and V. G. Vaidya, "Parallelization tools," in Second International Conference on MetaComputing, 2011.
2   "www.openmp.org".
3   Aldea, Sergio, Alvaro Estebanez, Diego R. Llanos, and Arturo Gonzalez-Escribano,"*A New GCC Plugin-Based Compiler Pass to Add Support for Thread-Level Speculation into OpenMP.*", Euro-Par 2014 Parallel Processing, . Springer International Publishing, 2014, pp. 234–245
4   Cintra, M., Llanos, D.R. " Toward efficient and robust software speculative parallelization on multiprocessors," Proceedings of PpoPP, 2003 , pp. 13–24.
5   Estebanez, A., Llanos, D.R., Gonzalez-Escribano, A., "New Data Structures to Handle Speculative Parallelization at Runtime", Proceedings of HLPP, 2014 .
6   Aldea, S., Llanos, D.R., González-Escribano, A.Chapman, B.M., Massaioli, F., Müller, M.S., Rorro, M. (eds.), "Support for thread-level speculation into OpenMP,"*IWOMP 2012. LNCS,vol. 7312,Springer, Heidelberg (2012)*, pp. 275–278 .

Liu[1] and Baiocchi[2]

Overview and Evaluation of an IoT Product for Application Development

Abstract: This paper presents the results of examining an emerging Internet of Things (IoT) product from Texas Instruments. The product examined is CC3200. Some experiences of using the CC3200 are also reported. It is believed that the CC3200 and its software development tools are excellent products for effective IoT application development.

Keywords: Internet of Things, development tools, hardware, IoT nodes.

1 Introduction

The Internet of Things (IoT) is one of the new technologies that has received broad attention. Industry and academia have growing interest in this technology. In industry, "heavy-weight" companies have developed IoT hardware and software products [1][2][3][4]. In academia, universities have started teaching IoT courses [5][6][7][8].

An IoT system is complex. However, logically speaking, an IoT system consists of only two parts – the local part and the cloud part [9]. The logical part consists of one or more monitoring nodes each of which contains a device, resource, and controller service. The monitoring node is also called a "thing" or "IoT device". The cloud part stores data received from the "things", presents data when requested, performs data analysis, and coordinate activities among the "things".

Two main players in manufacturing hardware for the "thing"s are Intel [10] and Texas Instruments (TI) [11]. The authors have chosen to evaluate TI's products for IoT teaching [12] and research because TI also manufactures numerous lines of electronic products for instrumentation and industrial control which are potentially needed by low-level hardware development for the "things". The product selected is TI's IoT chip CC3200.

---

1 Institute of Technology, University of Washington Tacoma, Tacoma, USA
E-mail: xingliu8@uw.edu
2 Institute of Technology, University of Washington Tacoma, Tacoma, USA
E-mail: baiocchi@uw.edu

Although there are a large number of documents and online resources that describe different aspects associated with the CC3200, so far the authors have not found a concise but inclusive overview for the product. It took the authors a couple of weeks to gain a fairly good understanding on different aspects of the CC3200. The authors then believe an overview with sufficient details as presented in this paper would be beneficial to new users of the product.

In the following, the paper reviews the characteristics of a "thing" in an IoT system in Section II, followed by a detailed review of the CC3200 in Section III. Section IV introduces the authors' experiences in using the CC3200. Section V summarizes the findings and concludes the paper.

2 Characteristics of "Things" in IoT

A "thing" in an IoT system is substantially more powerful than a traditional sensor/control node. Similar to a traditional sensor/control node, the "thing" is able to perform sensing, monitoring and control. However, a "thing" is also able to process data locally and communicate with the cloud or other "things" to exchange data and commands over the Internet and the Web.

2.1 Hardware Requirements for A "Thing"

In order to possess the above functions and capabilities, a "thing" should have wired and/or wireless I/O interfaces to connect sensors and actuators, interfaces for network connections, interfaces for memory storage, and interfaces for audio and video devices. Typical hardware found in an IoT "thing" is shown in Figure 1.

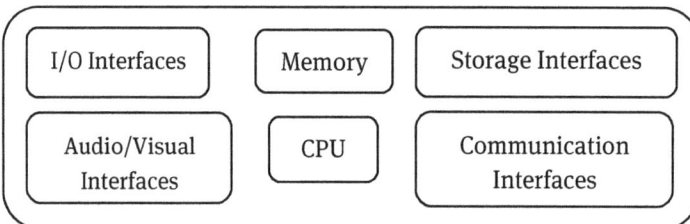

Figure 1. Hardware requirements for a "thing"

Only part of the above hardware elements are found in traditional microcontrollers which are frequently used to build sensors or control devices. Therefore, a traditional microcontroller is not sufficient to work as a "thing" in an IoT system by itself, unless it is supplemented by many other external devices. However, the disadvantage is a "thing" constructed this way will be less energy efficient and less reliable. This is not acceptable for most IoT applications.

2.2 Software Requirements for A "Thing"

An IoT "thing" is a smart object with substantial computing power. Therefore, it should be embodied with rich software resources. It should have moderately powerful software for local data processing and actuation. It should have software to accomplish data communication speedily with the cloud and other remote "things", either at the network level or at the web level. It should be equipped with security software to maintain the integrity of the "thing" and the IoT system. Because of the large number of functions supported, a "thing" should be able to run real-time tasks of different priorities concurrently. Figure 2 illustrates the software elements expected to exist inside a "thing".

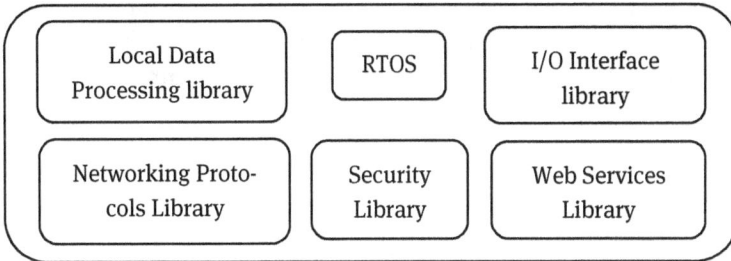

Figure 2. Software requirements for an IoT "thing"

Overall, a "thing" requires adequate hardware and software to collect data from the physical world, process it and deliver it to the cloud or share the data with other "things", or receive data and commands from the cloud or other "things". Communications between the cloud and the "things" should be fast and secure.

Achieving all of the above is not an easy task. Figures 1 and 2 indicate that a "thing" essentially can be considered a very powerful computer with strong sensing, control, and communication capabilities. It has most gadgets a normal computer would have, except with a smaller amount of memory because data will mainly be stored in the cloud. On the other hand, because a "thing" is frequently a sensor node and is likely to be powered by batteries, power consumption is also a great concern. Ideally the entire functions of "thing" is implemented in an integrated circuit (IC) chip, except the sensors and motor drives which have to be external to the IoT chip.

Among the companies that make IoT IC chips, Intel and TI are the main players. The authors chose to evaluate TI's IoT chip CC3200 because of the reasons mentioned previously. In addition, the authors have experience in using some other wireless products from TI. All of these helped with the decision-making for selecting a TI product to evaluate. The product purchased is the CC3200 SimpleLink LaunchPad.

The following sections examine the detailed functions of TI's IoT chip CC3200 and its development tools.

3 Examination of Texas Instrument's CC3200

TI designed the CC3200 as its benchmark solution for IoT. Careful considerations have been made in the design in terms of what to include on the chip. It appears that the design philosophy of CC3200 is that it should have everything a traditional microcontroller has, but it should also be Internet-ready and Web-ready, along with careful power management.

3.1 The Overall CC3200 Architecture

Essentially, CC3200 is a single IoT chip of the size of 9mm x 9mm. It has three main building blocks: an application microcontroller of the traditional sense, another microcontroller dedicated to Wi-Fi networking, and a power management system, as shown in Figure 3.

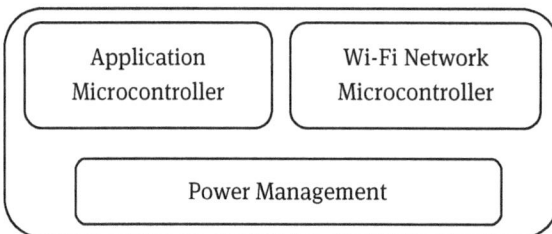

Figure 3. CC3200 building blocks

3.2 The Application Microcontroller of CC3200

The application microcontroller of CC3200 is an ARM Cortex-M4 microprocessor that runs at a frequency of 80 MHz. It has 256 KB of RAM, supports external serial Flash for the bootloader and ROM (which holds the peripheral drivers), and has an SD/MMC interface.

The application microcontroller supports a wide range of I/O interfaces. It has two UARTs, one SPI, one I2C, four timers with PWM, four ADC channels available to users, up to twenty-seven GPIO pins, one multichannel audio serial port, and one 8-bit parallel camera interface. The application microcontroller also has a hardware crypto engine which supports AES, DES, 3DES, SHA2, MD5, CRC, and checksum.

No networking or Web functions are provided by the application microcontroller. They are completely off-loaded to the network microcontroller which is discussed below.

3.3 The Network Microcontroller of CC3200

The CC3200 network microcontroller is co-located on the same IC chip as the application microcontroller. It is another ARM processor and is dedicated to networking and Web services. The network microcontroller implements the TCP/IP stack, other Internet protocols and the security protocols TLS and SSL.

The CC3200 does not support wired networking, but has extensive support for wireless networking. It implements the driver, the MAC, the baseband, and the radio for Wi-Fi.

The radio works with IEEE 802.11 b/g/n. The network microcontroller can work in three modes in Wi-Fi: station, AP and Wi-Fi direct. The radio has a transmission power between 14 dBm and 18 dBm, and a receiving sensitivity between -95 dBm and -74.0 dBm. UDP and TCP data rates are 16 Mbps and 13 Mbps respectively.

The CC3200 has an embedded HTTP server and can work as an HTTP client. More importantly, the CC3200 implements a light weight machine to machine connectivity protocol named MQTT (Message Queue Telemetry Transport), allowing speedy communications with other devices and the cloud services in IoT systems.

3.4 The Power Management System of CC3200

The CC3200 has a power management system which works with a wide range of voltages from 2.1 V to 3.6 V, or with pre-regulated 1.85 V.

Designed for IoT applications, the CC3200 has advanced low-power modes. Its hibernate mode current is 4 µA. Its low-power deep sleep mode current is 250 µA, while the maximum idle connected current is 825 µA.

Reception and transmission in the CC3200 consumes the most power, with a receiving current of 59 mA and transmission current of 229 mA at 54 Mbps with OFDM. This suggests that the system designer should minimize wakeup time, transmission time and frequency when the CC3200 is used in an IoT application.

3.5 Software Libraries for the CC3200

The CC3200 has a rich set of libraries for application development. Its peripheral control library, packaged as driverlib, provides the API functions for accessing the I/O ports on the application microcontroller. Its socket layer API provides functions for UDP and TCP sockets. The WLAN API helps with scanning and accessing points, adding and removing access point profiles, and security.

The Netapps library is a large library. It has functions for DNS resolution, pinging, and address resolution. It also provides code for HTTP client and server, SMTP, TFTP (Trivial File Transfer Protocol), XMPP (Extensive Messaging and Presence Protocol), JSON parsing, and MQTT (Message Queue Telemetry Transport) client and server.

In addition, the CC3200 has a power management library that allows power-aware applications to be developed.

The TI RTOS library is TI's own embedded real-time operating system for CC3200. In the meantime, the CC3200 also works with third-party software such as FreeRTOS.

3.6 Development Tools for the CC3200

Code Composer Studio (CCS) is the main IDE for CC3200 IoT application development. Apart from the key functions typical IDEs have, CCS has integrated the libraries for a broad range of TI products including wireless, microcontrollers and embedded processors.

There is also a cloud-based CCS IDE. The cloud-based CCS has most functions the desktop version has, but without the capabilities of viewing registers and memory, advanced debugging and tracing, and high performance probes.

In addition, programmers can use the IAR Workbench or GCC and Cygwin (for Windows) to develop applications without having to use CCS.

4 Experiences of Using the Development Tools

4.1 Code Composer Studio and Debugging Tools

The authors have been using the CC3200 SimpleLink LaunchPad and Code Composer Studio in teaching an embedded and real-time system course since January 2016. Experiences indicate that CCS has an environment very similar to other popular IDEs used by software developers in the industry such as Visual Studio. What make CCS special are its integration with the CC3200 and its tools for debugging, monitoring and analyzing the status of the hardware. For example, developers can find out whether their code handles external interrupts correctly by using the Pin Connect View tool. There is also the Port Connect tool which can be used to simulate port reading and writing.

Non-intrusive debugging and analysis of system activities can be performed in real-time using the Trace Analyzer when a program is running on the CC3200. Hardware tracing does not require changing any code in the application program.

CCS also includes a System Analyzer which can be used to perform analysis when a real-time operating systems (RTOS) is used. The System Analyzer does data analysis and visualization of the CPU, thread loads and durations, and execution sequence. Data can be collected without disrupting the running program too. Data can also be recorded and played back later.

4.2 UniFlash and PinMux

In order for developers to download binary code for applications to the external Flash memory attached to CC3200 via a serial connection, TI has developed a free standalone tool called UniFlash. UniFlash has a graphical user interface (GUI). Developers can use the GUI to decide which binary files to be downloaded to the Flash memory.

CC3200 has only a limited number of pins on the chip to interface with external signals. Therefore some pins have to be configured for multiple purposes. Even for the same pins, specifying the input/output data directions requires lengthy coding. TI has developed the PinMux utility tool which makes configuring the pins easy. PinMux has a GUI for users to specify configuration parameters such as pin multiplexing settings and thus avoid potential conflicts. The users only have to specify the peripheral signals which require external pins. Then PinMux generates a C header file and a C source file which should be included in CCS projects. The compiled code will configure the pins at run time.

Both UniFlash and PinMux are very easy to learn and use.

4.3 Code Examples and Clarity

There are no comprehensive tutorials available for CC3200 application development. In order to be able to carry out development, developers have to read through numerous user guides and sample codes. Essential guides include the Quick Start Guide, Getting Started Guide, Technical Reference Manual, (hardware) User Guide, Programmer Guide, circuit schematics, and TI RTOS Getting Started Guide. A user guide for FreeRTOS is needed if developers opt to use FreeRTOS instead of TI RTOS. These guides together amount to about 850 pages.

Instead of providing tutorials, TI has supplied 63 coding examples for programming the application microcontroller and the network microcontroller. There are brief application notes for all these 63 examples. Another 77 examples are provided for users who want to develop applications based on the TI RTOS. All the 140 examples come with the CC3200 SDK when it is installed.

Users do not have to load each example individually into CCS. Examples can all be imported through the "Project" → "Import CCS Projects" menu. TI has also provided a "Resource Explorer" under the "View" menu which allows users to load, build, and execute any individual example of their choice from the 140 examples through a web interface on a browser.

Luckily for developers, these coding examples are well documented.

4.4 RTOSes

To facilitate multi-tasking/multi-threading application development, CCS and CC3200 support two embedded real-time operating systems: FreeRTOS and TI RTOS.

FreeRTOS is an open source software written in C which only implements a minimum set of RTOS functions including preemptive multi-tasking, memory management, queues, semaphores, mutexes and inter-task communication through queues. However, FreeRTOS does not support networking, external hardware access and filesystem. Although the source code of FreeRTOS comes with the CC3200 SDK, documentation is not free. Developers have to purchase them from the owner of the software via freertos.org.

TI RTOS is more powerful. It has everything FreeRTOS has as a RTOS. But TI RTOS also has support for networking, USB, FAT file system, and device drivers. The real-time kernel of TI-RTOS is called SYS/BIOS. A tool named XDCtools is used for configuring and building SYS/BIOS based applications.

On the other hand, TI RTOS has a Unified Instrumentation Architecture (UIA) for data collection. In CCS, UIA is used with the System Analyzer to present data.

The authors have covered FreeRTOS along with CC3200 in the course being taught. It seems that CCS does not compile some FreeRTOS functions if they are used directly in the source code. Instead, the wrapped up counterparts of those functions found in the osilib had no problem. Students were reporting that CCS Cloud did not compile the FreeRTOS examples either. More investigation is needed on these issues in order to find out what the real problems are.

4.5 Observations from Classroom Teaching

As stated previously, the CC3200 has been used in teaching a 4th year embedded real-time system programming course. The course is designed for students to gain programming experience in developing software that interfaces with external devices based on concepts such as task scheduling, multi-threading or multitasking in a RTOS, and critical sections. The I/O interfaces CC3200 has, and the RTOS software CC3200 supports, have enabled the authors to achieve the desired learning outcomes.

Although the amount of reading is challenging, the students have managed to solve the problems assigned to them. The initial course project presentation conducted recently indicated that the students have gained confidence in building projects that are beyond the expectations of the instructors.

Conclusions

From the hardware perspective, Texas Instruments' IoT product CC3200 has all functions required by an IoT "thing" in an IoT system. It has wired and/or wireless I/O interfaces to connect sensors and actuators and interfaces for data communication and networking, interfaces for memory storage, and interfaces for audio and video devices.

From the software perspective, the CC3200 is supported by rich libraries to work with wide range of peripherals, connect with the Internet and Web servers.

CC3200's support for TI RTOS and FreeRTOS makes it an excellent platform for a smart "thing" in an IoT system.

The numerous development tools created by Texas Instruments enable rapid IoT application development. The code examples serve as a good starting point for new projects.

The design of the CC3200 SimpleLink LaunchPad had indeed the new users in mind. Most students appear to enjoy using the product. CC3200 has stimulated students' interest in writing software that interacts with hardware including sensors, and has given students great opportunity to gain skills for IoT application development.

Overall, TI's CC3200 is a top-notch IoT product for implementing intelligent IoT nodes in an IoT system. It is also a suitable platform for teaching IoT courses.

References

1   Microsfot.com, "Tap into the Internet of Things with the Azure IoT Suite", [Online]. Available:http://www.microsoft.com/en-ca/server-cloud/internet-of-things/azure-iot-suite.aspx. [Accessed: 24-Dec-2015].
2   Aws.amazon.com, 'AWS IoT', 2015. [Online]. Available: https://aws.amazon.com/iot/. [Accessed: 24-Dec-2015].
3   Intel.com, 'The Internet of Things (IoT) Starts with Intel Inside', 2015. [Online]. Available: http://www.intel.com/content/www/us/en/internet-of-things/overview.html. [Accessed: 24-Dec-2015].
4   Ti.com, 'TI's SimpleLink™ Wi-Fi® Family: Connect More: Anywhere, Anything, Anyone', 2015. [Online]. Available: http://www.ti.com/ww/en/simplelink_embedded_wi-fi/cc3200.html. [Accessed: 24-Dec-2015].
5   Uw.edu, 'Certificate in Internet of Things', 2015. [Online]. Available: http://www.pce.uw.edu/certificates/internet-of-things.html. [Accessed: 24-Dec-2015].
6   Harvard.edu, "CS 144r/244r: Secure and Intelligent Internet of Things", 2014. [Online]. Available: https://www.eecs.harvard.edu/htk/courses/. [Accessed: 24-Dec-2015].

7   Calpoly.edu, "CSC 520: Internet of Things", 2015. [Online]. Available: http://users.csc.calpoly.edu/~foaad/IOTS15.pdf. [Accessed: 24-Dec-2015].
8   Ncsu.edu, "CSC 591: Internet of Things: Applications and Implementation", 2016. [Online]. Available: http://www.csc.ncsu.edu/courses/. [Accessed: 24-Dec-2015].
9   V. Madisetti and A. Bahga, *Internet of Things (A Hands-on-Approach)*, VPT, 2014.
10 Intel.com, '*The Internet of Things (IoT) Starts with Intel Inside*', 2015. [Online]. Available: http://www.intel.com/content/www/us/en/internet-of-things/overview.html. [Accessed: 24-Dec-2015].
11 Ti.com, 'TI Internet of Things Overview', 2015. [Online]. Available: http://www.ti.com/ww/en/internet_of_things/iot-overview.html. [Accessed: 24-Dec-2015].
12 X. Liu and O. Biocchi, "An IoT Course for A Computer Science Graduate Program", International Conference on Communication, Management and Information Technology ICCMIT'16, Cosenza, Italy, April 26-29, 2016.
13 Ti.com, "CC3200 SimpleLink™ Wi-Fi® and Internet-of-Things solution, a Single-Chip Wireless MCU", [Online]. Available: http://www.ti.com/product/cc3200. [Accessed: 20-Feb-2016].

A.Senthamaraiselvan[1] and Ka.Selvaradjou[2]

A TCP in CR-MANET with Unstable Bandwidth

Abstract: In recent years, Cognitive Radio (CR) plays the vital role in solving the problem of spectrum scarcity which arise due to the impact of unlicensed users operating under the licensed user's bandwidth without affecting their performance. All through communication process, to make an efficient use of available resources, characteristics such as Bottleneck bandwidth, Interference and Round trip time need to be modified and adaptively update by TCP in its congestion window (CWND). Here we propose TCP CR-MANET to whelm the problem of efficient resource utilization by computing and administrating the licensed users Bottleneck bandwidth, Interference and Round trip time in the available buffer space of the relay nodes. TCP CR-MANET is implemented in NS2 simulator and experimentally analyzed throughput's outcome by tuning the above said characteristics.

Keywords: Cognitive Radio, Congestion control, Spectrum Sensing, Transport Protocol, and Mobile ad hoc network.

1 Introduction

Cognitive radio technologies have created the deep impact towards the growth of wireless communication, due to its ability in handling spectrum utilization effectively by redistributing the Unused spectrum in dynamically changing environments. The unlicensed bands, mostly that fall under 900MHz and 2.4GHz, are getting more and more congested [1]. The radio spectrum demand has been increased dramatically and still few more available spectrums can be allocated. However, according to Federal Communications Commission's (FCC) report [2], the same spectrum bands are underutilized due to existing amount of

1 Research Scholar, Department of Computer Science and Engineering, Pondicherry Engineering College, Puducherry, India.
E-mail:senselvana@pec.edu
2 Professor, Department of Computer Science and Engineering, Pondicherry Engineering College, Puducherry, India.
E-mail:selvaraj@pec.edu

idle spectrum holes at spatial and temporal measurements. Cognitive Radio Technology has the potential to ameliorate the scarcity of wireless resources. In this paper, we cautious to pin-point that Cognitive Radio Mobile Ad Hoc Network (CR-MANET), internally neither consists of federal party to obtain the spectrum usage information from the neighborhood nor external third party provision (spectrum broker) that empowers the distribution of the offered spectrum resources.

In classical ad hoc networks, the mobility of relay nodes and the ambiguity residing with wireless channels are the two key factors that affect the reliable distribution data from source to destination [3]. CR-MANET can be deployed in various aspects of Intelligent Transport Systems (ITS) applications [4]. The main challenges of transport layer in a classical wireless ad hoc networks are [5] Congestion, Packet drops based on channel related problems, Packet losses on mobility. In case -1: The RTT value was increasing based on the increased queuing delay of relay nodes. When RTT value goes beyond the given limit, the relay nodes fail to forward packets, likely this event degrades the performance of TCP. In case 2. In the network, a packet drops due to channel related problems or channel induced, likely of fading and shadowing performance of channel. In case 3. Relatively Packet losses occurred in the network, when there was mobility related losses or Permanent losses [6]. The source node would mistakenly consider the above mentioned cases as congestion event. All these losses are taken as inducing factors that are applicable to CR-MANET. In CR-MANET, we rely an intermediate nodes, which periodically piggyback the spectrum information with Acknowledgement (ACK) and also update's Primary User's arrival on time, explicitly informing the source. Our protocol ensures prospectively channel switching event by adhering to the momentous updation in bandwidth of the interfered link. Thus we propose a TCP congestion window (CWND) which levers rapidly to the transformation in the environment. Hence the objective of TCP CR-MANET is to facilitate window based methodology of the classical TCP, and increased applicability.

2 MOTIVATION

Here we analyze the one of the problems of the preceding approaches of transport protocols like TCP NewReno in Cognitive Radio ad hoc network which drive us to have enhanced performance with the proposed system TCP CR-MANET. With respect to Cognitive radio ad hoc network, each node is furnished with a Radio frequency transceiver. The key deeds of the Cognitive Radio Networks are 1. Spectrum sensing, 2. Impact of primary user activity, 3. Spectrum change. Primary Users are to be established as Poisson arrivals, by having an "on" time as $\left(\frac{1}{\alpha}\right)$ and as well as an "off" time as $\left(\frac{1}{\beta}\right)$ respectively to all kinds of provided channels of the network.

2.1 Spectrum sensing

The Secondary User or Cognitive User and the intermediate nodes has to do a periodic check over the current channel according to pre-defined sensing time for identifying a presence of licensed users. When a sensing time t^s is zero, the sensing is disabled on that period, we observed that the congestion window keep increasing until it reaches the maximum capacity of the channel. TCP sender side obtains a multiplicative delay reach the maximum retransmission timeout (RTO), RTO event could be triggered which results the degradation of the end to end performance [8]. Sensing time plays the vital role in deriving the optimized performance [9] via 1) thorough recognition of primary user and 2) effective utilization of the channel, which are the two contradictory goals that need to be achieved by diagnosing a better blending factor. The transport layer acclimatizes the current rate of sensing state and decides optimal setting of sensing time [10].

2.2 Impact of Primary User Activity

A primary user's (PU) activity is periodically detected during spectrum sensing or data transforming. In case, on arrival of the primary user, if secondary user's (SU/CR user)operation affects the current channel, the system will be in search of various unoccupied channel in the spectrum. When the current channel's spectrum sensing is of periodic and well defined interval, two activities will be performed as Available Channel set Discovery at various spectrum bands, Harmonize with the subsequent hop neighbors to derive mutually adequate channels in the set.The transport protocol [11] to differentiate these states, based on the value of "on" and "off" stage (α and β) of Primary User activity, that comprises four different patterns as follows.

1. High Activity$\left(\frac{1}{\alpha} \leq 1, \frac{1}{\beta} > 1\right)$, 2. Low – Activity $\left(\frac{1}{\alpha} > 1, \frac{1}{\beta} \leq 1\right)$.

3. Short term Activity $\left(\frac{1}{\alpha} > 1, \frac{1}{\beta} > 1\right)$, 4. Long term Activity $\left(\frac{1}{\alpha} \leq 1, \frac{1}{\beta} \leq 1\right)$.

The packet losses that would appear due to congestion and from the one occurring due to primary user's interference. The proposed event handler is triggered in some constraints, as shown in Table 1.

Table 1: Event Occurrences

Information Type	Low PU Activity	High PU Activity
Low Bandwidth Capacity	PU Interference, Congestion, ARTT.	PU Interference.
High Bandwidth Capacity	PU Interference.	PU Interference, ARTT

2.3 Spectrum change state

The effective utilization of the spectrum resources by Cognitive Radio Technology. Secondary users have opportunistically transmitting in the license bands on limited duration. Primary users is modeled as position arrival with "on" and "off" time respectively. TCP must be regulating accordingly the new available bandwidth [8][12]. TCP must be scaled the CWND to meet the channel condition.

3 TCP CR-MANET

The CR-MANET nodes employs a single radio transceiver, to regulate channels between licensed and unlicensed spectrum. The channels in different spectrum bands may have disparate channel bandwidths. Channels are operated under spectrum band and the data transfer is being carried over using OFDMA with CSMA/CA technique and processed by priority queue at Medium Access Control (MAC) layer. This MAC protocol [13] is based on the following assumptions: 1) A fixed MAC super-frame size for SU. An SU decides to access the channel based on SU scans. 2) The PU and SU to use N orthogonal (interference-free to each other) channels. An SU can access 1 to N channels simultaneously. 3) An SU is able to (and must) vacate the channels in use whenever interrupted by the PU. 4) Each channel contains at least one PU transmitter-receiver pair. 5) The MAC layer contains the information such that historical channel utilization of the PU and real-time channel availability. 6) The MAC layer can distinguish the PU's signal from noise.

3.1 Network Modeling

The chain topology network is simulated in CR-MANET. The Chain Topology wireless nodes are constructed and symbolized in the cognitive radio scenario, the transmission is initiated among the CR users when the PU starts transmitting the packets, if the channel used by CR user is the licensed channel of PU, it switches to another available channel, allocating the current channel to PU in order to maintain the performance of primary user.

3.2 Connection Establishment

The three-way handshake protocol are used to establish the connectivity of CR-MANET. The source sends a Synchronization (SYN) packet to the Sink. A relay nodes in the routing path appends the following information to the SYN packet: ID, Timestamp, and the tuple. On the receiver part,after getting the SYN packet, it sends a SYN-ACK message to the source.

3.3 Spectrum Sensing

CR-MANET uses ED (Energy Detection) as the sensing technology. Each SU performs sensing and data transmission processes in an asynchronous time-division manner. Therefore, the TCP sender regulates the timing and duration in a routing path based on sensing schedules [13]. The node requires different schedules for sensing, it may happen that a node receives messages from more than one TCP sender [14]. Sensing time t^s of each node is calculated using following equation (1)

$$t_i^s = \frac{1}{W\gamma^2}\left[Q^{-1}(P_f) + (\gamma + 1)Q^{-1}\left(\frac{P_{off}P_f}{P_{on}}\right)\right]^2 \tag{1}$$

with the parameters bandwidth (W), standard function (Q^{-1}), Probability of missed PU detection (P_f), SNR (γ), Probability of PU on period (P_{on}), and Probability of PU off period (P_{off}). The PU interference occurs, when PU attempts to access a channel used by an SU. According to the characteristics of CR-MAC, an SU must not occupy the channel whenever the PU accesses it, in turn which may lead to have "PU-Interference Loss" event [15]. In a PU-interference loss event, the TCP client may face a time-out condition, if the MAC fails to recover the collided packets within deadline. Such an event probably leads to have a decreased TCP throughput. The PU's historical channel utilization, is recognized and maintained as "PU-Activity" information, by acquiring the knowledge about PU's activity, the SU senses (and decides to access) channels periodically in each MAC super-frame. Proximate arrival of PU may certainly push the channel to have PU-interference loss, subsequently the collided packets can be retransmitted in the beginning of the next super-frame rather than waiting for other retransmission mechanisms. Starvation is more serious when MAC super frame is longer which causes throughput decline .On the other hand, to aid process of retransmission the MAC super-frame is made short, to recover the collided packets .By forcing the collided packets to be retransmitted at the beginning of the MAC super-frame, the starvation cases can be relaxed. However, forcing the packets to retransmit causes duplicate packets. To maintain the congestion window, ACKs of re-transmitted packets are not taken into count and thus framed as duplicate ACKs.

3.4 Channel Switching

The channel switching is detected periodically in spectrum band, for PU arrival in the network and SU switching to the available bandwidth. The RTT-adjustment function [14] is triggered in the cases of bandwidth changes and propagation delay.   RTT = Propagation Delay + 1/ Bandwidth.
The $RTT_{PATH} = RTT_{VL} + RTT_{SL}$, Where RTT_{VL} and RTT_{SL} are the RTT of the CR link and the stable bandwidth link, respectively.

$$RTT_{VL} = t_{VL} + l_{VL}/ bw_{VL}, RTT_{VL} = t_{SL} + l_{SL}/ bw_{SL} \qquad (2)$$

Where t_{VL} propagation delay for RTT_{VL}, t_{SL} propagation delay for RTT_{SL}, l_{VL}, l_{SL} packet length of CR link, and packet length of stable link.

$$RTT_{PATH} = C + \frac{L}{BW_{VL}} \qquad (3)$$

Where ($C = t_{VL} + t_{SL} + l_{SL}/ bw_{SL}$), $L = l_{VL}$, We can use the least squares method to train C and L in Equation (3). Giving n entries of the corresponding RTT-bandwidth information pair where (BW_i, RTT_i) represents the ith entry, C and L can be derived from:

$$L = \frac{\sum_{i=1}^{n}(BW_i - \overline{BW})(RTT_i - \overline{RTT})}{\sum_{i=1}^{n}(BW_i - \overline{BW})^2} \qquad (4) \qquad C = \overline{RTT} - LX\,\overline{BW} \qquad (5)$$

Where $\overline{BW} = \frac{1}{n}\sum_{i=1}^{n} BW_i$, $\overline{RTT} = \frac{1}{n}\sum_{i=1}^{n} RTT_i$, $RTT(L,C,BW) = C + \frac{L}{BW}$ (6)

The nodes are switching their channels, to evaluate communication characteristics link bandwidth BW=W, link delay (L^T). The TCP sender aware these characteristics and RTT then compute easily the bottleneck bandwidth (W'_b). In channel switching, the sender must update CWND and RTT from equation (7) and (8). L'^T is the link delay before channel switching. A relay node change its channel, if primary user communication is detected and also measure ssthresh.

$$RTT_{new} = RTT_{old} + L^T - L'^T \quad (7) \qquad CWND = \alpha. W'_b \ RTT_{new} \qquad (8)$$

4 IMPLEMENTATION

When a relay node changes its channel, on account of PU communication detection, its bandwidth and link delay ($L'^{T}_{i;i+1}$) can also be changed. This change
is drastic when the bottleneck bandwidth or RTT changes. Therefore, appropriately updates its congestion control parameters cwnd and ssthresh, using the
feedback information received from the relay node is essential. The bottleneck
node is located in two cases, the first case is that the bottleneck node is placed
on a path from the TCP sender to bottleneck node, the buffer will be empty by
the end of channel switching when the TCP sender stops sending packets during channel switching. The second case is that the bottleneck node is just the
switching node and forward to the TCP sender. The TCP sender receives the
feedback message after channel switching, then it calculates the RTT_{new} by using equation (9) and bottleneck bandwidth W'_{b}. Estimation of the CWND [17]
and RTT process as shown in Fig1.b. The network consists of the Primary user
and the CR users where the CR users actively participate in the packet transformation when the PU is idle and handoff the channel when the primary user
enters the packet transferring mode.

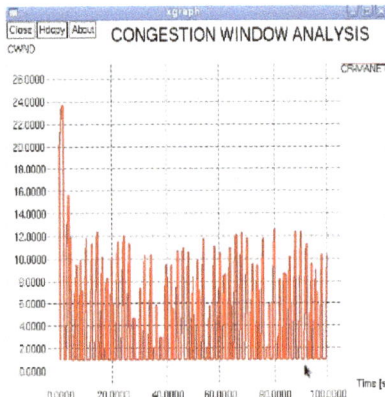

Fig.1. a. Congestion window Analysis   b. Algorithm to update the cwnd

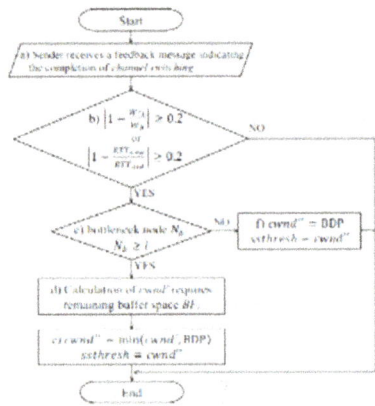

$$RTT_{new} = L_{1,2}{}^{T} + .. + L_{i-1,i}{}^{T} + L'^{T}{}_{i,i+1} + ... \qquad (9)$$

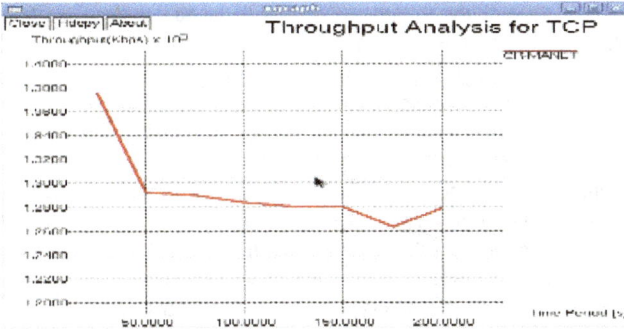

Fig.2. Throughput Analysis for CR-MANET

Fig.2. clearly states the throughput analysis of CR-Manet by tuning different parametrical setup with respect Bottleneck Bandwidth,congestion window and RTT.

Conclusion

TCP CR-MANET experimentally proved to support optimal resource utilization by incorporating buffer space of relay node in computation and administration of Bottleneck bandwidth, Interference and Round trip time of the communication channel.TCP CR-MANET protocol implementation involves beside connection with the fundamental link and network layers, specifically during on events, such as channel switching, mobility and Interference.TCP sender should examine both the bottleneck node and buffer resource in the network and updates the CWND when either the bottleneck bandwidth and RTT parameters were modified in terms of channel switching.Accordingly experiment results show that TCP CR-MANET give way for efficient resource utilization while resolving spectrum scarcity of the communication networks.

References

1   I.F. Akyildiz, W.Y. Lee, and K. Chowdhury, "CRAHNs: Cognitive Radio Ad Hoc Networks," Ad Hoc Networks J., vol. 7, no. 2, pp. 810-836, Elsevier, 2009.

2   FCC, Second Memorandum Opinion and Order, ET Docket No. 10-174, 2010.

3   M. D. Felice, K. R. Chowdhury, W. Kim, A. Kassler, L. Bononi, "End-to-end Protocols for Cognitive Radio Ad Hoc Networks: An Evaluation Study," Elsevier Performance Evaluation, Vol. 68, No. 9, pp. 859-875, 2011. 492, 2011

4   A.O. Bicen and O.B. Akan, "Reliability and Congestion Control in Cognitive Radio Sensor Networks," Ad Hoc Networks J., vol. 9, no. 7, pp. 1154-1164, Elsevier, 2011.

5   J. Liu and S. Singh, "ATCP: TCP for Mobile Ad Hoc Net-works," IEEE Journal on Sel. Areas of Comm., Vol. 19, No. 7, pp. 1300-1315, 2001.

6   T. Melodia, D. Pompili, and I.F. Akyildiz, "Handling Mobility in Wireless Sensor and Actor Networks," IEEE Trans. Mobile Computing, vol. 9, no. 2, pp. 160-173, 2010

7   K.R. Chowdhury, M. Di Felice, and I.F. Akyildiz, "TP-CRAHN: A Transport Protocol for Cognitive Radio Ad Hoc Networks," Proc. IEEE INFOCOM, pp. 2482-2491, 2009.

8   A.M.R. Slingerland, P. Pawelczak, R.V. Prasad, A. Lo, and R. Hekmat, "Performance of Transport Control Protocol over Dynamic Spectrum Access Links," Proc. Second IEEE Int'l Symp. New Frontiers in Dynamic Spectrum Access Networks (DySPAN), 2007.

9   M. Di Felice, K.R. Chowdhury, and L. Bononi, "Modeling and Performance Evaluation of Transmission Control Protocol over Cognitive Radio Ad Hoc Networks," Proc. 12th ACM Int'l Conf. Modeling, Analysis and Simulation of Wireless and Mobile (MSWIM '09), pp. 4-12, 2009

10 D. Sarkar and H. Narayan, "Transport Layer Protocols for Cognitive Networks," in Proc. of IEEE INFOCOM Computer Communication Workshop, pp. 1-6, 2010,

11 K. Sundaresan, V. Anantharaman, H. Y. Hsieh, and R. Sivakumar, "ATP: A Reliable Transport Protocol for Ad Hoc Net-works," IEEE Trans. On Mobile Computing, Vol. 4, No. 6, pp. 588-603, 2005.

12 Kaushik R. Chowdhury, Marco Di Felice, and Ian F. Akyildiz, "TCP CRAHN: A Transport Control Protocol for Cognitive Radio Ad Hoc Networks" IEEE Transactions On Mobile Computing, Vol. 12, No. 4, pp 790 – 803, 2013.

13 T. Issariyakul, L. S. Pillutla, and V. Krishnamurthy, "Tuning radio resource in an overlay cognitive radio network for TCP: Greed isn't good," Communications Magazine, IEEE, vol. 47, pp. 57-63, 2009.

14 C.Y.Chun, C.F.Chou, E.H.Wu, G.H.Chen, " A Cognitive TCP Design for a Cognitive Radio Network with an Unstable-Bandwidth Link" IEEE Transaction on Computers, Vol.64, Issue No.10, 2015.

15 N. Parvez, A. Mahanti, and C. Williamson, "An Analytic Throughput Model for TCP NewReno," IEEE/ACM Trans. Networking, vol. 18, no. 2, pp. 448-461, 2010.

16 A. Al-Ali and K. R. Chowdhury, "TFRC-CR: An Equation-based Transport Protocol for Cognitive Radio Networks," Elsevier Ad Hoc Networks Journal, Vol. 11, No. 6, pp. 1836-1847,2013.

17 Senthamaraiselvan, Ka. Selvaradjou," Ameliorate pursuance of TCP in CR-MANET utilizing accessible resources "International Conference on Computing, Communication and Security (ICCCS), 2015.

Morande Swapnil[1] and Tewari Veena[2]

Impact of Digital Ecosystem on Business Environment

Abstract: As of today, many of the information technology giants have been designing great hardware, some of them are working to make software and some are creating value based digital services. Few computer systems. That said, without Digital Ecosystem units has become next to impossible to fulfil user promise and lead digital industry. The overall user experience works best when tied up closely, pushing the boundaries of hardware systems. This research would study the tangible and non-tangible benefits that a Digital Ecosystem can provide to revolutionize financial margins of businesses. Further, it will also propose an optimal approach on a cross platform aspect of information management and analyse the overall effect of the Digital Ecosystem towards mapping their own business model.

Keywords: Business Intelligence, Digital Ecosystem, IoT (internet of things), User Experience, Workflow

1 Introduction

'Digital Ecosystem' is defined as the interdependence of Computer software, hardware and applications, and Digital services that play a role to achieve certain Objectives. The concept of Digital Ecosystem was proposed to describe a business community that relied on Information technology practices to achieve the objectives of higher market penetration, more customers and greater social inclusion in our digital lives. With the help of today's Internet based economy Digital Ecosystem can help push the product to the consumer. The emergence of business intelligence (Bi) can gather data from digital devices and further serve the consumer's interests. It is expected that the user experience would be easier and enhanced for organization with this approach. There are many startups who have not realized the effectiveness of this concept and few market players

---
1 Majan University College, P.O. Box 710, Postal Code 112, Ruwi, Muscat, Sultanate of Oman
Email: swapnil.morande@majancollege.edu.om, 00968-24730464
2 Majan University College, P.O. Box 710, Postal Code 112, Ruwi, Muscat, Sultanate of Oman
Email: veena.tewari@majancollege.edu.om, 00968-24730455

are yet to consider the impact of such Digital Ecosystem on their business practices to work out respective financial margins. With given research it can be stated that Digital Ecosystem is a part of services economy where innovative products can get customer excited to provide a customized experience.

2 Literature Review

With the technological advances, companies have developed innovative technologies by recognizing that software is tightly coupled with, and a vital element of, all software and hardware platforms. According to Forbes (2015) as technology continues to move at a breakneck pace — with social, mobile, analytics, cloud and other technologies driving the rapid evolution of digital businesses — pioneering enterprises are rewriting the digital playbook. They're stretching their boundaries by tapping into a broad array of other digital businesses, digital customers and even digital devices at the edge of their networks to create 'Digital Ecosystems' that are re-shaping entire markets and changing the way we work and live.

This shift is best highlighted in the rapidly growing Industrial Internet of Things (IoT) — i.e., the interconnection of embedded computing devices within the existing Internet based infrastructure — as companies are using, these connections to offer new services, reshape experiences and enter new markets through these Digital Ecosystems. According to PR Newswire (2011) Mobile cloud has the great potential to change, a lot bigger than what can be conceived today. The size of the mobile cloud market is poised to reach over 45 billion by the year 2016. With centralized data management (contacts, calendar, Bookmarks, emails and notes along with other application based settings) mobile cloud ecosystem is rapidly evolving. And with new, complex and secure products available from companies like Apple, Google and Microsoft, other cloud product companies as well as enterprises want to capitalize on the opportunities. According to CIO magazine (2015) a recent Accenture global survey of more than 2,000 IT and business executives found that four out of five respondents believe that the future will see industry boundaries dramatically blur as platforms reshape industries into connected ecosystems. Sixty percent of respondents said they plan to engage new partners within their respective industries and 40 percent plan to leverage digital partners outside their industry.

Thor Olavsrud (2015) documents that organizations are working with manufacturers to make all of its connected home products compatible with smart home system and healthcare equipment. Some organizations are working to create a platform through which an ecosystem of applications would help enable collaboration and workflow between stakeholders. In his article in CEO, he further notes - "As the digital transformation continues, today's organizations are building Digital Ecosystems that tap an array of other digital businesses, digital customers and digital devices at the edge of their networks."

There are companies like CloudSigma who are customer centric and deal with cloud IaaS (Infrastructure as a Service) platform have been busy developing Media Services based ecosystem exclusively for production companies, Such Digital Ecosystem provides high performance compute, storage and data transfer while content distribution driving the content driven services. According to Murat Ayranci (2016) into the digital world; companies have created Digital Ecosystems in which, more or less, everything exists and everything more or less suits everything. In doing so, they have made the life of their users easier while filling their pockets. It is apparent that such Ecosystems are driving the service economy for these companies. Also, Paul Daugherty (2015) - chief technology officer of Accenture - says Accenture's annual outlook of global technology trends — Accenture Technology Vision 2015 — found that these visionary companies recognize that as every business becomes a digital business, together they can effect change on a much bigger stage, collaborating to shape experiences, and outcomes, in ways never before possible. As a result, these leading enterprises are shaping a new type of service economy known as the "We Economy."

3 Motivation

– To explore the influence of Digital Ecosystem on the overall performance of service organizations.
– To understand the opportunities for an extended Digital Ecosystem, in relation to service organizations.

4 Problem Domain

Digital Ecosystem is a convergence of recent broadband adoption and technological advances that have pushed the Internet, telecommunications and Media industries into a flux. Digital Ecosystem can be represented with an example illustrated as below where several products from the company that belong to one of the categories such as hardware and Software which further drives service economy.

5 Problem Definition

Hence such Ecosystem based on digital products and with emerging competition across market players is making content delivery a lot appealing. Hence proposed research would work to answer following questions-
- Find out what is the influence of Digital Ecosystem.
- Investigate market opportunities that can be extended using Digital Ecosystem.

6 Statement

This research is an exploratory in nature and attempts to summarize the available literature in relation to the objectives of the research. In addition to that, as the authors have also expressed their own views in terms of their own experience, expertise and learning in this area. This is being done to support the various arguments, supporting the objectives.

7 Innovative content

The concept of Ecosystem can be experienced with the help of simulation website at – Echosim.io which provides first hand experience of amazon ecosystem based on Artificial intelligence integrated with services - Philips Hue, Pandora, WeMo, IFTTT, Wink, BBC, Audible etc.

8 Problem representation

According to Moore's law the exponential advancement in technology is heading towards high performance, high capacity computing. Keeping that in mind every organization can stand a chance to become a market leader while selling their products. However to augment technological development with a company's own product line to deliver better user experience digital ecosystem pals a greater part! Lack of ecosystem can play out as one of the strongest competitive advantages while playing in digital space.

9 Solution methodologies

9.1 Digital Content Management

Leading computing companies are currently employing the principal of Digital Ecosystem where hardware, software and services are built together from the ground up. We know that success of an iPad where Apple Inc. used software like IOS and services like iCloud to bring things under singular ecosystem. Apple's digital supply chain completely works on online content where hardware products make content consumption easier to make the entire purchase process very effective. The same can be leveraged while working with any other product within apple's ecosystems such as Apple iPhone or Apple TV. According to Keynote presentation by Apple Inc. in 2015 a less significant do feature such as Hand-Off can also be extremely valuable while working with Digital Ecosystem. The communication established with other systems and integration with peripheral devices can be very strong.

Figure 1. Apple's Digital Ecosystem

As shown in above Figure 1, Customers buy one of Apple's legendary products and use iTunes to purchase the content - where iCloud will works as a data synchronization system - for customers locking customers within Apple's Product Ecosystem. It does not stop here the similar ecosystem models are being repeated by leading companies such as Amazon, Microsoft and Google.

9.2 Unified Communication

Unified communications is a technology geared towards the user regarding communication via smart phones, PDAs and other types of mobile devices. It provides access to voice, Instant Messaging, Location data along with other business applications. This form of unified communications is mainly used for supporting personal productivity, however, in this research I may be able to draw innovative models for service/product delivery. With the help of enhanced artificial intelligence system Google takes it to another level with soon to launch a product called Google Home. Beyond doubt it would use Google's supremacy in the 'online search' field to deliver amazing services. Google Home leverages Google search. Because Google assistant is connected to the best search engine available, Home is going to provide much smarter results for queries. A portion of the Google Home demo shows the device notifying the user with the help of Digital Ecosystem. When the user says "OK, Google," Home warned the user their flight was delayed by 30 minutes. Google assistant is supported by Google notifications.

Manage everyday tasks effortlessly

Figure 2. Google Home – An element of Digital Ecosystem

In the recent worldwide developer conference of Google I/O (2016) company demonstrated the device, however, yet to reveal how its ecosystem, including Google Play and other apps will mesh with its digital assistant.

9.3 Cloud computing

Cloud computing is a means of providing computer facilities via the Internet, but that is only half of the picture. The other half is that it is also a means of accessing those same computer facilities via the Internet from different positions. In Cloud computing based service economy users will freely make use of these mobile devices and with the aid of ever expanding 2G/3G/4G (LTE) data services to find themselves running in for impulse purchases of content including Applications, Music, Movies, Television shows and E-books. Thus Cloud computing means tapping software, hardware or storage over the high speed Internet, then using and paying for it on an as-needed basis. Microsoft Inc. (2015) is for the first time in history making the computing device by itself - the Surface – which is expected to integrate seamlessly into the Microsoft product ecosystem - such as Windows OneDrive and MS-Office 365 and its integration with Windows Phones looks quite acceptable as of now. The same can be observed with 'Windows Continuum'.

Figure 4. Microsoft's Digital Ecosystem

As per the keynote presentation made by Microsoft Inc. in 2015, a state of the art feature used by a Microsoft branded Lumia 950 XL phone that connects to a Microsoft Display Dock and use it with an external monitor, a keyboard, and a mouse. Office apps and Outlook scale up to create a big screen-optimized work environment that makes you more productive. It's a PC-like experience that's powered by your phone. It is apparent that Microsoft has further plans to integrate Xbox gaming platform into their ecosystem enticing more customers. The same was reflected during the last conference by Microsoft in the year 2015 when Microsoft focused on an element of artificial intelligence called 'Microsoft Cortana.'

10 Justification of the Results

Digital Services discussed within a given research include cloud applications as well as online services would facilitate not only content management, but also provide a glimpse of technologies such as 'Cloud Computing" and Internet of Things (IoT). Finally, this research outlines the framework on how services can be delivered with the help of Digital Ecosystem in order to enhance customer loyalty.

Conclusion

Today the digital world exists with Multiple Operating Systems, Multiple Platforms, Multiple Computing Devices, and Multiple User Interfaces with multiple interactive approaches, yet these types of Cross platforms can generate issues with users. This is where given research can make things easier to understand following across various platforms-
- To understand the concept of Digital Ecosystem
- To understand various components of Digital Ecosystem
- To understand the benefits of deploying Digital Ecosystem

Future work

Digital Ecosystem has made information exchange and collaboration a lot easier. In service economy these cross platform devices such as modern tablet PCs and smartphones would leverage high speed internet to scale consumer experience. This evolution in digital services would further be supported by leading companies with existing Hardware and productive applications. Future devices will also include breakthrough technologies to provide an immersive experience as compared to previous generations with integration of Cloud based Unified information exchange services.

References

1   Briscoe, Gerard, and Alexandros Marinos. "Digital Ecosystems in the clouds: towards community cloud computing." Digital Ecosystems and echnologies, 2009. DES '09. 3rd IEEE International Conference on. IEEE, 2009.
2   "What Actually Is a Digital Ecosystem? - Kobil." Kobil. 29 June 2015. Web. 21 May 2016. Save to EasyBib
3   Olavsrud, Thor. "How Digital Ecosystems Are Creating the 'we Economy'" CIO. 02 Feb. 2015. Web. 21 May 2016.
4   Chang, Elizabeth, and Martin West. "Digital Ecosystems A Next Generation of the Collaborative Environment." iiWAS. 2006.
5   Mobile cloud computing industry outlook report: 2011-2016. (2012, Jan 30). PR Newswire Retrieved from http://search.proquest.com/docview/918562973?accountid=33562
6   Schmitt, G. (2009). How will crowdsourcing trend shape creativity in the future? Dvertising ge, 80(14), 13-13. Retrieved from http://search.proquest.com/docview/208411951?accountid=130240
7   "On Innovations." Washington Post. N.p., n.d. Web. 24 ug. 2012. <http://m.washingtonpost.com/national/on-innovations/microsoft-surface-consumer-tech-gets-a-third-ecosystem/2012/06/20/gJQ QwrqV_story.html>.
8   "Digital Ecosystem Convergence between IT, Telecoms, Media and Entertainment: Scenarios to 2015." The World Economic Forum. N.p., n.d. Web. 20 ug. 2012.

Narayan Murthy[1]

A Two-Factor Single Use Password Scheme

Abstract: The need for password protection is a vital part of cybersecurity. In order to keep information secure, companies must integrate technologies that will keep the data of their employees and customers safe from attackers. In practice, several techniques for password authentication exist. Some of the techniques in industry and research include: a crypto-biometric model; an authentication scheme using the camera on a mobile device; the noisy password technique; and one-time password schemes with a QR-code ([4], [5], and [8]). Traditional static passwords simply aren't enough to protect against today's dynamic threats. Hackers continue to find new ways to steal the credentials to enter the system. Clearly, a password is the weakest link in the security chain of a company; and hackers can crack most of the simple passwords in less than a minute.

Recent trends in password authentication involve two factors. In the US, the FDIC has published a financial institution letter in which they recommend a two-factor authentication system for financial transactions [2]. You wouldn't want your bank to allow access to your account with just one factor ([1] and [2]). The Web site [9] gives a list of financial institutions which do not yet support a two-factor authentication scheme. In this short paper, which addresses a combination of one-time usable passwords and a two-factor mechanism, we propose an authentication scheme in which the first factor is a password and the second factor involves modifying the password to make it more secure.

Keywords: Password, two-factor password, single-use password

1 Introduction

The most commonly used authentication method still relies on the use of passwords; and many passwords are formed using alphanumeric characters and special characters. In order to decrease the vulnerability of passwords to brute-force and dictionary attacks, many organizations impose conditions on creating passwords. Examples of these enforcements are a minimum required length,

---
1 Department of Computer Science, Pace University, NY, USA
Email: nmurthy@pace.edu

use of both uppercase and lowercase letters, including numerals and special characters and so on. This leads to a dilemma on the part of the users. They have to find ways to make up sufficiently strong and yet easy-to-remember passwords. Some organizations further require that passwords be changed regularly. But, in spite of making passwords complicated, there are sophisticated password cracking tools, which crack passwords by brute-force methods. Sometimes it may take a very long time, but can be done.

The next incarnation of password authentication is the one-time password (OTP). A one-time password is a password that is valid for only one login session or transaction. By making a password valid for only one use, the risk of a hacker reusing the password is eliminated. After one use, the new OTP may be sent to the mobile phone number of the account owner via text, email or even the telephone.

Researchers have provided various variations of the one-time password method. For example, the paper [6] suggests a one-time password method using a crypto-biometric model. The paper [4] presents the design of a challenge-response visual one-time password authentication scheme that is used in conjunction with the camera on a mobile device. The paper [7] focuses on the different aspects of one-time passwords and the ways of creating this system, in which a new noisy password technique is discussed. A new one-time password algorithm is discussed in [6], which implements a system that allows users to protect their accounts with a one-time password that adds minimal additional complexity over a simple reusable password system. This algorithm, as the authors acknowledged, does not offer the degree of security provided by most other one-time password systems, but can provide additional security when compared to reusable passwords [7]. There are also several papers which present one-time password schemes with QR-code [for example 6].

The next step in making a password mechanism stronger is the two-factor authentication (2FA). As the name indicates, a two-factor authentication is one in which the user is asked to present two forms of proof that he/she is who he/she claims to be. A form of proof in addition to a password is required. Some organizations use a hardware device which gives a new "PIN" number whenever the device is clicked. Users use this PIN as the second form of authentication. The paper [3] proposes a technique which uses wearables as the 2FA device, and allows authentication information to be transferred seamlessly and automatically from the device to the Web application.

2 Two-Factor Single Use Passwords

In this short paper we introduce a new password scheme that combines one-time and two-factor password methods, which we call the "two-factor single use" scheme. This is the basic idea: when your account is created, the company/bank sets up a password; let us call it the base password. The base password is not fixed for life. More on that later. In addition to the base password, the user is asked to select another small set of characters (say, 3-4); lets us call them "drops". The user then uses the new password, which is the character string resulting from inserting drops in the base password. The user can insert the drops anywhere in the base password. To authenticate the user, the organization server should verify if the user has sent the base password and predefined drops inserted in the base password. This can be easily verified by the server by checking that additional characters in the submitted password are the same as the predefined drops.

Let us look at the following example:

Base password: abcdef. And Drops: {x,y,z}

The password is the string that is the result of inserting x,y,z in abcdef. That is, xabcdefyz.

Now, the base password is not fixed for life, but the drops are (until the user explicitly changes them).

When the user uses the password xabcdefyz once, the password xabcdefyz becomes invalid. The company/bank sends, via email or any other communication mechanism, the user a different base password. This is done immediately and automatically. Let us say the new base password is pqpprstyu; then the new password is now xpqpprsyztyu.

The user still has to remember the drops. But these can be short and something that the user can remember easily. The user, however, probably wants to avoid simple commonly used words like cat or pen. This is because, after you insert them in the basic passwords, these words may be easily read by humans: cpqpprsattyu

Suppose a hacker steals the new password (base password + drops) by packet sniffing, and decrypts to see the clear text. This will not help the hacker in any way, because the life of the base password has already expired (it is used once). He really cannot reuse it.

The question, however, is how easily can the hacker extract the drops from the hacked password? Because the new password is a string, which contains characters from the base password and drops, no one can identify the drops by looking at the password string once or twice. If on the other hand the hacker sniffs the password several times, can he identify the drops? How many attempts are needed to establish the drops? This question needs further analysis.

3 Pros

– One-time passwords are innately more secure than reusable passwords. Additionally, multifactor authentication is innately more secure than single factor. The proposed two-factor single use scheme is stronger than other one-time usable and other two-factor methods.
– Having to only remember a short "drop" makes it easier on the users since the longer base password does not have to be recollected.
– Even if the base password is compromised, the separate 'drop' portion not being present keeps the account from being compromised and vice versa. Also, in order to retrieve the base password sent to the user, another device or email account would have to be hacked to obtain it.
– The base passwords being long and random (the user just copies them and pastes), are not susceptible to brute-force and dictionary attacks. For the same reason, they prevent shoulder-surfing vulnerabilities.
– The base password is valid only for one attempt making it difficult for hackers to reverse engineer and try to track the base password.

4 Cons

– With enough tries, a hacker may be able to identify drop characters.
– Base passwords are necessarily long and random, which users might find cumbersome. (User can copy and paste the base password. So the length and random ness should not really pose inconvenience.)
– The requirement of having to get a new password for a new session might have the user feeling inconvenienced and that more time has to be spent gaining access to their account.
– User can forget the drops.

5 Future work

As mentioned, with enough tries, a hacker may be able to identify drop characters. The question then is how many attempts are needed to establish the drops? It really depends on the total number of characters used in forming the base password, the length of base passwords, and the number of characters in the drops. To answer these questions several statistical studies have to be done.
Is it necessary to ask the user to change drops? If so, how frequently?

Is it better to send the new base password to the user only when requested instead of after the current base expires? Does this "waiting" cause inconvenience?

Is there a need to impose a "timer" on the time from which the base password is sent and entered? This additional imposition gives hackers a smaller window of time to work with.

Conclusion

Password security is a major goal of information security. Therefore, by enforcing a more restrictive authentication, such as, the two-factor password methods as a possible solution for a desired human behavior may be one answer to this complex issue of securing users' passwords. The proposed Two-Factor Single Use Password Scheme eliminates the burden of remembering complex passwords for users. The added security features offer greater security but might be perceived as an inconvenience to the user. We must then ask ourselves, is the security of our data and information more important to us than spending an extra 30 seconds to enforce greater security? A trustworthy system always leads to inconvenience.

References

1   Authentication in an Internet Banking Environment
    https://www.fdic.gov/news/news/financial/2005/fil10305.html
2   Authentication in an Internet Banking Environment, http:
    //www.ffiec.gov/pdf/authentication_guidance.pdf
3   Chen, A. Q., 2015, Two Factor Authentication Made Easy, Engineering the Web in the Big
    Data Era - 15th International Conference, ICWE 2015, At Rotterdam, The Netherlands.

4 Chow, Y., Susilo, W., Au, M.H., and Moesriami, B.A., A Visual One-Time Password Authentication Scheme Using Mobile Devices,
https://www.researchgate.net/publication/283837037_A_Visual_One-Time_Password_Authentication_Scheme_Using_Mobile_Devices

5 Harris, J.A., 2002, OPA: a one-time password system, Conference: Parallel Processing Workshops, 2002. Proceedings. International Conference. DOI: 10.1109/ICPPW.2002.1039708 · Source: IEEE Xplore

6 Liao, K.C., 2009 A One-Time Password Scheme with QR-Code Based on Mobile Phone, Conference: International Conference on Networked Computing and Advanced Information Management, NCM 2009.

7 Mahmoud, H.A., September 2009, Noisy password scheme: a new one time password system, Conference: Proceedings of the 9th WSEAS international conference on signal, speech and image processing, and 9th WSEAS international conference on Multimedia, internet & video technologies.

8 Mahto, D., and Yadav, D.K., Security Improvement of One-Time Password Using Crypto-Biometric Model,
https://www.researchgate.net/publication/278782992_Security_Improvement_of_One-Time_Password_Using_Crypto-Biometric_Model

9 Two Factor Auth (2FA) https://twofactorauth.org/

Dr.Ramesh k[1]

Design & Implementation of Wireless System for Cochlear Devices

Abstract: The latest cochlear Implant Naida CI Q70 with advanced wireless feature has got its speech processor and transmitter in the form of reliable wire communication architecture. This architecture creates problem for deaf in terms of maintenance and cost for its complex structure. Hence we propose a wireless architecture for speech processor to communicate with transmitter.

A band pass digital auditory filter is introduced in transmitter circuit which separates the frequency from overlapping and reduces unwanted noise. The whole optimized communication architecture design is to reduce the noise and gives a convenient structure for deaf to provide a reliable and improved hearing.

Keywords: Cochlear implant (C.I), digital auditory filterTransmitter, speech processor, finite impulse responsefilter(FIR), Signal to noise ratio (SNR).

1 Introduction

A cochlear implant [6] is a surgical treatment for hearing loss that works like an artificial human cochlea in the inner ear , which sends sound from ear to brain. Normally hair cells stimulate the hearing nerve, which transmits sound signals to the brain. When hair cells stop functioning the hearing nerve remain without stimulation and a person cannot hear. Cochlear implant has internal and external parts. Microphone, speech processor and transmitter as external devices. Stimulator and electrodes which are held inside the skull as internal device. Microphone picks up the sound from the environment. Speech processor which selectively filters sound to audible speech, it splits the sound into channels and sends the electrical sound signals through a cable to the transmitter. A transmitter is a coil held in the position by a magnet placed behind external ear. Transmitter sends sound to stimulator inside the skull by electromagnetic induction. Stimulator is connected with electrode cables which are connected to cochlea. Cochlea cell sends electric impulse to brain. [1].

---
1 Department of PG studies in Computer Science, Karnataka State Women's University Vijayapur, India
E-mail:rvkkud@gmail.com,rameshk@kswu.ac.in

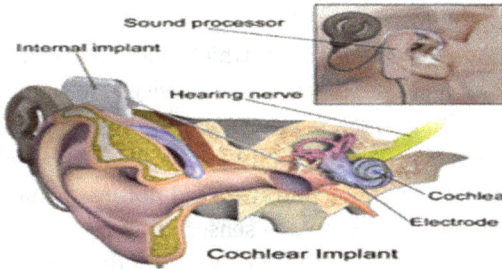

Figure 1. Cochlear device with its components Transmitter, Speech Processor, Electrode and Microphone

The clinical cochlear implant (CI) has good speech recognition under quiet conditions,but noticeably poor recognition under noisy conditions [1]. For 50% sentence understanding[2,3], the required signal to noise ratio (SNR) is between 5 and 15 dB for CIrecipients, but only –10 dB for normal listeners. The SNR in the typical daily environmentis about 5–10 dB, which results in <50% sentence recognition for CI users in anormal noise environment.Most previous studies on recognition improvement have focused on the coding strategy, design of the electrode array, and stimulation adjustment of pitch recognition, as well as on the virtual electrode technique [8,9] and optical CIs [10]. More recent efforts have focused on the microphone array technique [11,12]. This array beam forming method promises[7] to be more effective for situations in which the desired voice and ambient noise originate from different directions, the usual work environment for CI devices. Speech-enhancement methods include single and multichannel techniques. Spectral estimation methods are the most widely used single-channel techniques. Typical single channel approaches, such as the spectral subtraction [13,14], Wiener filtering, and subspace approach, are based on estimations of the power spectrum or higher order spectrum, assume the noise to be stationary, and use the noise spectrum in the non speech frame to estimate the speech-frame noise spectrum. Algorithm performance sharply weakens when the noise is non-stationary, or under typical situations with music or ambient speech noise. The microphone array technique considers the signal orientation information and focuses on directional speech enhancement. Specifically, the generalized sidelobe canceller and delay beam forming use multiple microphones to record signals for spatial filtering. For CI devices, the generalized sidelobe canceller is overly complicated and requires too many microphones, conditions that exceed the capabilities of current CI devices. Delay beam forming technologies, such as the first-order differential microphone (FDM) and adaptive null-forming method (ANF) , are adopted in hearing aids. These methods need only 2 microphones, which is an appropriate set-up for the CI size constraint and real-time processing.CI devices are similar with the hearing aids in size constraint and the requirement of front-end noise suppression. So, for CI speech enhancement, one simple solution for CI speech enhancement is to directly utilize the microphone-array–based noise reduction methods from the present hearing aids, in which the sensor-array techniques have been more widely used. However, the difference between CI devices and hearing aids is prominent, and a direct application of these algorithms to CI speech processingis not appropriate. Firstly, the principle is very different. CI devices transfer the acoustic signal to electrical stimulation into the cochlea wirelessly, and then the electrical pulses are used to directly stimulate the acoustic nerve to yield the auditory perception. But the hearing aids only need to change the corresponding gains in different sub bands for multi-frequency signal loss. In brief, the hearing aid is only an amplifier with adjustablegain in different frequency band. Secondly, the application of the microphone array technique is different. Many algorithms for speech application were borrowed from the narrowband methods in radar and antenna. Algorithms for front-end enhancement are indispensable to match the CI speech strategy. Thirdly, the solution for low frequency roll-off may be different. The hearing aids need to calibrate and preset the sub band

gain based on user's hearing loss. Therefore, in the hearing aid, one solution is to directly present the sub band gains in the filter banks in the processor by both taking the hearing loss and signal loss in microphone array algorithm into account. However, for CI devices with the modulated electrical pulse directly stimulate the cochlear nerves, we only need to adjust the algorithm loss. Finally, the signal distortion is different speech processor.

2 Functional Mechanism of Wired Communication between speech processor and transmitter

Cochlear implant replaces the normal inner ear byTransforming acoustic sound signals into electric stimuli and sends to the auditory nerve which is called as cochlea. Microphone picks up the sound and sends to Speech processor splits the auditory sound signals into different frequencies. Here speech processor consists of filter bank to split the speech spectrum into signals of Microphone picks up the sound and sends to speech processor. Speech processor splits the auditory sound signals into different frequencies. Here speech processor consists of filter bank to split the speech spectrum into signals of Microphone picks up the sound and sends to speech processor. Speech processor splits the auditory sound signals into different frequencies. Here speech processor consists of filter bank to split the speech spectrum into signals of various band width. Different frequency spectrum reaches transmitter through a wire.

3 Proposed wireless archtecture design between speech processor and transmitter

Wireless communication of sound signal is implemented between speech processor and transmitter by placing radio frequency transmitter and radio frequency receiver in speech processor and transmitter respectively. Sound signal transmit from speech processor is converted into digital data by using ADC(Analog to Digital Conversion). Radio frequency transmitters can be used to transfer digital data wirelessly to the transmitter. Transmitted signal from speech processor will undergo through a band pass auditory filter [4] toreduce noise and unwanted frequencies. Received signal in the transmitter get transfer to stimulator by electromagnetic induction.

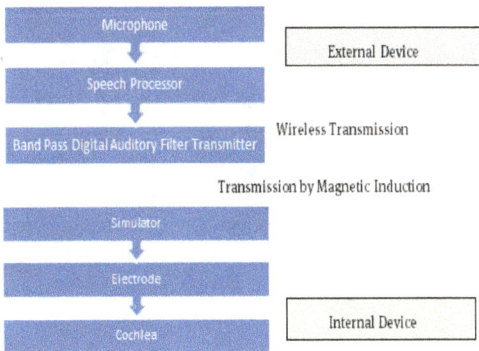

Figure 2. Function of finite impulse response filter

Filter design and its structure of finite impulse response is discussed in [4]. Filter design and its structure of finite impulse response simulations are proven to be efficient in terms of timing and suitable [4]. F.I.R. filters and avoids inter-action of speech signal of different channels. It send stable frequency to trans-mitter. F.I.R. filter[5] avoids the overlapping of signal.

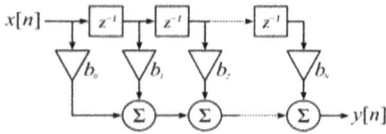

Figure 3. FIR filter of order N (direct form)

A direct form discrete-time FIR filter of order N. The top part is an N-stage delay line with N + 1 taps. Each unit delay is a z–1 operator in Z-transform nota-tion.

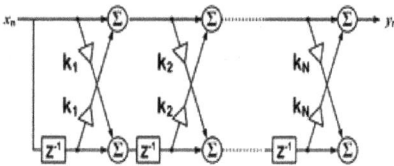

Figure 4. FIR filter of order N (lattice-form)

A lattice-form discrete-time FIR filter of order N. Each unit delay is a z–1 op-erator in Z-transform notation.For a causal discrete-time FIR filter of order N, each value of the output sequence is a weighted sum of the most recent input values:

$$y[n] = b_0x[n] + b_1x[n - 1] + \cdots + b_Nx[n - N] = \sum_{i=0}^{N} b_i * x[n - i]$$

where:

- $x[n]$ is the input signal,
- $y[n]$ is the output signal,
- N is the filter order; an nth order filter has $(N + 1)$ terms on the right-hand side
- b_i is the value of the impulse response at the i [th]instant for $0 \leq i \leq N$ of an nth order FIR filter. If the filter is a direct form FIR filter then b_i is also a coefficient of the filter.

This **computation** is also known as discrete **convolution**.

The $x[n-1]$ in **these** terms are commonly referred to as *taps*, based on the structure of a **tappeddelay line** that in many implementations or block diagrams provides the delayed inputs to the multiplication operations. One may speak of a *5th order/6-tap filter*, for instance.

The impulse response of the filter as defined is nonzero over a finite duration. Including zeros, the impulse response is the infinite sequence:

$$h[n] = \sum b_i \cdot \delta[n-1] = \{ b_n \quad 0 \le n \le N, 0 \ otherwise$$

If an **FIR** filter is non-causal, the range of nonzero values in its impulse response can start before $n = 0$, with the defining formula appropriately generalized.

4 Results and Discussion

Table 1: table shows the output frequencies through wireless transmission. Results shows the input frequencies will be same as input. These results confirm no loss of frequencies during wireless transmission

s.no	Example	Approximate frequency ranges(dB)	Input sound wave frequency(dB)at microphone	Output sound wave frequency at transmitter(dB)
1	Audible sound threshold	3-5 db	5	5
2	Normal breath	10-12 db	11	11
3	Normal conversion	30-35 db	33	33
4	Busy street	60-70 db	64	64
5	Subway or person shouting	80 -85 db	81	81
6	Loud stereo	90-95db	94	94
7	Table saw, auto horn	100-105 db	102	102
8	Elevated train, thunder	120-125 Db	120	120

Conclusion

The proposed wireless architecture for CI devices is based on introduction of band pass digital auditory filter in transmitter circuit which separates the frequency from overlapping and reduces unwanted noise. The whole optimized communication architecture design is to reduce the noise and gives a convenient structure for deaf to provide a reliable and improved hearing.Our results demonstrate the input frequencies and output frequencies through wireless transmission will be same. These results confirm no loss of frequencies during wireless transmission.

The whole exercise is to investigate the technique that aims to suppress the noise and improve the speech recognition of CI devices. In we construct the Algorithms and hardware Circuit which includes the Speech Processor and transmitter in single circuit and an algorithm that reduces the SNR ratio. In future we will carry the experiments to evaluate the Circuit and algorithm performance in a real working environment for CI users.

References

1   G Boll SF: Suppression of acoustic noise in speechusing spectral subtraction. IEEE Trans Acoust Speech Signal Process 1979, 27(2):113–120.M. Slaney,
2   "An Efficient Implementation of the Patterson-Holdsworth Auditory Filter Bank," AppleComputer Library, Apple Technical Report #35,1993.
3   Philipos C. Loizou, "Signal Processing Techniques for CochlearImplants", IEEE Engg. in medicine and Biology,1999,pp 34-45
4   Rekha ,V. Dandur, M.V Latte, S.Y.Kulkarni, and M.K. Venkatesh, "Digital filter for Cochlear Implant Implemented on a Field-Programmable Gate Array", PWASET,Vol 33, Sep 2008, pp 468-472.
5   Rajalakshmi,K., Kandaswami,A. "VLSI Architecture of digital Auditory Filter For Speech processor of Cochlear Implant ," International Journal of Computer APPLICATIONS(0975-8887).
6   Cochlear Implant Naida CI Q70
7   Spriet A, Van Deun L, Eftaxiadis K, Laneau J, Moonen M, Van Dijk B, Van Wieringen A, WoutersJ: Speechunderstanding in background noise with the two-microphone adaptive beamforming BETM in the Nucleus Freedom cochlear implant system. Ear Hear 2007, 28(1):62–72.
8   Rekha ,V. Dandur, M.V Latte, S.Y.Kulkarni, and M.K. Venkatesh, "Digital filter for Cochlear Implant Implemented on a Field-Programmable Gate Array",PWASET,Vol 33, Sep 2008, pp 468-472.

9 Chung K, Zeng FG: Using hearing aid adaptive directional microphones to enhance cochlear implant performance. Hear Res 2009, 250:27–37.

10 Rajalakshmi,K., Kandaswami,A. "VLSI Architecture of digital Auditory Filter For Speech processor of Cochlear Implant ," International Journal of Computer APPLICATIONS(0975-8887).

11 http://en.wikipedia.org/wiki/Cochlear_implant

12 Hanukumar. V., SeethaRamaiah. P., "Digital Speech Processing Design for FPGA Architecture for Auditory Prostheses", Journal of Computer ScienceAnd Engineering, Volume 6, Issue 1, March -2011.

13 Ramesh K, H M Guruprasad, Kishore M" New   Algorithms for Beam Formation and its Comparison" International Journal of Computer Applications (0975– 8887)Volume 51– No.3, August 2012.

Gurunadha Rao Goda[1] and Dr. Avula Damodaram[2]

Software Code Clone Detection and Removal using Program Dependence Graphs

Abstract: Code cloning or the act of copying code fragments and making minor, non–functional alterations, is a well known problem for evolving software systems leading to duplicated code fragments or code clones. A Clone Detection approach is to find out the reused fragment of code in any application to maintain. Since clone detection was evolved, it provides better results and reduces the complexity. In many existing system, it mainly focuses on line by line detection or token based detection to find out the clone in the system. So it makes the system to take long time to process the entire system. The proposed methodology comprises of 1. Cone detection by template conversion and metrics computation, 2. Program Dependence Graphs, 3. Dependence graph Matching. Various semantics had been formulated and their values were used during the detection process. This proposed method gives less complexity in finding the clones and gives accurate results.

Keywords: Clone detection, Textual Analysis, Metrics computation, Abstract syntax Trees, Precision and Recall, Program Dependence Graph.

1 Introduction

Software systems provide vital support for the smooth running of an organization's business. It is the responsibility of maintainers to keep the system up-to date and functioning correctly [3]. The success of free software is evident from the large and growing number of hardware devices that include free software components. Devices such as routers, televisions, set-top boxes and media players are commonly based on software such as the Linux kernel, the Samba file/print server and the Busy Box toolset [5]. Reusing code fragments by copying and pasting with or without minor adaptation is a common activity in software development. As a result software systems often contain sections of code that are very similar, called code clones [4]. A code clone is a code portion in

---
1 Tata Consultancy Services,Chennai, India
Email: Gurunadharaogoda@Gmail.Com
2 Vice-Chancellor, Sri Venkateswara University,Tirupati, Andhra Pradesh,India

source files that is identical or similar to another. Clones are introduced because of various reasons such as reusing code by 'copy-and-paste', mental macro (definitional computations frequently coded by a programmer in a regular style, such as payroll tax, queue insertion, data structure access, etc), or intentionally repeating a code portion for performance enhancement, etc [1]. Identifying software clones and understanding how software changes between releases are two important issues for maintainers where a text-based approach is likely to be useful. Maintenance of large software systems under pressure often leads to a phenomenon referred to as software cloning [2].

A clone detection system should have ability to select clones or to report only helpful information for user to examine clones, since large number of clones is expected to be found in large software systems [1]. Although some researchers argue not to remove clones because of the associated risks, there is a consensus that clones need to be detected at least. The rest of the paper is described as follows. A section 2 brief about the literature survey. The concept of textual and metric analysis is described in Section 3 and the proposed methodology is explained with necessary equations and diagrams in Section 4. The Results obtained in the proposed method is discussed in Section 5 and Section 6 concludes the work.

2 Related Work

A brief review of some recent researches is presented here.

R. R. Brooks *et al.* [6] proposed that, in cloning attacks, an adversary captures a sensor node, reprograms it, makes multiple copies, and inserts these copies, into the network. Shinji Kawaguchi *et al.* [7] proposed that, code clones decrease the maintainability and reliability of software programs, thus it is being regarded as one of the major factors to increase development/ maintenance cost. They have introduced SHINOBI, a novel code clone detection/modification tool that was designed to aid in recognizing and highlighting code clones during software maintenance tasks. Nam H. Pham *et al.* [8] proposed that, Model-Driven Engineering (MDE) has become an important development framework for much large-scale software. Previous research has reported that as in traditional code-based development, cloning also occurs in MDE. However, there has been little work on clone detection in models with the limitations on detection precision and completeness. Kodhai. E *et al.* [9] proposed that, clone detection has considerably evolved over the last decade, leading to approaches with better results but with increasing complexity.

Kodhai.E *et al.* [10] proposed that, a clone detection approach was to found out the reused fragment of code in any application to maintain .Various types of clones are being identified by clone detection techniques. Since clone detection was evolved, it provides better results and reduces the complexity.

3 An Efficient Clone Detection Process

In the previous work the clones were only detected and not removed. This disadvantage is overcome in the present work. The primary intension of our proposed work is software clone removal. The proposed methodology comprises of the following steps 1. Cone detection by template conversion and metrics computation, 2. Program Dependence Graphs, 3. Unification.

3.1 Preprocessing and input Selection

All the source code uninteresting to the comparison phase is filtered out in this phase. This phase also includes file integration, source code standardization and the normalization. File integration involves the grouping of all the files of the same project into a single large file for external parsing. This phase includes file integration, source code standardization and the normalization. File integration involves the concatenation of all the files of the same project into a single large file for external parsing.

3.2 Template Conversion

Template conversion is the process of transformation of the input source code into a pre-defined set of statements or conversion into a standard intermediary form. For example, renaming of data types, variables, function names etc as shown in fig. 2. This type of format used in textual analysis is called 'template'. The textual comparison of the selected candidates while detecting the type-2 cloned methods where as per the definition, function identifiers, variable names, types etc., are edited during the cloning process and mere textual comparison would not suffice. Once the template conversion is over, the source file and the template file is stored in the database for applying metrics. This transformation can vary from very simple e.g., just removing the white space and comments to very complex e.g., generating PDG representation and/or extensive source code transformations. Metrics-based methods usually compute an attribute vector for each comparison unit from such intermediate representations.

3.3 Metric Computation

A set of 12 existing method level metrics are used for the detection of type-1, type-2, type-3 and type-4 clone methods.

Table I shows the metric values for the code fragment in fig 2. After computing the metric values, the method pairs with equal or similar set of values are identified by comparing the records in the database. The short-listed set of candidates is then textually compared to be confirmed as clone pairs.

Table 1: Metric values for fig. 2

Sl. No.	Metrics	Value
1.	No. of lines of code	18
2.	No. of arguments passed	4
3.	No. of local variables declared	6
4.	No. of function calls	1
5.	No. of conditional statements	1
6.	No. of looping statements	1
7.	No. of return statements	2

3.4 Finding Clone Types and Clone Pairs

By taking up a line by line comparison of the standardized and normalized source code for type-1 clone method the identification of the potential clone pairs is done. That is identical code fragments are selected except for variations in whitespace, layout and comments. For type-2 clone comparison of templates are done. Here syntactically identical fragments except for variations in identifiers, literals, types, whitespace, layout and comments are taken. In the fragments there is some modifications except there is some similarities means it must be declared as type-3 by matching template with the exact code. Copied fragments with further modifications such as changed, added or removed statements, in addition to variations in identifiers, literals, types, whitespace, layout and comments can be said as type-3 clones.

It's declared as type-4 clone when the fragments are completely different but produce similar output. If the functionalities of the two code fragments are identical or similar and referred as Type IV clones. That is when two or more code fragments that perform the same computation but are implemented by different syntactic variants are said to be type-4 clone.

3.5 Program Dependence Graph

The clones are identified Program Dependence Graph. The inputs for our clone detection are the two table instances and a set of matching node pairs are the resultant outputs. Initially, the program instances are converted into graph format, called the program dependency graph. To generate the dependency graph for a particular table instance, two properties are needed to be found.

Properties needed to generate PDG

1. Entropy
2. Mutual Information

These two values are found separately for each table instances for making the dependency graph.

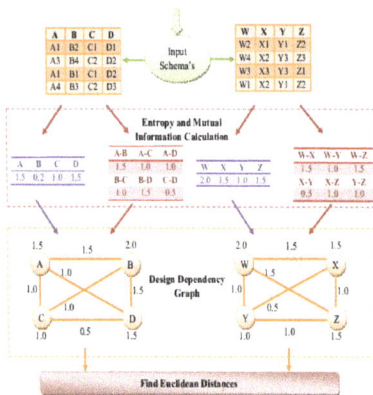

Fig 1: Block diagram of the proposed work

3.5.1. Entropy

For every attribute in both the tables, the property entropy is calculated separately. This entropy value is further helpful to get the mutual information between two attributes.

Let A be an attribute and each A has the probability distribution $p(a)$. Then, the entropy is defined as follows:

$$E(A) = -\sum p(a) \log p(a) \tag{1}$$

This entropy property that is obtained for every attribute in both the tables using eqn. (1) is independent of the actual values of the attributes.

3.5.2. Mutual Information

This property is also independent of the actual values of the attributes. In our work, after finding the entropy for every individual attribute in both the table instances, we need to find the correlation between the two attributes of the same table instance. We have considered two table instances as the input for our work. In order to find the correlation between the two attributes, it is not enough to consider one table, but it is needed to find the mutual information between both the table instances. In this proposed work, the mutual information between any two attributes of a table instance is measured by the correlation measure, Mean-Square Contingency Co-efficient (MSCC).

Let A and B be any two attributes with the domain sizes D_1 and D_2 respectively and let the probability distributions of A and B be $p(a)$ and $p(b)$, respectively with the joint probability distribution $p(a,b)$. Then, the mutual information using MSCC is defined as follows:

$$MI(A,B) = \frac{1}{\min\{D_1,D_2\}-1} \sum_{a=1}^{D_1} \sum_{b=1}^{D_2} \frac{\left(p(a,b)-p(a).p(b)\right)^2}{p(a).p(b)} \tag{2}$$

Where, a and b indicate the domains in the attributes A and B respectively (up to the domains with domain sizes D_1 and D_2).

Mutual information value using MSCC measures the total amount of information that is captured by one of the attributes about the other attribute. Mutual information of an attribute A with itself is known as self-information, which is indicated as $MI(A,A)$ and it is same as the property entropy of A, ($E(A)$).

3.5.3. Designing the Dependency graph

For the two input table instances S_1 and S_2 as in fig. 2 (a) and 2(b), we can find the two properties, namely, entropy and mutual information as described in the sections 4.1.1 and 4.1.2. The dependency graph is an undirected graph with nodes and vertices. The nodes and the vertices are represented using entropy and mutual information respectively. Every node in the dependency graph represents the entropy property of an attribute and every vertex between any two nodes represents the mutual information property between the corresponding attributes. With the aid of Square matrix SM, the dependency graph for a schema of table instances in fig. 2 (a) and 2 (b) can be designed as in fig. 2(c) and 2(d).

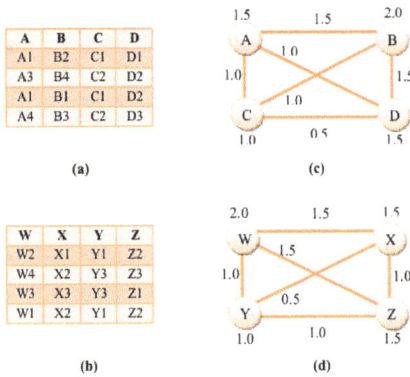

A	B	C	D
A1	B2	C1	D1
A3	B4	C2	D2
A1	B1	C1	D2
A4	B3	C2	D3

(a)

(c)

W	X	Y	Z
W2	X1	Y1	Z2
W4	X2	Y3	Z3
W3	X3	Y3	Z1
W1	X2	Y1	Z2

(b)

(d)

Fig. 2: Dependency graph Design (a) Schema-1 of table instances - S_1 (b) Schema-2 of table instances - S_2 (c) Dependency graph-1 (G_1) for S_1 (d) Dependency graph-1 (G_2) for S_2

Representation of dependency graph

Let S be a schema of table instances with N number of attributes. Also consider a_i to be the ith attribute in S, where $1 \le i \le N$. Then, we can design the dependency graph of S using Square matrix SM as follows:

$$SM = (sm_{ij}), \qquad where, sm_{ij} = MI(a_i, a_j) \ and \ 1 \le i, j \le N \qquad (3)$$

In the process of dependency matching, mutual information has a special role to find the matching node pairs between the two schemas of table instances. So, the main consideration is on the mutual information property to obtain the result of schema matching. Thus, among the two-step schema matching method, the first step is carried out by generating dependency graph for both the given input schemas of the table instances.

3.6 Dependency Graph Matching

After the generation of dependency graph using the entropy and mutual information properties, the dependency between the two dependency graphs of both the input schemas of the table instances is found. In this second step of Schema Matching method, Euclidean distance between the pair of nodes of two dependency graphs has a major role.

3.6.1 Distance measure – Euclidean Distance

Dependency between the two dependency graphs of both the schemas of the table instances is measured by means of Euclidean Distance. For this distance measure, the mutual information property is helpful to find the dependency between the graphs.

Let G_1 and G_2 be the two dependency graphs of same size and let A_{ij} and B_{ij} be the mutual information between the nodes i and j of the dependency graphs G_1 and G_2 respectively. Also consider m as a matching index by which a node in G_1 maps the node in G_2. The matching index is represented as $m(node\ in\ G_1) = matching\ node\ in\ G_2$. For the two graphs G_1 and G_2, the Euclidean distance measure representation for the dependency graphs is,

$$ED(G_1, G_2) = \sqrt{\sum_{i,j} A_{ij} - B_{m(i)m(j)}} \qquad (4)$$

From eqn. (4), the dependency between the two dependency graphs can be found and the matched node-pairs are determined using our proposed work.

4 Results And Discussion

The proposed software clone detection system has been implemented in the working platform of JAVA (version JDK 1.6.The main aim of the proposed method is to identify all the four clone types in the source code. This can be achieved by the combining both textual analysis and metrics.

Performance Measure

$$\Pr ecision,\ P = \frac{Number\ of\ clones\ \text{correctly}\ found}{\text{Total Number of clones found}}$$

$$\mathrm{Re}\,call,\ \mathrm{R} = \frac{Number\ \ of\ \ clones\ \ found\ \ correct}{Total\ \ \text{number of clones in the source code}}$$

The precision and recall of the proposed method will evaluate the proposed system's efficiency. The following graph describes the comparison of performance measure.

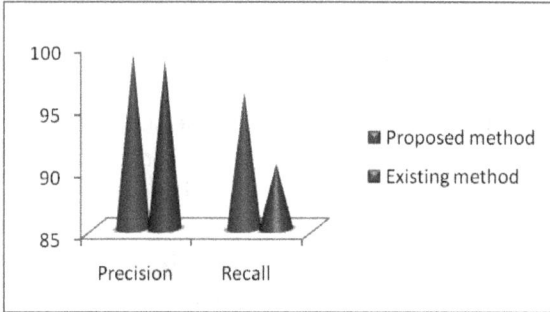

Fig. 9: Comparison of Precision and Recall

From the Fig. 9, we observe that our proposed method detects the clones available in the source files in an efficient mannar. We compare the proposed work with the already existing clone detection tool suffix tree it will give less precision and recall rate when compared to our proposed method. The measures for the above graph is given in table II.

Table 2: Performance Measure

Methods	Performance Measure	
	Precision	Recall
Proposed method	99	96
Existing method	98.5	90.25

Conclusion

The paper has proposed a light-weight technique to detect functional clones with the computation of metrics combined with simple textual analysis technique. With the usage of metrics the existing exponential rate of comparison overhead. Potential clones are compared line-by-line to determine whether two potential clones really are clones of each other. The early experiments prove that this method can do at least as well as the existing systems in finding and classifying the function clones in Java. The Precision and Recall plot describes the efficiency of the proposed work.

References

1 Toshihiro Kamiya, Shinji Kusumoto and Katsuro Inoue, "CCFinder: A Multi-Linguistic Token-based Code Clone Detection System for Large Scale Source Code," IEEE Transactions on Software Engineering, Vol. 28, No. 7, pg. Software Engineering, Jul 2002.
2 J Howard Johnson, "Substring Matching for Clone Detection and Change Tracking,," In Proc of the International Conference on Software Maintenance (ICSM), Victoria, British Columbia, pp. 120–126, Sep 1994.
3 Elizabeth Burd and John Bailey, "Evaluating Clone Detection Tools for Use during Preventative," In Proc. of the Second IEEE International Workshop on Source Code Analysis and Manipulation (SCAM'02), Montreal, Canada, Oct 2002.
4 Chanchal K. Roy, James R. Cordy and Rainer Koschke, "Comparison and Evaluation of Code Clone Detection Techniques and Tools: A Qualitative Approach," Science of Computer Programming, Vol. 74, No. 7, Feb 2009.
5 Rainer Koschke, Raimar Falke and Pierre Frenzel, "Clone Detection Using Abstract Syntax Suffix Trees," In Proc. of the 13th Working Conference on Reverse Engineering, Benevento, pp. 253 - 262, Oct 2006.
6 R. R. Brooks, P. Y. Govindaraju, M. Pirretti, N. Vijaykrishnan and M. Kandemir, "Clone Detection in Sensor Networks with Ad Hoc and Grid Topologies," International Journal of Distributed Sensor Networks, Vol. 5, pp. 209–223, 2009.
7 Shinji Kawaguchi, Takanobu Yamashinay, Hidetake Uwanoz, Kyhohei Fushida, Yasutaka Kamei, Masataka Nagura and Hajimu Iida, "SHINOBI: A Tool for Automatic Code Clone Detection in the IDE," In Proc. 16th Working Conference on Reverse Engineering, pp. 313 - 314, Oct 2009.
8 Kodhai. E, Kanmani. S, Kamatchi. A, Radhika. R and Vijaya Saranya. B, "Detection of Type-1 and Type-2 Code Clones Using Textual Analysis and Metrics," In Proc. of the 2010 International Conference on Recent Trends in Information, Telecommunication and Computing, Washington, DC, pp. 241-243, 2010.
9 Kodhai.E, Perumal.A, and Kanmani.S, "Clone Detection using Textual and Metric Analysis to figure out all Types of Clones," In Proc. of the International Joint Journal Conference on Engineering and Technology (IJJCET 2010), pp. 99 - 103, 2010.
10 Nam H. Pham, Hoan Anh Nguyen, Tung Thanh Nguyen, Jafar M. Al-Kofahi and Tien N. Nguyen, "Complete and Accurate Clone Detection in Graph-based Models," In Proc. of the 31st International Conference on Software Engineering, Washington, DC, 2009.

Dileep Kumar G.[1], Dr. Vuda Sreenivasa Rao[2], Getinet Yilma[3] and
Mohammed Kemal Ahmed[4]

Social Sentimental Analytics using Big Data Tools

Abstract: Today's consumers are heavily involved in social media by having
accounts on multiple social media services. Social media gives users a platform
to communicate effectively with friends, family, and colleagues, and also gives
them a platform to share common interests, hobbies, experiences, current issues, news, opinions, religion or politics. Social networks like Twitter and Facebook manage hundreds of millions of interactions each day. Historically, this
unstructured big data has been very difficult to analyze using traditional data
warehousing technologies. New cost effective solutions, such as Hadoop, are
changing this and allowing data of high volume, velocity, and variety to be
much more easily analyzed by employing big data analytics strategies. Hadoop
is a massively parallel technology designed to be cost effective by running on
commodity hardware. The use of social networking services in an enterprise
context presents the potential of having a major impact on the world of business
and work. This unstructured conversation of social media can give businesses
valuable insight into how consumers perceive their brand, and allow them to
actively make business decisions to maintain their image. The objective of this
paper is to perform social media sentiment analysis big data tools.

Keywords: Datasets, Big Data, Social Media, Sentiment Analysis, MapReduce.

1 Introduction

With the emergence of social media, the performance of Social media analytics
tools has become increasingly critical. Social media analytics refer to the analy-

---

1 Adama Science and Technology University, Adama, Ethiopia
E-mail: dileep.gdk@gmail.com
2 Adama Science and Technology University, Adama, Ethiopia
E-mail: vudasrinivasarao@gmail.com
3 Adama Science and Technology University, Adama, Ethiopia
E-mail: getinetyilma@gmail.com
4 E-mail: esmael2004@gmail.com

sis of structured and unstructured data from social media channels. Social media is a broad term encompassing a variety of online platforms that allow users to create and exchange content. Social media can be categorized into the following types: Social networks (e.g., Facebook and LinkedIn), blogs (e.g., Blogger and WordPress), microblogs (e.g. Twitter and Tumblr), social news (e.g., Digg and Reddit), social bookmarking (e.g., Delicious and StumbleUpon), media sharing (e.g., Instagram and YouTube), wikis (e.g., Wikipedia and Wikihow), question-and-answer sites (e.g., Yahoo! Answers and Ask.com) and review sites (e.g., Yelp, TripAdvisor). Also, many mobile apps, such as FindMyFriend, provide a platform for social interactions and, hence, serve as social media channels [1].

Social media analytics is a nascent field that has emerged after the advent of Web 2.0 in the early 2000s. The key characteristic of the modern social media analytics is its data-centric nature. The research on social media analytics spans across several disciplines, including psychology, sociology, anthropology, computer science, mathematics, physics, and economics. Marketing has been the primary application of social media analytics in recent years. This can be attributed to the widespread and growing adoption of social media by consumers worldwide [2], to the extent that Forrester Research, Inc., projects social media to be the second-fastest growing marketing channel in the US between 2011 and 2016 [3].

User-generated content (e.g., sentiments, images, videos, and bookmarks) and the relationships and interactions between the network entities (e.g., people, organizations, and products) are the two sources of information in social media. Based on this categorization, the social media analytics can be classified into two groups:

i. Content-based analytics: Content-based analytics focuses on the data posted by users on social media platforms, such as customer feedback, product reviews, images, and videos. Such content on social media is often voluminous, unstructured, noisy, and dynamic. Text, audio, and video analytics, as discussed earlier, can be applied to derive insight from such data. Also, big data technologies can be adopted to address the data processing challenges.

ii. Structure-based analytics: Also referred to as social network analytics, this type of analytics are concerned with synthesizing the structural attributes of a social network and extracting intelligence from the relationships among the participating entities. The structure of a social network is modeled through a set of nodes and edges, representing participants and relationships, respectively. The model can be visualized as a graph composed of the nodes and the edges. We review two types of network graphs, namely social graphs and activity graphs [4]. In social graphs, an edge between a pair of nodes only signifies the existence of a link (e.g., friendship) between the corresponding entities. Such graphs can be mined to identify communities or determine hubs (i.e., the users with a relatively large number of direct and indirect social links). In activity networks, however, the edges represent actual interactions between any pair of nodes. The interactions involve exchanges of information (e.g., likes and comments). Activity graphs are preferable to social graphs, because an active relationship is more relevant to analysis than a mere connection.

Sentiment Analysis is the field of study that analyzes people's opinions, sentiments, appraisals, attitudes, evaluations and emotions towards entities such as organizations, products, services, individuals, topics, issues, events, and their attributes as presented online via text, video and other means of communication. These communications can fall into three broad categories: positive, negative and neutral. SA has many other names such as opinion mining, opinion extraction, sentiment mining, subjectivity analysis, customer complaint, affect analysis, emotion analysis, review mining, review analysis etc [5].

In the current commercial competition, designers, developers, vendors and sales representatives of new information products need to carefully study whether and how do their products offer competitive advantages. Twitter, with over 500 million registered users and over 400 million messages per day, has become a gold mine for organizations to monitor their reputation and brands by extracting and analyzing the sentiment of the tweets posted by the public about them, their markets, and competitors [6].

Questions Sentiment Analysis might ask includes
- Is this product review positive or negative?
- Is this customer email satisfied or dissatisfied?
- Based on a sample of tweets, how are people responding to this ad campaign/product release/news item?
- How have bloggers' attitudes about the president changed since the election?

The paper is organized as the following: Section II presents literature survey, Section III discusses various sentiment analysis techniques, Section IV presents methodology of sentiment analysis. Section V is definitely the use of MapReduce for analyzing sentiments and finally Section VI presents the experimental results.

2 Related Work

Up to now, there have been many studies related to sentiment analysis on Hadoop. Khuc et al. [7] described a large-scale distributed system for real-time sentiment analysis on Hadoop. The system consists of two components: a lexicon builder and a sentiment classifier, which are capable of running on a large-scale distributed system using a MapReduce framework and a distributed database system. Bautin et al. [8] introduced the TextMap Access system, Lydia, which provides ready access to a wealth of interesting statistics on millions of people, places, and things across a number of web corpora using the Hadoop system. Lydia consists of five primary components: spidering, NLP markup, sentiment analysis, entity analysis and aggregation, and visualization.

Mukherjee and Bhattacharyya [9] investigated the utility of linguistic features for detecting the sentiment of Twitter messages and showed that part-of-speech (POS) features might not be useful for sentiment analysis in the microblogging domain. They found using hashtags to collect training data proved useful. They investigated the role of the hashtag and parts of speech in sentiment analysis. However, they also did not consider the hardware aspects related to data collection and processing. They also presented a lightweight method for using discourse relations for polarity detection of tweets. The method incorporates discourse information in the bag of words model to improve accuracy.

There are other studies on sentiment analysis. In particular, some studies deal with sentiment analysis on Twitter, such as the following. Go et al. [10] described a distant supervision-based approach for sentiment analysis classification. They use hashtags in tweets to create training data and a multiclass classifier to decide the polarity of the sentence. Barbosa and Feng [11] proposed a method for sentiment analysis in Twitter. The method used POS-tagged n-gram features and hashtags.

A variety of open source projects for big data processing are in progress with the Hadoop ecosystem [12]. The database management system that is widely used for big data processing with the Hadoop system is Not-Only Structured Query Language (NoSQL) [13]. It stores big data and retrieves some data using the consistency model, which is less restrictive than traditional relational databases. It is a very flexible database because it provides horizontal and vertical scalabilities, does not use the join operation between tables and table schema, and serves up fast response for reads and writes. Currently, a number of database models based on NoSQL are being introduced in academia and industry: BigTable from Google [14], Amazon DynamoDB from http://amazon.com/ [15],

HBase from the Apache Software Foundation [16], Cassandra [12], and MongoDB [16] are typical.

In particular, MongoDB has no schema and provides both regular expression search and searches for whether the sentence flexibly includes a particular value. Because NoSQL can be scaled with the horizontal extension method (horizontal scalability), a big data processing system can be extended by adding the existing systems in parallel at a low cost and not upgrading the system with an expensive central processing unit (CPU).Therefore, it is possible to process a large amount of data in parallel, compared to a conventional relational database management system, using the MapReduce techniques, filtering, data clustering operation, statistics, and data extraction.

A multimedia data processing system and algorithms are proposed [17] to analyze the sentiments of users from large amounts of unstructured text data generated by SNSs. The proposed method is composed of a parallel HDFS system based on the Hadoop ecosystem and on four MapReduce functions. The proposed method stably processes data loading according to the increase in the number of data items. The system load is distributed to each node by parallel processing. When the proposed sentiment analysis functions have processed the data effectively, the system load is not concentrated on a single node but is evenly distributed among all nodes.

3 Sentiment Analysis Techniques

Sentiment Analysis can be performed [18] in three ways: 1) Sentiment Analysis based on technical point of view, 2) Sentiment Analysis by using view of the text and 3) Sentiment Analysis by rating level, depicted in figure 1.

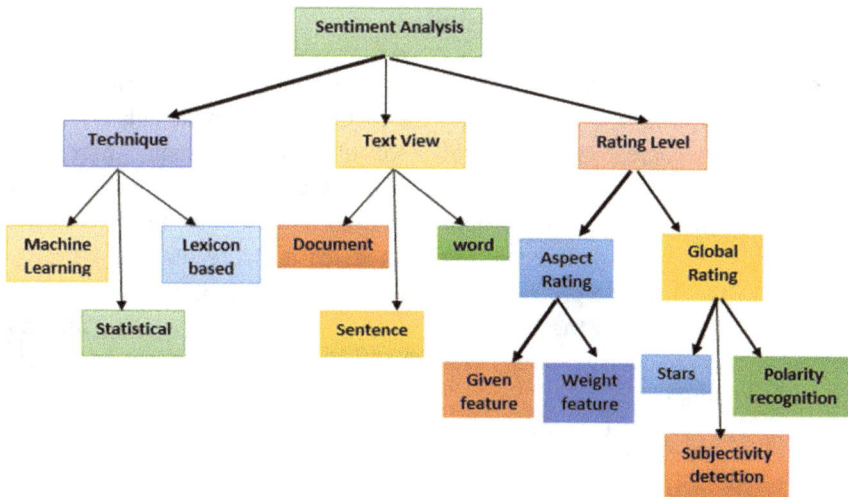

Figure 1. Sentiment Analysis Classification

3.1 Technical point of view

The machine learning, lexicon-based, statistical and rule-based approaches can be applied to sentiment analysis [19]. The machine learning method uses several learning algorithms in order to determine the sentiment by training on a known dataset. Figure 2 illustrates the two learning types of machine learning and algorithm categories [20]. The lexicon-based approach involves calculating sentiment polarity for a review using the semantic orientation of words or sentences in the review. The semantic orientation" is a measure of subjectivity and opinion in text. The rule-based approach looks for opinion words in a text and then classifies it based on the number of positive and negative words. It considers different rules for classification such as dictionary polarity, negation words, booster words, idioms, emoticons, mixed opinions etc. Statistical models represent each review as a mixture of latent aspects and ratings. It is assumed that aspects and their ratings can be represented by multinomial distributions and try to cluster head terms into aspects and sentiments into ratings [19].

3.2 Text View and Rating Level

The classification [18] can be by document level, sentence level or word/feature level. Document-level classification aims to find a sentiment polarity for the whole review, whereas sentence level or word-level classification can express a sentiment polarity for each sentence of a review and even for each word. Our study shows that most of the methods tend to focus on a document-level classification. We can also distinguish methods which measure sentiment strength for different aspects of a product and methods which attempt to rate a review on a global level. Most of the solutions focusing on global review classification consider only the polarity of the review (positive/negative) and rely on machine learning techniques.

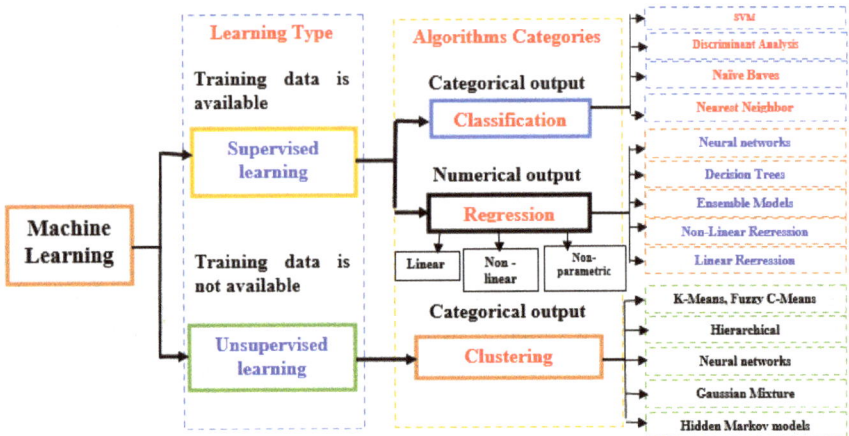

Figure 2. Machine Learning Overview

4 Sentiment Analysis Methodology

Social networks like Twitter and Facebook manages hundreds of millions of interactions each day. It has been a typical task to process unstructured data using traditional data mining and warehousing technologies. So, there is a need for new cost effective solutions such as Hadoop, which allowing data of high volume, velocity and variety to be much more easily analyzed. Hadoop is a massively parallel technology designed to be cost effective by running on commodity hardware. Fortunately, there exists several tools and technologies which are suitable for the nature of unstructured data poses by text which contains useful insight for social sentiment analysis, illustrated in Table1.

Table 1: Big Data Tools

Tool	Usage
HDFS (Hadoop Distributed File System	HDFS divides the data into smaller parts and distributes across several clustering nodes. It also enables the underlying storage for Hadoop cluster.
Apache Flume, SQL Server Integration Service, Facebook's Scribe, Apache Kafka	These allows us to download posts and load into Hadoop.
MapReduce	Performs Data Transformations in Hadoop cluster
Pig	A high-level data-flow language and execution framework for parallel computation.
Hive	A data warehouse infrastructure that provides data summarization and ad hoc querying.
HBase	It is column oriented integrated database management system with Hadoop.
Zookeeper	A high-performance coordination service for distributed applications.
Cassandra	It is a distributed database system to handle big data distributed across many utility servers.
Jaql	It converts high-level queries into low-level queries.
Avro	A data serialization system.
Oozie	Workflow management tool, runs jobs parallel.
Sqoop	Data movement between SQL and Hadoop
Mahout	It provides a machine learning library on top of Hadoop, with the goal to provide machine learning algorithms that are scalable for large amounts of data.
RStudio	Present analyzed data in a graphical manner.

Sentiment Analysis Process:
The social media posts can be downloaded and fetched into Hadoop using tools like SQL Server Integration Services, or specially built tools like Apache Flume. Social networks like Twitter and Facebook manage hundreds of millions of interactions each day. Because of this large volume of traffic, the first step in analyzing social media is to understand the scope of data needed to be fetched for analysis and then load it into Hadoop's HDFS. Quite often the data can be limited to certain hash tags, accounts, and key words.

The second step is to transform it into a format that can be used for analysis. Data transformation in Hadoop is done using MapReduce. MapReduce jobs could be written in a number of programming languages, including .Net, Java, Python, and Ruby, or can be system generated by tools such as Hive (a SQL like language for Hadoop that many data analysts would be immediately comfortable with) or PIG (a procedural scripting language for Hadoop).

A simple example of MapReduce could map social media posts to a list of words and a count of their occurrences, and then reduce, that list to a count of occurrence of a word per day. In a more complex example we can use a dictionary in the map process to cleanse the social media posts, and then use a statistical model to determine the tone of an individual post.

The third step is once MapReduce has done its magic, the meaningful data now stored in Hadoop can be loaded into an existing enterprise business intelligence (BI) platform or analyzed directly using powerful self-service tools like PowerPivot and PowerView. Customers utilizing SQL Server as their enterprise BI platform have a variety of options to access their Hadoop data, including: Sqoop, SQL Server Integration Services, and Polybase (SQL Server PDW 2012 only).

Having social media data loaded into an existing enterprise BI platform allows dashboards to be created that give at a glance information on how customers feel about a brand. Imagine how powerful it would be to have the ability to visualize how customer sentiment is affecting top line sales over time! This type of powerful analysis allows businesses to have the insight needed to quickly adapt, and it's all made possible through Hadoop. The figure 3 illustrates the complete process.

Besides commercial and off-the-self BI tools, there are several other tools which can utilized for big data analytics such as R, Apache Mahout, Alchemy API, Python NLTK, Semantria, Lucene and GATE.

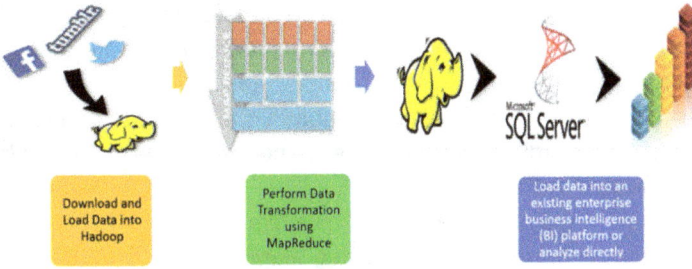

Figure 3. Sentimental Analysis Process

5 Using MapReduce for Sentiment Analysis

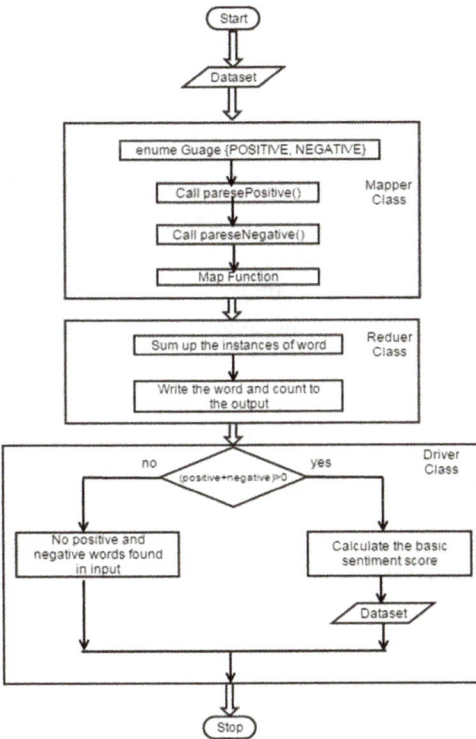

Figure 4: The Complete Process

MapReduce allows us to take unstructured data and transform (map) it to something meaningful, and then aggregate (reduce) for reporting. All of this happens in parallel across all nodes in the Hadoop cluster.

To perform sentiment analysis as shown in figure 4, first count up the positive words and negative words in a given data set. Then divide the difference by the sum to calculate an overall sentiment score.

sentiment = (positive - negative) / (positive + negative)

We have a developed an application using MapReduce. The application consists 3 classes: SAMapper, SAReducer and SADriver. Most of the work has done in the Mapper class. The Mapper class begins with an enumeration used by the custom counters to store the number of positive and negative words from the input given.

enum Gauge{POSITIVE, NEGATIVE}

The setup method in Mapper class processes the command line arguments. If -skip is true, it calls parseSkipFile. It then calls the parsePostitive( ) and parseNegative( ) methods to populate the hash sets used to compare and identify words in their respective lists. The parsePositive method cycles through the list of positive terms and creates an entry for each word. The parseNegative method does the same with the negative terms.

The map method has two additional counters which would filter and capture the positive and negative terms. The reduce method in SAReducer class assembles the results and returns them to SADriver. Instead of returning the results immediately, though, SADriver stores the result in a variable. This gives you a chance to work with the results and write to the console before ending the program.

6 Results and Discusion

The experiment environment for conducting the proposed study used Cloudera's quick start virtual machine is shown Table 2.

Table 2: Single Node Hadoop Configuration

OS	CPU	RAM	HDD
CentOS 6.4 x64	Intel Core i3 2.20 GHz	4 GB	500 GB

The proposed study has used Shakespeare's text from poems, comedies, histories and tragedies and classifies the text; creates and copies three files: stop-words.txt, pos-words.txt, and neg-words.txt to HDFS and uses this for processing.

After running the application successfully, SADriver appends the results of custom counters to the end of the list of standard counters. The sentiment and positivity scores appear after the standard output. A positivity score 50% or higher indicates that the words from the input tend to be mostly positive. In the case of Shakespeare, it falls just 1% short of that goal.

```
cloudera@quickstart:~/sentimentAnalysis
File  Edit  View  Search  Terminal  Help
        Map-Reduce Framework
                Map input records=173141
                Map output records=940003
                Map output bytes=8489898
                Map output materialized bytes=666641
                Input split bytes=965
                Combine input records=940003
                Combine output records=48303
                Reduce input groups=23834
                Reduce shuffle bytes=666641
                Reduce input records=48303
                Reduce output records=23834
                Spilled Records=96606
                Shuffled Maps =7
                Failed Shuffles=0
                Merged Map outputs=7
                GC time elapsed (ms)=6663
                CPU time spent (ms)=25460
                Physical memory (bytes) snapshot=1590550528
                Virtual memory (bytes) snapshot=12010168320
                Total committed heap usage (bytes)=1221029888
        Shuffle Errors
                BAD_ID=0
                CONNECTION=0
```

```
cloudera@quickstart:~/sentimentAnalysis
File  Edit  View  Search  Terminal  Help
        org.myorg.Map$Gauge
                NEGATIVE=42163
                POSITIVE=41184


**********



Sentiment score = (41184.0 - 42163.0) / (41184.0 + 42163.0)
Sentiment score = -0.011746074



Positivity score = 41184.0/(41184.0+42163.0)
Positivity score = 49%



**********
```

Conclusion and Future Scope

The paper studied various sentiment analysis techniques and proposed big data analytics methodology for sentiment analysis. The paper also demonstrated using MapReduce to perform sentiment analysis in a single node configuration.

A next step in this path would be to store real-time social media data into HBase, analyze in multimode configuration experimental testbed to perform comparative analysis and explore data using big data visualization tools.

References

1 Barbier, G., & Liu, H. (2011), "Data mining in social media". In C. C. Aggarwal (Ed.), Social network data analytics (pp. 327–352). United States: Springer.
2 He, W., Zha, S., & Li, L. (2013), "Social media competitive analysis and text mining: A case study in the pizza industry". International Journal of Information Management,33(3), 464–472.
3 VanBoskirk, S., Overby, C. S., & Takvorian, S. (2011). US interactive marketing forecast 2011 to 2016. Forrester Research, Inc. Retrieved from https://www.forrester.com/US+Interactive+Marketing+Forecast+2011+To+2016/fulltext/-/E-RES59379
4 Heidemann, J., Klier, M., & Probst, F. (2012), "Online social networks: A survey of a global phenomenon". Computer Networks, 56(18), 3866–3878.
5 Gundecha, P., & Liu, H. (2012). Mining social media: A brief introduction. Tutorials in Operations Research, 1(4)
6 [6] Bakliwal, A., Arora, P., Madhappan, S., Kapre, N., Singh, M., Varma, V., "Mining sentiments from tweets". Proceedings of the WASSA 12 (2012)
7 V. N. Khuc, C. Shivade, R. Ramnath, and J. Ramanathan, "Towards building large-scale distributed systems for Twitter sentiment analysis," in Proceedings of the 27th Annual ACM Symposium on Applied Computing (SAC '12), pp. 459–464, March 2012.
8 M. Bautin, C. B. Ward, A. Patil, and S. S. Skiena, "Access: news and blog analysis for the social sciences," in Proceedings of the 19th International World Wide Web Conference (WWW'10), pp. 1229–1232, April 2010.
9 S. Mukherjee and P. Bhattacharyya, "Sentiment analysis in twitter with lightweight discourse analysis," in Proceedings of the 24th International Conference on Computational Linguistics (COLING '12), pp. 1847–1864, December 2012.
10 A. Go, R. Bhayani, and L. Huang, "Twitter sentiment classification using distant supervision," Project Report, Stanford University, 2009.
11 L. Barbosa and J. Feng, "Robust sentiment detection on twitter from biased and noisy data," in Proceedings of the 23rd International Conference on Computational Linguistics (Coling '10), pp. 36–44, August 2010.
12 Hadoop, http://hadoop.apache.org/.

13 J. Han, E. Haihong, G. Le, and J. Du, "Survey on NoSQL database," in Proceedings of the 6th International Conference on Pervasive Computing and Applications (ICPCA '11), pp. 363–366, October 2011.

14 F. Chang, J. Dean, S. Ghemawat et al., "Bigtable: a distributed storage system for structured data," ACM Transactions on Computer Systems, vol. 26, no. 2, article 4, 2008.

15 S. Sivasubramanian, "Amazon dynamoDB: a seamlessly scalable non-relational database service," in Proceedings of the ACM SIGMOD International Conference on Management of Data (SIGMOD '12), pp. 729–730, May 2012.

16 K. Chodorow, MongoDB: The Definitive Guide, O'REILLY, 2nd edition, 2013.

17 Ilkyu Ha, Bonghyun Back and Byoungchul Ahn, "MapReduce Functions to Analyze Sentiment Information from Social Big Data", in International Journal of Distributed Sensor Networks, 2015.

18 Haseena Rahmath, Tanvir Ahmad, "Sentiment Analysis Techniques - A Comparative Study", IJCEM International Journal of Computational Engineering & Management, Vol. 17 Issue 4, July 2014, ISSN 2230-7893.

19 Anaïs Collomb, Crina Costea, Damien Joyeux, Omar Hasan and Lionel Brunie: A Study and Comparison of Sentiment Analysis Methods for Reputation Evaluation

20 Bogdan Batrinca, Philip C. Treleaven, "Social media analytics: a survey of techniques, tools and platforms", AI & Soc (2015) 30:89–116.

J. Prakash[1] and A. Bharathi[2]

Predicting Flight Delay using ANN with Multi-core Map Reduce Framework

Abstract: With the increase in airline transportation, it is necessary to concentrate on delays by Airline Company to improve their business. Passenger may book a flight ticket in a particular airline agency by its reputation in punctuality. They should improve their punctuality by predicting the delay time. To achieve this error rate of delay prediction should be minimized. This paper focuses on predicting and comparing a delay time of a flight with the minimized error rate i.e. more improved accuracy. The error rate is minimized by automating the prediction model using Artificial Neural Network and also by using ANN with multi core map reduce framework.

Keywords: ANN, Multi-core, Map reduce, Delay prediction.

1 Introduction

Flight Delay and cancellation is a major issue in United States over a period of time. The Federal Aviation Administration (FAA) considers a flight to be delayed when it is 15 minutes late than its scheduled time. A cancellation occurs when the airline does not operate the flight for a certain reason. The various causes for flight delay and cancellation are maintenance in aircraft, fueling, congestion in air traffic and weather. When flights are canceled or delayed, passengers may be entitled to compensation due to rules obeyed by every flight company [1]. Because of this delay, the airlines are forced to pay $22 billion yearly to FAA and passengers also get affected that they could miss connecting flight. To overcome these problems research is going in this area to identify the factors that cause flight delays, to predict whether an individual flight will be delayed and also if there is a delay, its magnitude is estimated [2]. Many researchers were used Ma-

1 PSG College of Technology/Department of Computer Science & Engineering, Coimbatore, India.
E-mail: prakashera36@gmail.com
2 PSG College of Technology/Department of Computer Science & Engineering, Coimbatore, India.
E-mail: newbharathi2011@gmail.com

chine Learning algorithm named Logistic regression to identify the factors that causes the flight delay, models such as SVM, Decision Tree, Naïve Bayes to determine whether the individual flight delay, and logistic Regression algorithm to predict the delay magnitude of flight. From the results of various researches, all the mentioned above techniques predicted the result with an accuracy of 50% approximately. In this paper, we have proposed a automated model to predict the flight delay using Artificial Neural Networks (ANN) with the improved accuracy and improvement in performance by analyzing the flight delay data from United States Department of Transportation, Bureau of Transportation Statistics using R and RHadoop.

Machine learning is a method of data analysis for automating analytical model building. Machine learning algorithms learn from the data iteratively. The iterative aspect of machine learning is important because as models are able to independently adapt when they are exposed to new data. They learn from previous computations to produce reliable, repeatable decisions and results. On a large scale, automated prediction model can analyze complex data and deliver faster and more accurate results. To achieve this, flight delay prediction model uses a hadoop framework for a model building[3].

2 Related Work

Man-Seok Ha, Jung-II Namgung and Soo-Hyun Park analyzed the flight data in "Analysis of Air-Moving on Schedule Big Data Based On CRISP-DM Methodology" and derives the most important factors using correlation for predicting the average arrival delay. This work also compares the machine learning algorithm such as ANN [4][5], Logistic Regression and Decision trees for predicting the average arrival delay. They concluded that ANN is the best model for predicting flight delay in comparison with logistic regression and decision tree [6][7].

Mahdi Pakdaman Naeini, Hamidreza Taremian, Hamidreza Taremian in the paper "Stock Market Value Prediction Using Neural Networks" used ANN for stock market prediction and concluded that error rate of prediction in ANN is less than linear regression model [8] [9].

Kushagra Sahu, Revati Pawar, Sonali Tilekar, Reshma Satpute in paper "Stock Exchange Iforecasting Using Hadoop Map-Reduce Technique" proved that prediction using multi node framework for large volume of data can be done effectively and also takes lesser CPU time[10].

Cheng-Tao Chu , Sang Kyun Kim, Yi-An Lin, YuanYuan Yu, Gary Bradski, Andrew Y. Ng , Kunle Olukotun proposed a multi core map reduce frame work for machine learning algorithms such as Locally Weighted Linear Regression, Naïve Bayes, Gaussian Discriminative Analysis, K – Means, Logistic Regression, Neural Network, PCA, SVM. They states that for a Neural Network in a training process the mapper performs partial gradient and reducer performs batch gradient [11].

3 Proposed System

This section discusses about the automation of model to predict flight delay using ANN and also applying map reduce framework for prediction model.

3.1 Data Preprocessing

Flight delay data set is taken from RITA, Bureau of Transportation Statistics. Preprocessing techniques like Missing Value Handling and Normalizing Data is performed. Missing values is handled by attribute mean and Flight Delay Data is normalized in the range of 0 to 1.

3.2 Automation of Prediction model

The prediction model is automated using Artificial Neural Network. The algorithm is shown in figure 1.

3.2.1 Structuring Input and Output for Neural Network

The first step is deciding on input layer and output layer for predicting airline delay. From the analysis of data set from RITA, Bureau of Transportation Statistics, the input to predict delay are CarrierDelay, WeatherDelay, NASDelay, SecurityDelay, LateAircraftDelay and the output is ArrivalDelay and DepartureDelay. Both input and output are represented in minutes.

3.2.2 Formation of Feed Forward Neural Network

Feed Forward Neural Network decides on random number of Hidden layers and number of neurons in each hidden layer. Also Bias will be allocated for all hidden layers and output layer except input layer. A directed edge will flow from one neuron to all other neurons in the next layer and also from bias to all the neurons in that layer. Then a random weight is assigned to all directed edge. In this case, there will be five input neurons and 2 output neurons. Totally the neural network will have 3 hidden layers. In first hidden layer, it has 4 neurons, 3 neurons in second hidden layer and 2 neurons in third hidden layer. Neural network has 65 random weights assigned to it.

Step1: *Structuring Input and Output for Neural Network*

  Input(CarrierDelay(I1), WeatherDelay(I2), NASDelay(I3),

  SecurityDelay(I4), LateAircraftDelay(I5))

  Output (ArrivalDelay(O1), DepartureDelay(O2)).

Step2: Formation of Feed forward Neural Network

  No. of Hidden Layers ← random (n)

  No. of Neurons ← 2n-1, where n is no. of input

  Weights ← F(W1,W2,W3,.......,Wn)

Step3: Calculating Output for each neuron in hidden layers and Output layers using sigmoid function

  Call *NeuralNet()*

  For each neuron in Neural Network

  Calculate *Sum* ← $\sum_{i=0}^{i=n} Ii * Wi$

  Calculate *Output* ← $1/(1 + e^{-sum})$

  return *ArrivalDelay(O1), DepartureDelay(O2)*, Where I – Input ,

                                          O – Output.

Step4: Training Neural Network

  Calculate Mean Square Error

      $MSE \leftarrow \sum_{i=0}^{i=1} \frac{(IO_i - AO_i)^2}{n}$ Where IO is Ideal Output, AO is Actual Output

  Call *Backpropagation ()*

  For each weight in Neural Network

  Calculate Delta rate

          $\delta_k \leftarrow -E * F'(x)$, Where E=AO-IO (Output Layer)

          $\delta_k \leftarrow F'(x) * \sum w_{ki} * \delta_k$ (Hidden Layer)

          $\frac{\partial E}{\partial wik} \leftarrow \delta_k * Output$

  For each Weight in Neural Network

  Update Weights

  $\Delta W_t \leftarrow -\varepsilon * \frac{\partial E}{\partial wt} + \alpha \Delta W_t - 1$

Step 5: Repeat Step 3 and Step 4 until the error rate is less than threshold value.

Figure 1: Algorithm for Artificial Neural Network.

3.2.3 Calculating Output and Training process

For each neuron in neural network, the sum and output is calculated using sigmoid function f(x) starting from hidden layers till output layer. Then Mean Square Error is calculated for the actual output produced. After this phase the training process of neural network begins. For all the weights in neural network the delta rate and gradient is calculated, then the weight is adjusted.

The following Figure 2 shows the structure, Input and the output of the implemented ANN.

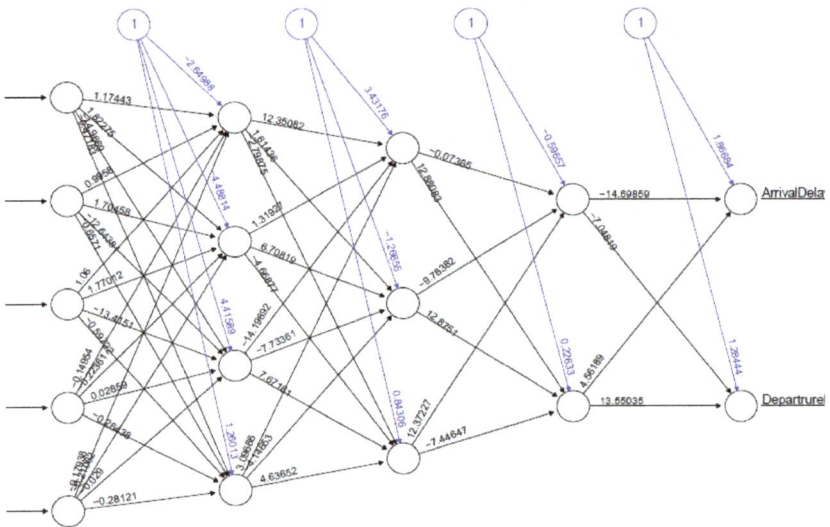

Figure 2. Structure of Artificial Neural Network

3.3 Map Reduce Framework for flight delay prediction model

A multi – core map reduce framework is adopted for Neural Network to predict the flight delay with faster and more accurate results. Large volume of flight data is submitted to multi core map reduce engine for automating flight delay prediction model. Master node splits the data into available mapper class. Then the engine performs process1 and process2 as shown in figure 3 in a iterative manner.

Input: Flight Data	
Master: Splits the flight data into four(No.of mapper in Multicore environment) by number of rows.	
Process1: Formation of Neural Network	**Process2: Training Neural Network**
Class Mapper()	Class Mapper()
{	{
For each neuron in neural network	For each weight in neural netowrk
Calculate *partial sum* and *output*	Calculate *partial Gradient*
}	}
Class Reducer()	Class Reducer()
{	{
For each neuron in neural network	For each weight in neural netowrk
Calculate *final sum* and *output*	Calculate *Batch Gradient*
Caluclate *MSE*	}
}	

Figure 3. Algorithm for Multi core Map reduce Prediction model using ANN

4 Results and Discussion

Bureau of Transportation Statistics contains flight delay data for the year 1987 to 2008. For result analysis, we have taken the flight delay data for the year 2008. This data is trained using ANN with and without map reduce frame work. The error rate, accuracy and performance for training and testing flight delay data is shown in following Table 1.

Table 1: Results comparison of ANN and ANN + Mapreduce.

		Training	Testing	Performance
Artificial Neural Network	Mean Square Error	47.2 %	38.9%	3.78 min
	Accuracy	52.8%	61.1.%	
Artificial Neural Network with multi core map reduce	Mean Square Error	23.2.%	13.7%	1.38 min
	Accuracy	76.8%	86.3%	

From Figure 5, we observed that the error rate is reduced by 24 % and accuracy is improved by 24% for training phase. From Figure 6, we observed that the error rate is reduced by 25.2% and accuracy is improved by 25.2%. From the observations it is cleared that the automated flight delay prediction model using

ANN with multi core map reduce predicts delay of airline with the more improved accuracy.

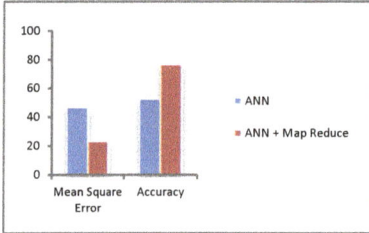

Figure 5: Comparison result of error and accuracy for training ANN.

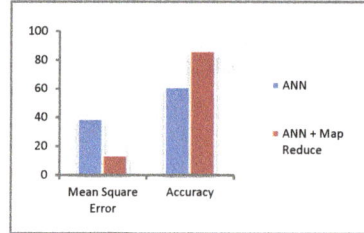

Figure 6: Comparison result of error and accuracy for testing ANN.

5 Conclusion

From the results and discussions, it is clear that the automated machine learning algorithm (ANN) delivers the result faster and accurate results i.e., the flight delay is predicted more accurately when that algorithm is adapted with multi core map reduce frame work. In the future work, intelligence is applied to automated prediction model using agent based framework to predict the delay with more improved accuracy.

References

1   Rapajic, Jasenka. "Beyond airline disruptions", p. 16, ISBN: 978-0-7546-7550-5,2009.
2   Choi, Candice, "When it comes to weather-related flight cancellations, airlines are off the hook", Boston.com (AP), 5 September 2011.
3   SAS Institute, http://www.sas.com/it_it/insights/analytics/machine-learning.html.
4   J.E. Dayhoff, J.M. DeLeo, "Artificial neural networks: Opening the black box". pp. 1615–1635, 2001.
5   Prof. Jeff Heaton,"Artificial Neural Network", http://www.heatonresearch.com/, 2009.
6   Man-Seok Ha, Jung-Il Namgung and Soo-Hyun Park, Analysis Of "Air-Moving On Schedule" Big Data Based On Crisp-Dm Methodology, Vol.10, NO.5, ISSN 1819-6608, March 2015.
7   Jessie Steinwig woods, https://jessesw.com/Air-Delays/Predicting flight delays, March 16, 2015.

8 Mahdi Pakdaman Naeini, Hamidreza Taremian, Hamidreza Taremian "Stock Market Value Prediction Using Neural Networks", International Conference on Computer Information Systems and Industrial Management Applications (*CISIM*), 2010.

9 A. Victor Devadoss, T. Antony Alphonnse Ligori, "Stock Prediction using Artificial Neural Networks", *International Journal of Data Mining Techniques and Applications*, Vol:02, Pages: 283-291, December 2013.

10 Kushagra Sahu, Revati Pawar, Sonali Tilekar, Reshma Satpute, "Stock Exchange Iforecasting Using Hadoop Map-Reduce Technique", *International Journal of Advancements in Research & Technology*, Vol. 2, Issue 4, 380 -382, ISSN 2278-7763, April-2013.

11 Cheng-Tao Chu , Sang Kyun Kim, Yi-An Lin, YuanYuan Yu, Gary Bradski, Andrew Y. Ng , Kunle Olukotun, "Map-Reduce for Machine Learning on Multicore", NIPS 2006: 281-288, 2006.

Dr.Ramesh K[1], Dr. Sanjeevkumar K.M[2] and Sheetalrani Kawale[3]

New Network Overlay Solution for Complete Networking Virtualization

Abstract: Software Enabled Networking Architecture (SENA) is a latest solution that supports the next major advances in enterprise communications. This is going to be the most exciting and disruptive data center networking technology and a new network paradigm that separates network control logic from the underlying network hardware. SENA can directly configure each connected device that makes up a network, administrators can dynamically establish multiple networks. They can also allocate bandwidth and route data flows for optimized performance using high-level control programs. By overlaying virtual networks onto physical networks, administrators can make existing infrastructure more adaptable to different workloads. The result is an agile, optimized, scalable network that is responsive to the needs of the business. SENA can provide a network overlay[1] solution that supplies a complete implementation framework for network virtualization and fully deployable data center expansion. The purpose of this work is to provide an overview of the functions and benefits of SENA, and outline its Components and the steps for implementation.

Keywords: Insert five to six keywords separated by commas

1 Introduction

Overlay networks[1] run as independent virtual networks on top of a physical network infrastructure. These virtual network overlays allow cloud providers to provision and orchestrate networks alongside other virtual resources. They also offer a new path to converged networks and programmability. However, net-

1 Department of PG studies in Computer Science, Karnataka State Women's University Vijayapura Karnataka, India
Email: rvkkud@gmail.com,rameshk@kswu.ac.in
2 Department of PG studies in Management, Karnataka State Women's University, Vijayapura Karnataka, India
Email: sanjeevkarnaik@gmail.com
3 Department of PG studies in Computer Science, Karnataka State   Women's University, Vijayapura Karnataka, India
Email: sheetalrani@kswu.ac.in

work overlays shouldn't be confused with other forms of SDN[2]. With overlay networks, engineers have a physical network that lies on top of their physical infrastructure. That virtual network overlay allows network resources to be dynamically provisioned similarly to virtual compute and storage. This form of virtualization allows the network to be part of the overall cloud orchestration framework.SDN The research was published in a 2011 paper describing Distributed Overlay Virtual Ethernet (DOVE).IT is founded on this host-based overlay technology, which achieves advanced network abstraction that enables application-level network services in large-scale multitenant environments. The technological breakthrough that SENA represents demonstrates not to just accommodating major shifts in technology, but leading them by pioneering new technology models. SENA is a direct outcome of this initiative.

2 The SENA solution for agile data center [4] connectivity

2.1 The Problem

In most large scale data centers, network administrators strive to wire the network one time then operate and maintain it without change.The fact is, changing the underlying physical infrastructure to support new business application requirements is hard to do and typically takes days to weeks to complete. This is a central problem and data center managers must resolve. When compute and storage the resources can be provisioned rapidly but network connectivity cannot, it can negatively impact the business agility.

2.2 Proposed idea

SENA can help data center managers increase business agility by enabling rapid provisioning of virtual network services without disrupting existing physical assets. The software does not require any changes to existing networks to operate, a valuable attribute that simplifies adoption. The only requirement to implement SENA is a simple one. The physical network infrastructure on which the software is overlaid must be capable of providing IP address-based connectivity. Every enterprise data center network supports this capability.

SENA is a multi-hypervisor, server-centric solution comprising multiple components that overlay virtual networks onto any physical network that provides IP connectivity. The software is designed to support multivendor data center environments. Although implementing the software does not require changes to physical infrastructure, the hypervisor must be updated. Specifically, implementing SENA ware requires an Virtual Switch (an upgrade to Distributed Virtual Switch) to be resident in VMware. SENA VMware Edition is packaged for easy installation using VMware install and update tools.

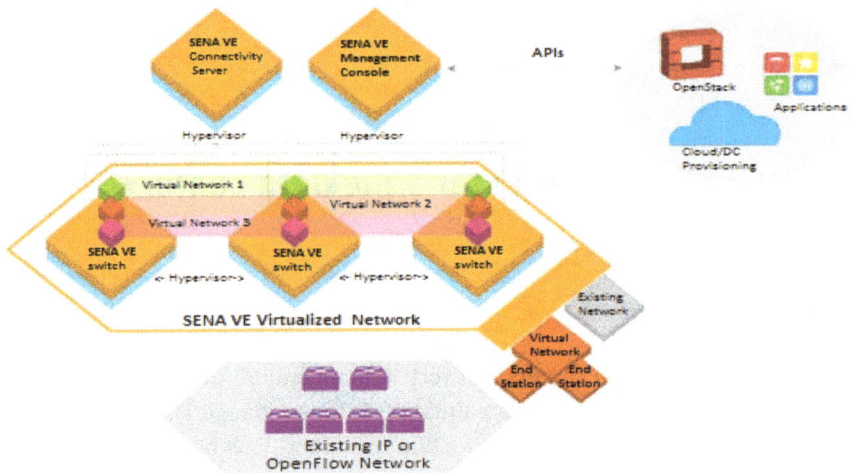

SENA VE Connectivity Server

SENA VE Management Console

APIs

OpenStack

Applications

Hypervisor

Hypervisor

Cloud/DC Provisioning

Virtual Network 1

Virtual Network 2

Virtual Network 3

SENA VE switch

SENA VE switch

SENA VE switch

‹ Hypervisor›

‹ Hypervisor›

SENA VE Virtualized Network

Existing Network

Virtual Network

End Station

End Station

Existing IP or OpenFlow Network

Figure 1. SENA is a multi-hypervisor virtual network overlay that uses existing IP infrastructure

2.3 Components of the SENA solution

The SENA solution is made up of four software components that work in combination to provide effective host-based network virtualization

    i.   **An SENA Virtual Switch** is software that resides in the hypervisor. It serves as the start and end point of each virtual network[8]. The SENA Virtual Switch provides Layer 2 and Layer 3 network virtualization[2] over a UDP overlay, and implements the data path of the virtual network. The virtual switch also performs control plane functions to support virtual machine (VM) address auto discovery, VM migration and network policy configuration

ii.   **connectivity service** disseminates VM addresses to the virtual switches participating in an SENA virtual network. The connectivity service software is deployed as a cluster of virtual appliances.

iii. **A management console** is the centralized point of control for configuring VE. It configures each virtual network, controls policies and disseminates policies to the virtual switches. It also helps administrators to manage individual virtual networks. The software resides on a server as a virtual appliance.

iv.   **VLAN- and IP-based gateways** enable SENA VE to establish interoperability with networks and servers that are external to the SENA VE environment. For Layer 2 networks, SENA VE provides VLAN-based gateways. For Layer 3 networks, the software provides IP-based gateways.

3 SENA Working Principle

SENA efficiently overlays virtual networks onto existing networks, thus decoupling application connectivity from the physical network infrastructure. This enables a "wire once" physical network that can support multiple SENA virtual networks which can be flexibly managed and controlled through highly available clusters of SENA Connectivity servers and the SENA Management Console. This architecture separates the control plane from the data plane, a central tenet of SENA. SENA operates by adding a distinct header to packets sent by VMs. Each SENA data transfer is just an ordinary IP packet sent to the existing switches in the data center network and the switches can use existing IP forwarding routes and tables. Devices continue to operate at line rates. The SENA solution builds on the network that is already in place, and provides the flexibility to create and manage virtual networks on demand.

3.1 What SENA does for networks and its virtualization does for compute

SENA is a logical extension of the virtualization trend that has become the dominant feature in the data center[2,4]. The software extends the efficiency and productivity advantages achieved with server virtualization to the process

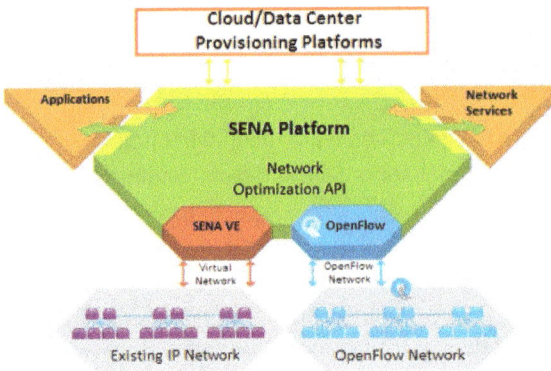

Figure 2. SENA is a core component and is a platform for Software Defined Networking. It abstracts theunderlying network and presents it to applications as either a service or as an infrastructure

of network provisioning and management[5]. These advantages allow data centers[2,4] to be more as in the following:

Efficient, because SENA improves resource use. It allows secure, dedicated virtual networks to be created quickly and easily, without requiring changes to the underlying physical infrastructure.Agile, because SENA cuts network provisioning time from days to minutes. With SENA, you can establish secure virtual networks as easily as starting up VMs. Scalable, because SENA offers data center managers the scalability needed for current and future growth. Up to 16 million networks can be specified in the architecture. The first release supports 16,000 virtual networks.

4 Functions of SENA

4.1 High availability, the ultimate data center [4] imperative

Enterprise data centers[4] maintain uncompromising standards for high availability, which reflects the value that data center operations contribute to the enterprise. In many cases, the data center is one of the most valuable components in the business because the enterprise cannot function if the data center is down. SENA supports enterprise needs for high availability with customizable, redundant component design.Two or more active SENA Connectivity Servers control each virtual network. The number of SENA Connectivity Servers that can be assigned to individual virtual networks is user configurable. This ensures that the user can select the level of high availability needed for a given virtual network. This redundant design allows the state of each SENA Connectivity Server to be replicated in at least one other instance of the SENA Connectivity Server at all times. The SENA Management Console provides high availability in Active and Standby modes. One instance operates in Active mode, and the other functions in Standby mode. If an Active SENA Management Console experiences a failure or outage, automatic failover to the Standby SENA Management Console occurs. SENA Gateways also support redundancy, allowing failover in the event of an outage. In these ways, SENA is a high-performance, high availability solution.

4.2 Secure, scalable multitenancy for cloud [8] providers

What can be gained by adopting SENA if you are already using VLANs? With SENA you can create secure, scalable multitenant networks with individual network control. Each virtual network created with SENA can be managed individually using the application programming interface (API) the software provides. In addition, you get greater scalability with SENA: A traditional network is physically limited to 4,096 VLANs[10], and requires configuration of end-to-end VLANs on some or all physical devices in the network. With SENA, the maximum number of VLANs that can be supported increases from a physical limit of 4,096 networks to an architectural limit of 16,000,000. The first release of SENA—SENA VMware Edition—supports 16,000 virtual networks. Cloud providers need to support multiple customers with dedicated, reliable, secure and scalable networks, and SENA can help supply these services with increased cost effectiveness and efficiency. exposed to the physical network. SENA only exposes one network address per NIC. This greatly simplifies the process of creating and deploying virtual networks on demand.

4.3 Data center consolidation

Data center consolidation is a common practice among large enterprises today because of the increased economy and efficiency that can be gained. Consolidation can also be necessitated by mergers and acquisitions because the acquiring company wants to ensure that all customers receive the same service experience. The difficulty centers on combining IP addresses. Redesigning complete network schemas is an exceptionally complex and time-consuming challenge. SENA resolves this problem by reusing existing IP addresses. In fact, the network address of each VM in an SENA virtual network is not

4.4 Maximizing server ROI

VMs require real network connections. However, since it is much easier to create VMs than it is to network them, your network resources can be exhausted before you can use your servers to the fullest extent. Maximizing server use is a principal reason to implement SENA. With the software in place, VM[9] density can be increased to the limits of memory, and processor cycles and server virtualization can continue without concern for VM network bottlenecks. With SENA, you can establish a "wire-once" data center network environment with expansion capacity for future growth and increased virtualization.

4.5 Optimizing network provisioning with programmable APIs[11]

The SENA solution provides programmatic access to virtual network functions using RESTful APIs, which can provide web services to any client program able to transmit messages using the HTTP protocol. SENA also supports the OpenStack Quantum API, which is a network abstraction that allows OpenStack to use the underlying network as the infrastructure without requiring it to have knowledge of the underlying resources.

5 Benefits of the SENA solution

– The SENA solution offers data center managers many ways to expand services and control costs. Benefits of the software are following.
– Virtualizes existing IP networks with no change to the underlying physical network infrastructure.
– Automates network provisioning and simplifies administration, which can help to reduce operating expenses.
– Expedites data center consolidation by allowing existing network addresses to be retained

- Enables large-scale multitenancy with independent management and optimization of multiple virtual networks
- Improves server resource utilization and return on investment (ROI) by removing the network as a bottleneck to increased VM density
- Provides API-based programmatic access to virtual networks, which allows data center provisioning platforms and network services to use virtual networks as a service or as an infrastructure.

Conclusion

The proposed work helps you to build a software-defined environment In the era of Smarter Computing, entire data center infrastructures will become as programmable as individual systems are today. Compute, storage, network and middleware components will be tuned to the workload, endlessly scalable and adaptable to dynamic workload demands. The data center, in short, will be efficient, flexible, and purpose-built and aligned with the needs of the business. With SENA software, secure multitenant network virtualization and abstraction of physical assets are not merely capabilities your network will have in the future. They are benefits you can achieve with the network you have today.

References

1   Overlay Networks: An Akamai Perspective Ramesh K. Sitaraman1,2,, Mangesh Kasbekar1, Woody Lichtenstein1,and Manish Jain11Akamai Technologies Inc∗2University of Massachusetts, Amherst
2   Technical white paper Software-defined networking and network virtualization Unifying the virtual overlay and physical underlay
3   Balakrishnan, H., Lakshiminarayanan, K., Ratnasamy, S., Shenker, S., Stoica, I., Walfish, M. A layered naming architecture for the Internet. In Proc. ACM SIGCOMM '04 (Portland, OR, Aug. 2004), pp. 343–352.
4   Cisco. Nexus 3064 Switch Data Sheet, 2012.
5   DMTF. Virtual Networking Management White Paper, Version 1.0.0. DSP2025, February 2012.
6   Sridhavan, M., Greenberg, A., Venkataramaiah, N., Wang, Y., Duda, K., Ganga, I., Lin, G., Pearson, M., Thaler, P., and Tumuluri, C., "NVGRE: Network Virtualization Using Generic Routing Encapsulation," draft-sridharan-virtualization-nvgre-01 (work in progress), July 2012
7   Technology Challenges for Virtual Overlay Networks Professor Kenneth P. Birman Dept. of Computer Science, Cornell Univ.1

8   Network Virtualization – Overlay Networks in Cloud Environments, Part 2 by Yael Frank |
     Apr 7, 2016 Networking

9   http://anonym.to/?http://www.microsoft.com/learning/en/us/course.aspx?id=20413b

10 www.flukenetworks.com/vlan/optiview

11 H. Zimmerman. OSI reference model -- The ISO model of architecture for open systems
     interconnection. IEEE Transactions on Communications, 28, Apr. 1980, pp. 425-432.

Konda.Hari Krishna[1], Dr.Tapus Kumar[2], Dr.Y.Suresh Babu[3],
N.Sainath[4] and R.Madana Mohana[5]

Review upon Distributed Facts Hard Drive Schemes throughout Wireless Sensor Communities

Abstract: The most critical purpose of distributed knowledge Storage plans in wireless Sensor Networks is to proficiently disperse knowledge over the WSN. Conveyed understanding stockpiling can expect a relevant phase in progressing knowledge accessibility, safety, vitality proficiency and method lifetime of remote sensor programs. Countless scientists had been proposed different methods to retailer understanding in a dispersed way. This paper portrays a evaluation on dispersed know-how stockpiling plans in wireless Sensor Networks. We've characterized these plans into for probably the most section two classes notably completely pletely circulated information stockpiling and information driven ability. At that point, these plans are further arranged into four courses beneath the imperatives topology, security, load-adjusting and unwavering great. Favourable occasions and weaknesses of every plan too concentrated on and we made the examination of each plans with amazing obstacles.

And aside from this we presents "correlation and replication based distributed data storage Protocol", an efficient correlation and replication established information storage protocol for colossal scale Wi-Fi sensor networks with cell sink. Contrarily to related protocol "correlation and replication founded disbursed information storage Protocol" quite simply manages the data replication

1 Ph.D -Research Scholar- Lingaya's University & Assistant professor, Dept. Of. Computer Science & Engineering, Bharat Institute of Engineering and Technology, Hyderabad
Email: kharikrishna396@gmail.com, Mob No: 9490247527
2 Associate professor, Dean & HOD, Dept. Of Computer Science & Engineering, Lingaya's University, Faridabad
Email: Kumartapus534@gmail.Com, Mob No: 9818339510
3 Professor, P.G Dept of Computer Science, JKC College, Guntur
Email: yalavarthi_s@yahoo.com, Mob No: 9885194691
4 Associate professor, Dept. Of Computer Science & Engineering, Bharat Institute of Engineering and Technology, Hyderabad
Email: natukulasainath@gmail.com, Mob No: 9059812935
5 Associate professor & HOD, Dept. Of Computer Science & Engineering, Bharat Institute of Engineering and Technology, Hyderabad
Email: rmmnaidu@gmail.com, Mob No: 9440793154

among the Storage nodes within the community. The proposed protocol considers sensible correlation mechanisms for summarization of the info packets without affecting the data best and also it might probably overcome the node disasters that will occur in related protocols. "Correlation and replication based distributed data storage Protocol" can guarantee extended community lifetime and vigour efficient knowledge storage schemes compared with present protocols for distributed data storage. The proposed efficient correlation and replication schemes can keep the information assortment efficiency as equal as within the related protocols..

Keywords: Wi-Fi sensor programs; disbursed understanding stockpiling; information accessibility; protection; community lifetime; Vitality effectiveness.

1 Introduction

A WSN contains of spatially circulated independent sensors to display bodily or ecological stipulations, for instance, temperature, sound, weight, and so on. A WSN most of the time has subsequent to zero framework. It comprises of various sensor hubs cooperating to screen a field to gather expertise about the earth. At present such programs are utilized as part of countless present day and client purposes, for example, mechanical method watching and control, desktop health checking, et cetera. In such applications some of the large experiment is the position and how one can retailer the detected understanding. Knowledge stockpiling in WSNs principally falls into two classifications, to be detailed brought together information stockpiling and dispersed information stockpiling. Data accessibility, protection, inquiry preparing and know-how restoration, system lifetime, vitality proficiency are the real difficulties confronted by way of understanding stockpiling in faraway sensor systems. In this paper, we examine about one of a kind circulated information stockpiling the place information is put away on multiple hubs, usually in a reproduced form in faraway sensor programs regarded in figure 1.

There are two predominant procedures: know-how driven capability and utterly appropriated information stockpiling. In a wholly conveyed information stockpiling process, all hubs contribute simply as to detecting and striking away. In know-how driven potential procedure some recognized stockpiling hubs are in charge of gathering information. Each the plans make utilization of distinctive approaches for conveyed know-how stockpiling. Additionally, each

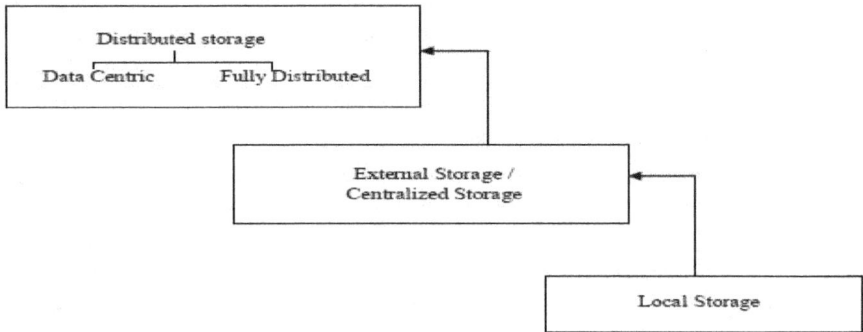

Fig 1: Information storing methods in WSN

system is described via distinct homes comparable to topology, protection, and load-adjusting and unwavering pleasant. This work introduces an efficient correlation and replication established entirely dispensed knowledge storage scheme which is able to overcome the disadvantages of existing protocols. The proposed technique is established on entirely allotted knowledge storage with mobile sink, in order that this protocol does not forget all sensors are accountable for sensing and storing. The important thing study undertaking of this paper is learn how to gain good stage of vigour effectively and community lifetime."

2 Problem statement

Expertise stockpiling in WSNs for essentially the most phase falls into two classes, to be specified unified information stockpiling and dispersed knowledge capability. Within the earlier case, understandings are detected, all set, collected and oversaw at a focal field, extra most likely than now not a sink. In the latest case, after a sensor hub has produced some knowledge, the hub shops the expertise locally or at some assigned hubs inside the approach, rather than quickly sending the expertise to a unified area out of the method. Expertise accessibility, safety, query getting ready and information recovery, process lifetime, vitality proficiency are the essential challenges confronted via expertise stockpiling in remote sensor systems. Considering sensor hubs are extra inclined to disappointment, knowledge misfortune will happen in WSNs. As an after effect of that information accessibility in WSN turns out to be low. Utilization of expertise replication accessories will stay away from such instances. Protection of expertise put away either in the neighbourhood or remotely is additionally a hassle in WSN. It is easy for the assailant to get to knowledge from traded off hubs if WSN failed to bolster any protection components. Embracing static sink hub in conveyed knowledge stockpiling for knowledge restoration will carry about to minimize method life time, due to the fact that static sink effect in vitality opening difficulty, in which hubs in the direction of the sink probably devour extra vitality considering of expertise transferring from one-of-a-kind hubs in the approach. Henceforth, separations would occur in the system. In our work a completely distributed information storage scheme for information storage is developed which introduce a Correlation mechanism in order that community lifetime, vigour affectivity, message overhead and information gathering affectivity will also be accelerated to a better degree".

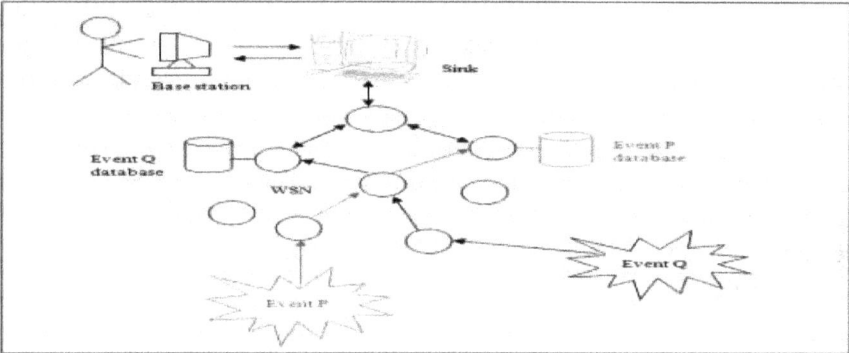

Fig 2: Dispersed safe-keeping

3 Conveyed knowledge Storage Schemes

In WSN, more than a few plans are utilized for circulated knowledge stockpiling. The DDS may also be particularly arranged into wholly conveyed understanding stockpiling [FDDS] and information pushed potential [DCS], which can be additional ordered into topology founded DDS, security established DDS, load-adjusting founded DDS and dependability based DDS.

4 Entirely distributed data Storage

On this approach, all hubs make a contribution in a similar way to detecting and putting away. All hubs try and store the sensor readings in the community and, then, appoint one-of-a-kind hubs in the WSN to retailer recently gathered expertise when their regional recollections are full. Totally dispersed knowledge stockpiling can be labelled into for probably the most phase 4 lessons all things regarded as **1) Topology established FDDS, 2) security established FDDS 3) Load-adjusting situated FDDS, and 4) Reliability established FDDS.**

4.1 Topology headquartered FDDS

On this methodology know-how stockpiling in far flung sensor programs rely upon the topology of the system. Most likely tree topologies are embraced. Community topologies are likewise provided in some wonderful cases. A few illustrations are given as takes after.

4.1.1 ProFlex

The principal goal of ProFlex is to gift disseminated expertise stockpiling for heterogeneous remote sensor systems with transportable sink. It's a probabilistic and adaptable expertise stockpiling plans. ProFlex develops distinct information replication constructions. On the point when contrast and related conventions, ProFlex has a helpful execution under message misfortune instances, diminishes the overhead of transmitted messages, and diminishes the event of the vitality opening dilemma. The convention is created from three phases: tree progress, value element dispersion and understanding dissemination. Tree topology is in charge of making numerous replication constructions. Features of curiosity of ProFlex comprise 1.Reduced message misfortune, 2.Diminish the overhead of transmitted messages, 3.Cut back event of vitality whole predicament and four. Applicable to giant scale WSN. Coverage to protection of expertise is the foremost problem of ProFlex.

4.1.2 SUPPLE

A bendy probabilistic data dissemination protocol for WSNs that considers

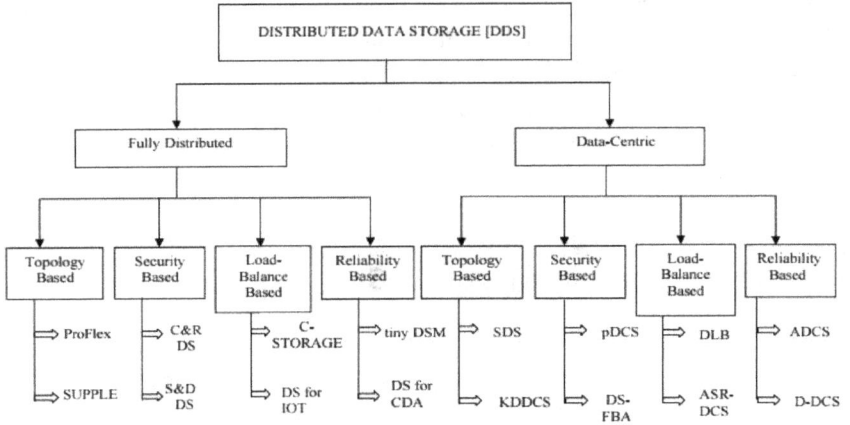

Fig 3: Distributed data schemes in WSN

static or cellular sinks. The Supple protocol has three phases: tree development, weight distribution, and information replication. This protocol used to be introduced to beat the drawbacks of Deep protocols. Apart from ProFlex SUPPLE makes use of single multiplication structure utilizing tree topology. The primary phase is a tree development initiated by way of a crucial sensor node of the sensing discipline. The primary sensor node is responsible for receiving and replicating the collected knowledge in the network. The second section assigns weights to nodes, which signify the chance of a node storing knowledge. Within the final segment, the sensor nodes send their data to the central node and this node replicates each and every knowledge to certain quantity of occasions utilizing the tree infrastructure and in line with its storage chance. Benefits of supple comprise **1.**Self organizing network, **2.**Low conversation overhead and **3.** Information availability. Hazards of supple are message overhead and excessive power consumption.

4.2 Safety headquartered fully distributed data storage

protection situated thoroughly dispensed knowledge storage participate in disbursed information storage through because security and privacy of data because the major constraint. A number of study papers are there which center of attention on security whilst information storing.

4.3 Load-balancing situated FDDS

These schemes participate in completely disbursed data storage founded on load-balancing using unique approaches. Such schemes deal with the drawback of low-reminiscence capability of sensor nodes in WSN.

4.4 Reliability based fully distributed data storage

Reliability centered thoroughly distributed information storage concentrate on how we are able to reap robustness in disbursed storage, so that data availability can also be expanded to an acceptable degree.

Table 1: Comparison of Distributed Data Storage Schemes in WSNs

Main Classification	Sub classification	Title	Data Availability	Security	Energy efficient	Network lifetime
Fully Distributed Data Storage [FDDS]	Topology Based	ProFlex	Y	N	Y	Y
		SUPPLE	Y	N	N	Y
	Security Based	C&R- DS	N	Y	N	N
		S&D –DS	N	Y	Y	N
	Load-Balancing Based	C-STORAGE	N	N	Y	Y
		DS for IOT	Y	N	N	Y
	Reliability Based	TinyDSM	Y	N	N	N
		DS for CDA	Y	N	Y	Y
Data centric Storage [DCS]	Topology Based	SDS	N	N	Y	Y
		KDDCS	N	N	Y	N
	Security Based	Pdcs	N	Y	Y	N
		DS-FBA	N	Y	Y	Y
	Load-Balancing Based	DLB	N	N	Y	Y
		ASR	Y	N	Y	Y
	Reliability Based	ADCS	Y	N	N	Y
		D-DCS	Y	N	Y	N

5 Knowledge Centric Storage

In expertise pushed capability process some famous stockpiling hubs, e.g., managed by means of a hash capability, are in command of gathering a precise kind of knowledge. DCS can be arranged into basically 4 lessons accordingly as **1)** Topology based DCS, **2)** safety centered DCS, **3)** Load-adjusting based DCS, and **4)** Reliability headquartered DCS.

5.1 Topology founded DCS

In topology based know-how pushed capability know-how stockpiling in far flung sensor techniques depend on the topology of the procedure. Most quite often tree topology is embraced.

5.2 Security based DCS

In safety based understanding driven potential, giving more significance to security and safeguard of understanding amid potential and inquiry preparing, just a few paper arrive to inform about how we are able to accomplish safety utilizing diverse programs.

5.3 Load-adjusting situated DCS

These plans participate in information pushed capacity in mild of burden adjusting making use of uncommon methodologies. Such plans deal with the quandary of low-reminiscence restrict of sensor hubs in WSN hence information accessibility can be multiplied.

5.4 Reliability situated DCS

Reliability centered information centric storage pay attention to how we can acquire robustness in allotted storage, so that knowledge availability can be extended to an appropriate stage.

6 "Effective Correlation and Replication established allotted data Storage Protocol"

In this work we endorse an efficient "correlation and replication situated distributed knowledge storage Protocol". "Correlation and replication situated disbursed information storage Protocol" can overcome the difficulties of ProFlex. Eventually "correlation and replication established distributed information storage Protocol" can warranty improved community lifetime and vigour affectivity. Also it might probably preserve the data availability and knowledge gathering affectivity same as the common variant. It makes use of cellular sink to accumulate information from the community in order that we can obtain higher knowledge collection affectivity. The proposed protocol introduces:

6.1 Power efficient Replication procedure

The reproduction management methods can provide very excessive data availability. Using excess number of replicas may have an effect on the vigour efficiency and community lifetime. As a consequence, "correlation and replication centered distributed knowledge storage Protocol" presents a replication scheme for distributed data storage in wireless sensor network which can warranty each information availability and community lifetime. "Correlation and replication established dispensed data storage Protocol" performed the goals by means of enhancing the value aspect distribution algorithm utilized in ProFlex. Right here we expect the partial view dimension every tree shall be equal to v, where v is the aggregation of storage node for every tree (outlined in ProFlex). Hence, number replicas will also be diminished since the partial view size is equal to the quantity of replicas created via the trees i.e., $r(v) = v$. As a result the number of complete number of replicas in the network is decreased than in ProFlex, for this reason storage nodes turn out to be power effective.

6.2 Information correlation centered on communiqué variety

Information correlation in "correlation and replication centered dispensed information storage Protocol" introduces an advanced approach for know-how summarization scheme. To accomplish this, a small amendment to the data correlation implemented in ProFlex is done. When a sensor node produces a packet, it forwards the information packet to the foundation of tree as in ProFlex. H sensor collects packets from its youngsters and shops the packets in its buffer rather of right away creating the replicas this method will continue except buffer size of H-sensor emerge as two times of the partial view buffer measurement of Storage nodes of the tree rooted on the corresponding H-sensor node.

When H-sensor buffer dimension reached to two time of the partial view of storage nodes, the H-sensor node summarizes the correlated packet sand sending to its own tree and neighbouring tree after growing the replicas as in information distribution algorithm. We assume the information packets within the buffer can correlate if the gap between the nodes produces the information are less than or equal to the square of the conversation range of node that send the info first, i.e., d= r2, the place r is the communication variety of node that ship the data to H-sensor node.

6.3 Performance analysis of "correlation and replication centered dispensed knowledge storage Protocol"

The first step of our performance evaluation is headquartered on the affect of reduction made in the quantity of replicas and there after we analyze the correlation mechanism used in CORREP. Right here all the simulations are carried out utilising the NS2 network simulator.

Conclusion

On this paper we targeting numerous circulated information stockpiling plans in faraway sensor techniques and we characterised these strategies into for probably the most phase two types to be particular completely dispersed knowledge stockpiling (FDDS) and expertise pushed potential (DCS). In FDDS all hubs make contributions in a similar way to detecting and placing away even as in DCS some recognized stockpiling hubs are in charge of gathering a distinct knowledge. In each and every of these groupings the methods can be once more ordered into topology based, safety centered, load-adjusting centered and unwavering pleasant situated dispersed knowledge stockpiling plans.

The topology founded knowledge stockpiling performs conveyed know-how stockpiling in gentle of the topology of the approach and the security headquartered understanding stockpiling receives some information stockpiling plots that bolster protection highlights. The heap adjusted founded appropriated know-how stockpiling makes use of framework like structural engineering to accomplish load adjusting and we will accomplish force in disseminated stockpiling via unwavering satisfactory headquartered conveyed expertise stockpiling. At long last we made an examination between special conveyed understanding stockpiling plans under distinctive imperatives corresponding to know-how accessibility, safety, vitality productivity and method lifetime. A component from this paper proposes "correlation and replication established dispensed knowledge storage Protocol", a totally allotted knowledge storage protocol for Wi-Fi sensor networks.

"Correlation and replication based distributed knowledge storage Protocol" is the protocol with excessive network lifetime and energy efficient compared with associated protocols ProFlex. Such a change to information connection was done taking into account the correspondence extent and incomplete perspective size of capacity hubs in the remote sensor systems can make strides the system lifetime. As future work, it is fascinating to include security and information consistency elements to "correlation and replication based data storage protocol" so that circumstances like assaulting of trading off hub by the gatecrashers can be evaded and in this manner secrecy and nature of detected information can be made strides.

References

1   C Viana, T.Herault, T.Largillier, S.Peyronnet, F.Zaïˮdi (2010), "Supple: a flexible probabilistic data dissemination protocol for wireless sensor networks", 13th ACM International Conference on Modeling.

2   Guilherme Maia, Daniel L. Guidoni a, Aline C. Viana b, Andre L.L. Aquino c, Raquel A.F (2013), "A distributed data storage protocol for heterogeneous wireless sensor networks with mobile sinks", Ad Hoc Networks 11,pp. 1588– 1602.

3   Krzysztof Piotroski, Peter Langendoerfer and Steffen Peter IHP (2009), "tiny DSM: A Highly Reliable Cooperative Data Storage for Wireless Sensor Networks", 978-1-4244-4586- IEEE.

4   M. Neenu, T. Sebastian (2013), "Survey On Distributed Data Storage Schemes in Wireless Sensor Networks", IJCSE, Vol. 4 No.6 Dec 2013-Jan 2014, pp.466-473.

5   M. Vecchio, A.C. Viana, A. Ziviani, R.Friedman (2010), "Deep: density-based proactive data dissemination protocol for wireless sensor networks with uncontrolled sink mobility", Elsevier Computer Communication33 (8) (2010).

6   Pietro Gonizzi, Gianluigi Ferrari, Vincent Gay b (2013), "Data dissemination scheme for distributed storage for IoT observation systems at large scale" Information Fusion (2013) .

7   Ren Wei, Ren Yi and Zhang (2010), "Secure, dependable and publicly verifiable distributed data storage in unattended wireless sensor networks", Science China Information Sciences.

8   Shen Yulong, Xi Ning, Pei Qingqi, Ma Jianfeng (2013), "Distributed storage schemes for controlling data availability in wireless sensor networks", Seventh International Conference on Computational Intelligence and Security.

9   Wen-Hwa Liao, Kuei-Plng Shih, Wan-Chi Wuaa (2009), "A grid-based dynamic load balancing approach for data centric storage in wireless sensor networks", Computers and Electrical Engineering 36 (2010) pp.19–30.

10 Z. Bar-Yossef, R. Friedman, G. Kliot (2008), "RaWMS – random walk based lightweight membership service for wireless ad hoc networks", ACM Transactions on Computer Systems 26 (2008) 5:1–5:66.

11 Dr. Shin young Lim.; Dr. Tae Hwan Oh.; Dr. Young B. Choi.; Security Issues on Wireless Body Area Network for Remote Healthcare Monitoring, 2010 IEEE International Conference on Sensor Networks, Ubiquitous, and Trustworthy Computing, 2010 IEEE International Conference on Sensor Networks, Ubiquitous, and Trustworthy Computing .

12 Kathy Dang Nguyen; Ioana Cutcutache.; Saravanan Sinnadurai.; Shanshan Liu.; Fast and Accurate Simulation of Bio monitoring Applications on a Wireless Body Area Network ; 2008 IEEE.

13 Min Shao, Sencun Zhu, Wensheng Zhang, and Guohong Cao (2007), "PDCS: Security and Privacy Support for Data-Centric Sensor Networks", ACM International Workshop on Wireless Sensor Networks and Applications.

14 Mohamed Aly, Kirk Pruhs, Panos K. Chrysanthis (2006), "KDDCS: A Load Balanced In Network Data Centric Storage Scheme for Sensor Networks", CIKM'06, November 5–11, 2006, Arlington, Virginia, USA.

15 Pietro Gonizzi , Gianluigi Ferrari , Vincent Gay b (2013), "Data dissemination scheme for distributed storage for IoT observation systems at large scale" Information Fusion xxx (2013) xxx–xxx.

16 Pooya Hejazi, Iran Hamed, Hassanzadeh Amin (2011), "An Adaptive Method for Structured Replication Data centric Storage in Wireless Sensor Networks", Proceedings of the 5th International Conference on IT & Multimedia.

17 Ren Wei, Ren Yi and Zhang (2010), "Secure, dependable and publicly verifiable distributed data storage in unattended wireless sensor networks", Science China Information Sciences.

18 Sepehr Babaei, Masoud Sabaei (2011), "Adaptive Data-Centric Storage in Wireless Sensor Networks", 978-1-61284-840- 2/11/$26.00 ©2011 IEEE.

19 Shen Yulong, Xi Ning, Pei Qingqi, Ma Jianfeng (2013), "Distributed storage schemes for controlling data availability in wireless sensor networks", Seventh International Conference on Computational Intelligence and Security.

20 Wen-Hwa Liao, Kuei-Plng Shih, Wan-Chi Wuaa (2009), "A grid-based dynamic load balancing approach for data-centric storage in wireless sensor networks", Computers and Electrical Engineering 36 (2010) 19–30.

21 Z. Bar-Yossef, R. Friedman, G. Kliot (2008), "RaWMS – random walk based lightweight membership service for wireless ad hoc networks", ACM Transactions on Computer systems 26 (2008).

Information About Author

KONDA. HARI KRISHNA received his M.TECH in computer science from Jawaharlal Nehru Technological University, Kakinada & A.P and pursuing Ph.D in LINGAYA's University, Faridabad. He is working as an Assistant Professor at Bharat Institute of Engineering & Technology in Dept. of Computer Science & Engineering. He published 18 Research Papers in Various International Journals of Reputed and His Research Area is Mining of applications in Wireless Sensor Networks. He is a good researcher & who has worked mostly on Wireless Sensor networks, Ad hoc Networks, Network security and Data mining.

Dr. TAPUS KUMAR, Working as a Professor, Dean & H.O.D in School of Computer Science & Engineering, Lingaya's University, Faridabad. He holds a Doctorate in Computer Science & Engineering. He has more than experience of 15 years in Academics & Administration. He has published various Research papers in various National & International Journals of Reputed.

Dr. Y. SURESH BABU, Working as a Professor in Dept of Computer Science, JKC COLLEGE, GUNTUR. He holds a Doctorate in Computer Science & Engg, Image processing as specialization with a combined experience of 23 years in Academics & Administration. He has published nearly 45 research papers in various National and International Journals of reputed.

N.Sainath, B.Tech from Jaya Prakash Narayana College of Engineering & M.Tech SE from Srinidhi Institute of Technology, Pursuing PhD from JNTU Hyderabad. Currently he is working as Associate Professor at Bharat Institute of Engineering & Technology. His areas of interest include Data mining, Network Security, Software Engineering, Sensor Networks, and Cloud Computing. He is enrolled for the memberships of IEEE, CSI, and ISTE. He has Published 17 papers in International Journals and has 11 International conference Proceedings and attended 15 workshops and 10 National conferences.

R. MADANA MOHANA, received his B.Tech, JNTU, Hyderabad, 2003 & M.E, Satyabama University, Tamil Nadu, 2006 & Ph.D (Thesis Finalization), from S. V. University, Tirupathi, 2008-2015. Currently he is working as Associate Professor at Bharat Institute of Engineering & Technology. He has more than experience of 10 years in Academics & Administration. He has published various Research papers in various National & International Journals of Reputed & attended workshops and conferences.

Mohd Maroof Siddiqui [1], Dr. Geetika Srivastava [2],
Prof (Dr) Syed Hasan Saeed [3] and Shaguftah [4]

Detection of Rapid Eye Movement Behaviour Sleep Disorder using Time and Frequency Analysis of EEG Signal Applied on C4-A1 Channel

Abstract: Sleep disorder is basically a disorder of the snooze patterns of living thing or a person. Some of the sleep disorders are serious capable to block with mental, normal substantial,   and emotional performance. This research article includes waveform of EEG signals and attribute of human being are realized. The results are drawn in the form of signal spectrum analysis by changing the domain in various sleeping stages. Sleep disorders like anxiety and depression can be diagnosed by analyzing the EEG patterns .Sleep disorders are the changes in chemical activities and electrical activities in the brain that can be observe by capturing the images and the brain signals. Paper includes Short Time frequency analysis of Power Spectrum Density (STFAPSD) approach which is applied on Electroencephalogram (EEG) Signals to Diagnose the Rapid Eye Movement Behavior Disorder (RBD). Results obtained after the observation can be easily used to detect the patient suffering from RBD and normal person.

Keywords: RBD, Analysis of EEG signal, PSD Estimation, Analysis of EEG Signal

1 Introduction

Normally there are two distinct sleep stages 1.Rapid eye movement (REM) 2. Non-Rapid eye movement (Non-REM) NREM is divided into four different stages they are stage 1, stage 2, stage 3, and stage 4. In REM sleep, breathing of person is

---

1 Research Scholar, Amity University, India
Email: maroofsiddiqui@yahoo.com
2 Department of Electronics and Communication, Amity University, India
Email: gsrivastava2@lko.amity.edu
3 Department of Electronics and Communication, Integral University, India
Email: s.saeed@rediffmail.com
4 Department of Electronics and Communication, Integral University, India

irregular, blood pressure increases and the brain is very active during REM stage of sleep and the electrical activity observed are similar to the activity of brain during wakefulness when EEG signals are observed. Normally in this stage temporary paralysis occurs i.e. there is a loss in tone of muscle.REM stage is usually associated with dream.REM stage is observed 20-25 % in whole sleep. It occurs because the paralysis of muscles that temporary occurs in REM stage is absent and it allows the person to act out as to his or her dream. Dream behavior involves, yelling, talking, kicking, sitting, punching arm flailing, jumping from bed and grabbing. These dreams are vivid and have large amount of movement during sleep. And there is a difference between night terror or sleep walking that in RBD the person can recall the dream easily. It is observed in middle aged to elderly people mostly in men. It occurs in less than 1 % of total population.

1.1 Causes

REM sleep is associated with dreams. It approximately occurs for one and half or for two hours from full sleep .During REM sleep muscles temporarily gets paralyzed while the brain will remain active. And if the brain does not work properly then the disorders are developed such as narcolepsy, sleep walking or RBD. In REM if the temporarily muscles are not paralyzed then the person will do the actions as they see in the dream, it may begin with actions, talk but after that it may lead to large actions like jumping, walking. It can harm the person or the bed partner too.

1.2 Diagnosis

Usually we all go through REM sleep phase every 1.5 to 2 hours of sleep from total sleep .Therefore these episodes can takes place 4 times in a night. And sometimes it occurs once in a week or in a month and episodes occurs normally in morning hours because REM is more frequent in morning.RBD can be diagnosed by the patient must be assessed at a sleep center which has the experienced and trained staff to deal with the patients. And the patient should be kept under the observation whole night which includes brain and muscle activity, monitoring of sleep, if any lack of paralysis of muscle is observed then it is also possible to search out the other causes of the problem.

1.3 Treatment

RBD can be treated successfully by the followingAfter the analysis of all the symptoms of the patient can be treated by the medication and this method is successful and effective. Low doses of benzodiazepine and clonazepam are effective in 90% of cases. These drugs used to relax the body and reduce the muscle activity during sleep. Sometimes clonazepam is not effective to the patient, then some melatonin or antidepressants can reduce the behavior at night.

The person should adopt predictable sleep-wake cycle. Alcohol consumption should be avoided.

2 Sleep data for the patient

The Table 1 below shows the data for the normal people after sleep. Sleep duration of 1 minute is used from entire sleep time of one night sleep observed and recorded at sleep centers Data is observed for different sleep stages like S0, S1, S2, S3, S4 and REM. time duration of sleep for normal person is 9-12 hours every day.

Table 1: Normal Patient's Sleep Data

S.NO.	SOURCE	PATIENT			SLEEP TIME DURATION OF S0 SLEEP STAGE		SLEEP TIME DURATION OF S1 SLEEP STAGE		SLEEP TIME DURATION OF S2 SLEEP STAGE		SLEEP TIME DURATION OF S3 SLEEP STAGE		SLEEP TIME DURATION OF S4 SLEEP STAGE		SLEEP TIME DURATION OF REM SLEEP STAGE	
		Name	GENER	AGE	Start Time	End Time	Start Time	End Time	Start Time	End Time	Start Time	End Time	Start Time	End Time	Start Time	End Time

1	2	3	4	5	6	7	8	9
physionet.org	physionet.org	physionet.org	physionet.org	physionet.org	physionet.org	physionet.org	physionet.org	physionet.org
N1	N2	N3.	N4	N5	N6	N7	N8	N9
Female	Male	Female	Female	Female	Male	Male	Female	Male
37	34	35	25	35	31	31	42	31
22:09:33	22:19:06	23:10:42	22:36:37	22:49:48	NA	NA	22:18:11	NA
22:10:33	22:20:06	23:11:42	22:37:37	22:50:48	NA	NA	22:19:11	NA
6:19:33	0:44:06	4:57:42	6:41:37	22:53:18	NA	NA	5:58:41	NA
6:20:33	0:45:06	4:58:42	6:42:37	22:54:18	NA	NA	5:59:41	NA
23:29:33	23:20:06	23:55:12	23:54:07	0:01:18	NA	NA	6:02:11	NA
23:30:33	23:21:06	23:56:12	23:55:07	0:02:18	NA	NA	6:03:11	NA
0:35:03	22:47:06	23:20:42	3:33:07	0:52:48	NA	NA	22:56:41	NA
0:36:03	22:48:06	23:21:42	3:34:07	0:53:48	NA	NA	22:57:41	NA
22:40:33	22:52:06	23:25:12	0:29:07	1:05:48	NA	NA	23:00:41	NA
22:41:33	22:53:06	23:26:12	0:30:07	1:06:48	NA	NA	23:01:41	NA
23:26:33	0:20:36	1:00:42	1:07:37	1:18:48	NA	NA	0:23:41	NA
23:27:33	0:21:36	1:01:42	1:08:37	1:19:48	NA	NA	0:24:41	NA

	10	11	12	13	14	15	16
	physionet.org	physionet.org	physionet.org	physionet.org	physionet.org	physionet.org	physionet.org
	N	N	N	N	N	N	N
	10	11	12	13	14	15	16
	Male	Female	Male	Female	Female	Male	Female
	23	28	29	24	35	34	41
	23:26:22	22:37:16	15:14:22	**	**	21:00:22	22:35:17
	23:27:22	22:38:16	15:15:22	**	**	21:01:22	22:36:17
	N/A	N/A	N/A	**	**	22:14:52	6:17:47
	N/A	N/A	N/A	**	**	22:15:52	6:18:47
	1:07:22	23:08:46	15:30:52	**	**	22:28:52	23:58:17
	1:08:22	23:09:46	15:31:52	**	**	22:29:52	23:59:18
	1:34:22	0:38:16	15:38:52	**	**	22:34:52	6:54:47
	1:35:22	0:39:16	15:39:52	**	**	22:35:52	6:55:47
	1:41:22	23:36:46	15:55:22	**	**	23:36:52	23:12:47
	1:42:22	23:37:46	15:56:22	**	**	23:37:52	23:13:47
	2:17:22	0:17:16	16:34:22	**	**	23:46:22	23:50:17
	2:18:22	0:18:16	16:35:22	**	**	23:47:22	23:51:17

Similarly, table 2 below shows the data for the RBD patients after sleep. Sleep duration of 1 minute is used from entire sleep time of one night sleep observed and recorded at sleep centres. Data is observed for different sleep stages like S0, S1, S2, S3, S4 and REM. time duration of sleep for RBD patients is also 9-12 hours every day but some difficulty during sleep or get interrupted during sleep.

Table 2: RBD Patient's Sleep Data

S.NO.		1	2	3	4	5	6
	SOURCE	physionet.org	physionet.org	physionet.org	physionet.org	physionet.org	physionet.org
PATIENT	Name	RBD 1	RBD 2	RBD 3	RBD 4	RBD 5	RBD 6
	GENER	Male	Male	Male	Male	Female	Male
	AGE	58	77	81	70	75	73
SLEEP TIME DURATION OF S0 SLEEP STAGE	Start Time	23:16:49	21:49:25	22:08:07	20:25:00	22:00:03	22:11:51
	End Time	23:17:49	21:50:25	22:09:07	20:26:00	22:01:03	22:12:51
SLEEP TIME DURATION OF S1 SLEEP STAGE	Start Time	23:18:49	1:59:25	22:54:07	2:38:00	N/A	22:19:21
	End Time	23:19:49	2:00:25	22:55:07	2:39:00	N/A	22:20:21
SLEEP TIME DURATION OF S2 SLEEP STAGE	Start Time	0:08:49	0:22:25	22:58:07	3:00:30	22:55:03	22:36:21
	End Time	0:09:49	0:23:25	22:59:07	3:01:30	22:56:03	22:37:21
SLEEP TIME DURATION OF S3 SLEEP STAGE	Start Time	0:42:29	5:26:25	0:42:07	3:06:00	23:22:03	23:17:51
	End Time	0:43:49	5:27:25	0:43:07	3:07:00	23:23:03	23:18:51
SLEEP TIME DURATION OF S4 SLEEP STAGE	Start Time	23:32:49	23:12:25	0:53:07	3:37:00	23:27:03	1:18:51
	End Time	23:33:49	23:13:25	0:54:07	3:38:00	23:28:03	1:19:51
SLEEP TIME DURATION OF REM SLEEP STAGE	Start Time	1:44:49	0:24:25	4:16:07	3:41:30	0:07:03	1:50:51
	End Time	1:45:49	0:25:25	4:17:07	3:42:30	0:08:03	1:51:51

7	8	9	10	11	12	13	14	15
physionet.org	physionet.org	physionet.org	physionet.org	physionet.org	physionet.org	physionet.org	physionet.org	physionet.org
RBD	RBD	RBD	RBD	RBD	RBD	RBD	RBD	RBD
7	8	9	10	11	12	13	14	15
Male	Male	Male	Male	Male	Female	Male	Male	Male
72	82	76	73	73	76	61	66	69
22:11:29	22:15:09	22:00:00	22:14:22	22:16:22	21:59:54	22:19:36	22:14:28	22:36:25
22:12:29	22:16:09	22:01:00	22:15:22	22:17:22	22:01:54	22:20:36	22:15:28	22:37:25
22:20:29	22:24:09	22:18:00	23:07:22	23:07:22	4:43:24	N/A	2:50:28	N/A
22:21:29	22:25:09	22:19:00	23:08:22	23:08:22	4:44:24	N/A	2:51:28	N/A
23:16:29	22:33:09	22:27:00	23:29:52	23:29:52	22:15:24	4:39:36	22:53:28	22:48:25
23:17:29	22:34:09	22:28:00	23:30:52	23:30:52	22:16:24	4:40:36	22:54:28	22:49:25
22:39:59	1:38:39	22:38:00	23:45:22	23:45:22	22:19:54	23:09:36	22:47:58	22:52:55
22:40:59	1:39:39	22:39:00	23:46:22	23:46:22	22:20:54	23:10:36	22:48:58	22:53:55
22:42:29	1:41:39	22:48:00	23:48:22	23:54:22	22:22:54	N/A	1:03:58	22:55:55
22:43:29	1:42:39	22:49:00	23:49:22	23:55:22	22:23:54	N/A	1:04:58	22:56:55
0:55:29	1:58:39	0:32:00	0:49:22	0:49:22	22:38:54	0:02:36	2:19:28	23:30:55
0:56:29	1:59:39	0:33:00	0:50:22	0:50:22	22:39:54	0:03:36	2:20:28	23:31:55

16	17	18	19	20	21	22
physionet.org	physionet.org	physionet.org	physionet.org	physionet.org	physionet.org	physionet.org
RBD 16	RBD 17	RBD 18	RBD 19	RBD 20	RBD 21	RBD 22
Male	Male	Male	Male	Male	Female	Male
65	70	59	65	72	73	70
22:21:16	22:06:31	22:54:52	22:17:54	22:53:33	22:24:17	23:40:42
22:22:16	22:07:31	22:55:52	22:18:54	22:54:33	22:25:17	23:41:42
22:44:46	N/A	N/A	22:36:54	22:59:33	N/A	8:35:42
22:45:46	N/A	N/A	22:37:54	23:00:33	N/A	8:36:42
23:28:16	22:18:31	23:02:52	2:19:54	23:05:33	2:48:17	1:33:42
23:29:16	22:19:31	23:03:52	2:20:54	23:06:33	2:49:17	1:34:42
0:02:16	22:29:31	23:16:52	23:48:54	23:14:33	2:21:17	1:40:42
0:03:16	22:30:31	23:17:52	23:49:54	23:15:33	2:22:17	1:41:42
0:16:16	23:41:31	23:19:52	23:13:54	23:20:33	0:12:47	4:53:42
0:17:16	23:42:31	23:20:52	23:14:54	23:21:33	0:13:47	4:54:42
2:12:16	23:55:31	4:31:52	23:32:54	0:33:33	1:00:17	0:11:42
2:13:16	23:56:31	4:32:52	23:33:54	0:34:33	1:01:17	0:12:42

3 Analysis of EEG Signal

3.1 Load EEG Data

According to the algorithm used, first step is to extract an EEG signal from the Polysomnography done at the sleep centers. The following figure shows all the channel of an EEG signal.

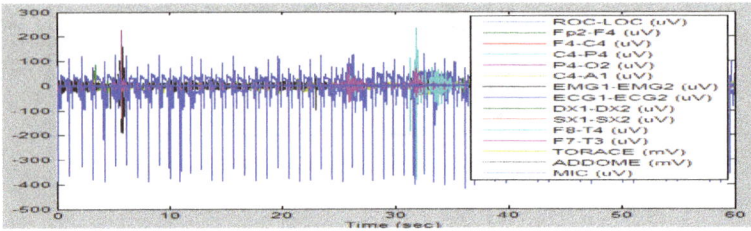

Figure.1. An EEG Signal with All Channels Present Together

3.2 Extraction Of Channel

Now the second step is to separate the channel from the extracted EEG signal. C4-A1

Considered all stages of the mentioned channels. These stages are S0, S1, S2, S3, S4, and REM.

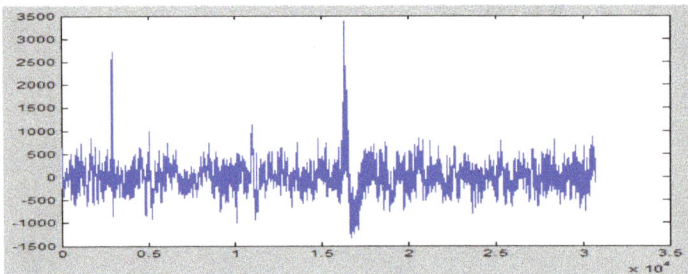

Figure. 2. An EEG Signal with C4-A1 Channel (REM stage) separated from the all channels considered together

3.3 Apply Low Pass Filter

In this step we basically label the X-axis. Time in second is labeled on the X-axis. This figure depicts the signal that is filtered using low pass filter. A low pass filter of cut off frequency 25 Hz is used.

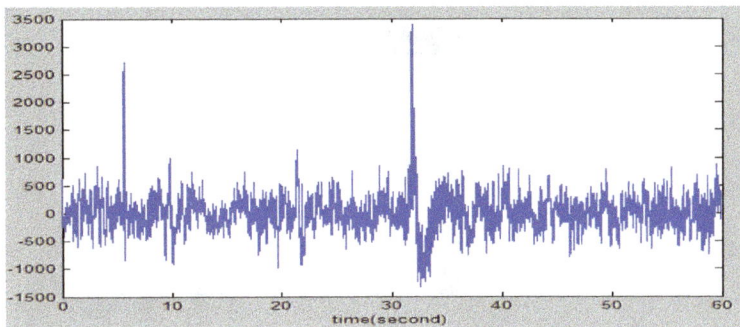

Figure. 3. An EEG Signal with C4-A1 Channel (REM stage) filtered using low pass filter with cut-off frequency of 25 Hz

3.4 Comparison between Filter and Non Filter

This shows a figure that depicts a comparison between zero-phase filtering and conventional filtering.

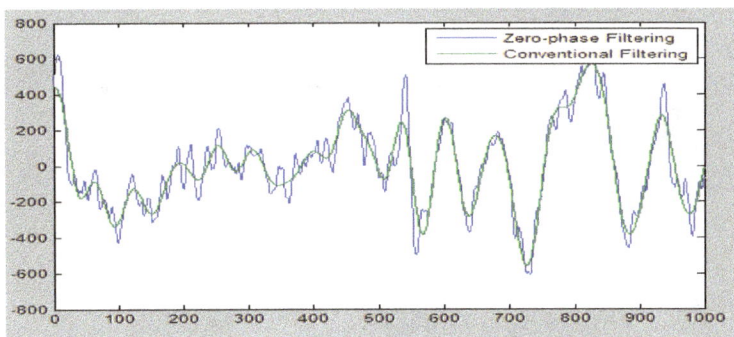

Figure. 4. Graph showing difference between an EEG signal with C4-A1 channel (REM stage) filtered by a conventional filter and zero phase filter

3.5 PSD Estimation

This figure shows the power spectral density estimate done using Welch method of the power spectral density estimation. The results are obtained using a Hamming window of the length 128 bin.

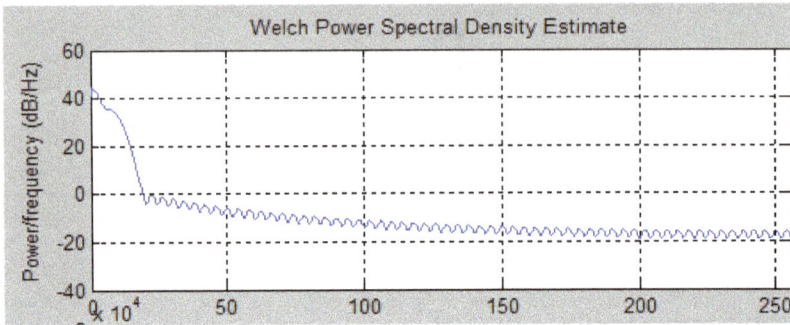

Figure. 5. Graph showing g the magnitude plot of power spectral estimation curve of An EEG Signal with C4-A1 Channel (REM stage). The area under curve shows the average power of delta, theta, alpha and beta waves

3.6 PSD Estimation Curve

This figure shows \ the magnitude of power spectral estimation curve of An EEG Signal with C4-A1 Channel (REM stage). The area under curve shows the average power of delta, theta, alpha and beta waves.

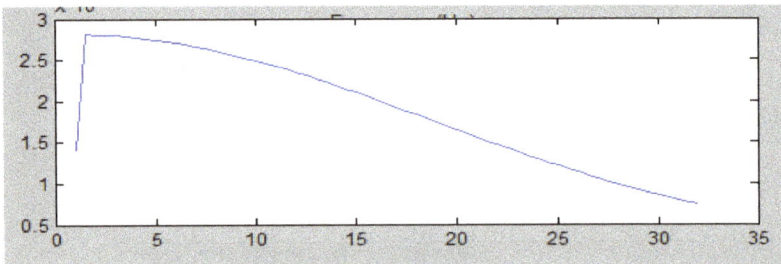

Figure 6. Graph showing g the phase plot of power spectral estimation curve of An EEG Signal with C4-A1 Channel (REM stage). The area under curve shows the average power of delta, theta, alpha and beta waves

Result

The research work is made with the help of the results obtained in analyzing an EEG signal. During the study, it is found that there is a difference in the normalized power of the normal patients and the patients suffering from Rapid Eye Movement Behavior Disorder. These differences have been shown in the conclusion tables discussed below.

Table 3 : Normalized Power of the Delta Wave of Normal Patient and RBD Patient for C4-A1 Channel and Stage REM

					CHANNEL- C4A1 / STAGE –REM										
NORMAL/ PATIENT	N1	N3	N5	N10	N11	N16	RBD 2	RBD 3	RBD 7	RBD 13	RBD 14	RBD 15	RBD 16	RBD 19	RBD 22
NORMAL-IZED POWER	0.53294	0.59256	0.55835	0.59473	0.59588	0.5039	0.37737	0.49499	0.46142	0.39149	0.48356	0.45753	0.44919	0.48904	0.4387
REMARK	HIGH						LOW								

The above table shows the normalized power of Delta wave for Normal patient and the RBD patient. The study is done for C4-A1 channel and stage REM. From the table it is observed that normalized power of Delta wave for Normal patient lie between 0.50 and 0.59 while those of RBD patient is between 0.37 and

0.49. Thus we can say normalized power of Delta wave in case of normal patient is high and that of RBD patient is low.

Table 4: Normalized Power of the Theta Wave of Normal Patient and RBD Patient for C4-A1 Channel and Stage REM

NORMAL/PATIENT	NORMALIZED POWER	REMARK
N2	0.35889	HIGH
N3	0.31803	HIGH
N5	0.33971	HIGH
N11	0.32161	HIGH
N16	0.31727	HIGH
RBD 4	0.26192	LOW
RBD 8	0.26133	LOW
RBD 10	0.2333	LOW
RBD 11	0.2333	LOW
RBD 12	0.27741	LOW
RBD 18	0.29287	LOW

CHANNEL - C4A1, STAGE – REM

The above table shows the normalized power of Theta wave for Normal patient and the RBD patient. The study is done for C4-A1 channel and stage REM. From the table it is observed that normalized power of Theta wave for Normal patient lie between 0.31 and 0.35 while those of RBD patient is between 0.23 and 0.29. Thus we can say normalized power of Theta wave in case of normal patient is high and that of RBD patient is low.

Table 5: Normalized Power of the Alpha Wave of Normal Patient and RBD Patient for C4-A1 Channel and Stage REM

NORMAL/PATIENT	N4	N5	N10	N11	N12	RBD	RBD	RBD	RBD	RBD	RBD	RBD	RBD 17	RBD	RBD
NORMAL/IZED POWER	0.098214	0.097007	0.10144	0.079779	0.080675	0.17376	0.13629	0.13382	0.15367	0.13278	0.17527	0.13994	0.13072	0.17212	0.16339
REMARK	LOW					HIGH									

CHANNEL- C4A1 — STAGE –REM

The above table shows the normalized power of Alpha wave for Normal patient and the RBD patient. The study is done for C4-A1 channel and stage REM. From the table it is observed that normalized power of Alpha wave for Normal patient lie between 0.07 and 0.10 while those of RBD patient is between 0.13 and 0.17. Thus we can say normalized power of Alpha wave in case of normal patient is low and that of RBD patient is high.

The above table shows the normalized power of beta wave for Normal patient and the RBD patient. The study is done for C4-A1 channel and stage REM. From the table it is observed that normalized power of beta wave for Normal patient lie between 0.0010 and 0.0049 while those of RBD patient is between 0.0076 and 0.014. Thus we can say normalized power of beta wave in case of normal patient is low and that of RBD patient is high.

Table 6 : Normalized Power of the Beta Wave of Normal Patient and RBD Patient for C4-A1 Channel and Stage REM

CHANNEL-C4-A1 / STAGE –REM

NORMAL/ PATIENT	NORMALIZED POWER OF BETA WAVE	REMARK
N1	0.006252	LOW
N2	0.00260	LOW
N4	0.001015	LOW
N5		LOW
N	0.00484	LOW
N11	0.002735	LOW
RBD	0.007939	HIGH
RBD	0.008297	HIGH
RBD	0.00764	HIGH
RBD	0.013705	HIGH
RBD	0.008951	HIGH
RBD	0.014279	HIGH
RBD	0.009545	HIGH
RBD 18	0.012282	HIGH
RBD	0.011128	HIGH
RBD	0.008016	HIGH

Conclusion

Normalized power of Rapid eye movement behavior disorder has been compared with the normalized power of normal patient. In this work REM stage is considered for the analysis to detect the disorder as REM stage is the dreaming stage and mostly disorder is observed in the REM stage of sleep. In REM stage the temporarily paralysis of muscles occur but in case of RBD patient, there is no occurrence of temporarily paralysis of muscles. Different waves of EEG i.e. delta, theta, alpha and beta of EEG waves are analyzed to compare with the normalized power of normal patient. In this paper the channel C4-A1 is considered to detect the RBD patient.

Normalized Power (Pnorm) of normal patients i.e. those person have no symptoms of sleep disorders has been detected and then compared with pathological cases of REM stage of C4-A1channel of EEG in all the band of EEG signals.

Normalized power is defined as the percentage of each EEG activity out of complete power. It is observed that it is improved indication of detection of features rather than of taking average power of each EEG activity. It is observed that not only RBD patient but also the patients suffering from anxiety and depression also detected by analyzing the EEG signals. By observing the result difference between RBD patient and normal person can be easily detected on the basis of Normalized Power.

References

1   V. Krajca, S. Petranek, K. Paul , M. Matousek, J. Mohylova, and L. Lhotska, "Automatic Detection of Sleep Stages in Neonatal EEG Using the Structural Time Profiles", Proceedings of the 2005 IEEE Engineering in Medicine and Biology 27th Annual Conference Shanghai, China, September 1-4, 2005

2   Siddiqui M. M, Srivastava G, Saeed S. H. Diagnosis of Nocturnal Frontal Lobe Epilepsy (NFLE) Sleep Disorder Using Short Time Frequency Analysis of PSD Approach Applied on EEG Signal. Biomed Pharmacol J 2016;9(1)

3   Siddiqui, Mohd Maroof, et al. "Detection of rapid eye movement behaviour disorder using short time frequency analysis of PSD approach applied on EEG signal (ROC-LOC)."Biomedical Research 26.3 (2015): 587- 593.

4   Alexandros T. Tzallas, Markos G. Tsipouras, and Dimitrios I. Fotiadis, "Epileptic Seizure Detection in EEGs Using Time–Frequency Analysis" IEEE transactions on information technology in biomedicine, vol. 13, no. 5, September 2009.

5   Siddiqui, Mohd Maroof, et al. "EEG Signals Play Major Role to diagnose Sleep Disorder." International Journal of Electronics and Computer Science Engineering (IJECSE) 2.2 (2013): 503-505.

6   Siddiqui M. M, Srivastava G, Saeed S. H. Detection of Sleep Disorder Breathing (SDB) Using Short Time Frequency Analysis of PSD Approach Applied on EEG Signal. Biomed Pharmacol J 2016;9(1)

7   Siddiqui, Mohd Maroof, et al. "Detection of Periodic Limb Movement with the Help of Short Time Frequency Analysis of PSD Applied on EEG Signals." Extraction 4.11 (2015).

8   Rajendra Acharya U., Oliver Faust, N. Kannathal, TjiLeng Chua, Swamy Laxminarayan. "Non-linear analysis of EEG signals at various sleep stages" Elsevier, Computer Methods and Programs in Biomedicine (2005) 80, 37—45

9   Liu, D.; Pang, Z.; Lloyd, S. R. "Neural Network Method for Detection of Obstructive Sleep Apnea and Narcolepsy Based on Pupil Size and EEG", IEEE Transactions on Neural Networks, Vol. 19, No. 2, (February 2008), pp. 308-318, ISSN 1045-9227.

10 E. Basar, T.H. Bullock "Induced Rhythms in the Brain", Boston, 1992.

11 Iasemidis, L. D., "Epileptic seizure prediction and control", IEEE Trans.Biomed. Engng., 50, 2003, 549–558.

12 Khandoker, A. H.; Palaniswami, M.; & Karmakar, C. K. "Automated scoring of obstructive sleep apnea and hypopnea events using short-term electrocardiogram recordings" IEEE Transactions On Information Technology In Biomedicine, Vol. 13, No.6, (November 2009), pp. 1057-1067, ISSN 1089-7771.

13 Malinowska, U.; Durka, P-J.; Blinowska K-J.; Szelenberger W. & Wakarow A. " Micro and Macrostructure of Sleep EEG", IEEE Engineering Medicine and Biology Magazine, Vol. 25, No. 4, August 2006, pp. 26-31, ISSN 0739-5175.

14 http://physionet.org/cgi-bin/atm/ATM

15 Rangayyan, R. M. (2002). Biomedical Signal Analysis: A Case-Study Approach, IEEE Press, ISBN 0-471-20811-6, USA

16 Khandoker, A. H.; Gubbi, J.; & Palaniswami, M. "Support vector machines for automated recognition of obstructive

17 Sleep apnea syndrome from ECG recordings" IEEE Transactions On Information Technology In Biomedicine, Vol. 13, No. 1, (January 2009), pp. 37-48, ISSN 1089-7771.

Komal Sunil Deokar[1] and Rajesh Holmukhe[2]

Analysis of PV/ WIND/ FUEL CELL Hybrid System Interconnected With Electrical Utility Grid

Abstract: Global energy crisis now-a-days is issue of concern for whole world .Existing electricity generation is prominently depends upon non-renewable fossil fuels,which are not only diminishing day by day but also causing harm to the environment. Fortunately to overcome this situation , renewable energy sources exists in nature freely and abundantly. Such sources includes, solar energy, wind energy, tidal energy, biogas, fuel cell .By using these renewables in conjunction with non-renewables estimated energy crisis can be reduced upto some extent. Concentrating on this view, in this paper hybrid model of PV/ Wind/ Fuel cell interconnected with electrical utility is analysed. For analysis, voltage, current , power output and power factor of each of subsystem i.e. PV, Wind And Fuel Cell power generation systems are observed when they are interconnected to utility grid. Results are checked at load site and at PCC. Voltage source inverter is used to convert dc ouput obtained from renewable generation systems into ac output supply. This VSI uses sinusoidal pulse width modulation technique. Model is analysed in Matlab as if it is working On-Line.

Keywords: Hybrid system, Renewables, Power generation, Voltage source inverter, Sinusoidal pulse width modulation

1 Introduction

Modern life style, highly equipped luxury homes, multi shopping complexes, fully or semi automated industries, these are the key features of the present human world. To achieve such lifestyle,huge amount of energy need has become thing of concern. Amongst all forms of energy, demand is high for electrical energy. Industrial and commercial consumers are being expensively charged

---

1 Bharati Vidyapeeth Deemed University's college of Engineering, Pune, India
Email: komaldeokar8@gmail.com
2 Bharati Vidyapeeth Deemed University's college of Engineering, Pune, India
Email: rajeshmholmukhe@hotmail.com

than domestic consumers for electricity but still facing energy shortage and hence production loss. Around 75% of total world energy is derived from burning fossil fuels but, the use of fossil fuels in undoubtedly associated with air pollution causing lowering environmental quality. Also one more thing is that whole world is in great fear of diminishing sources of fossil fuels. As fossile fuels are left in limited quantity, their cost also has increased & hence the cost of energy. To overcome all these problems, there is better option to switch towards maximum use of renewable, which are natural, abundant in quantity & clean sources of energy, which do not causes harm to environment. Renewable energy sources are – solar energy, wind energy, biogas energy, hydroelectric energy, tidal energy, fuel cell, OTEC plants. These are ecofriendly and reliable energy technology alternative for fossil fuels.[1] Since the past decade research is going on to develop economical & efficient models to derive energy from renewable. Renewable energy integration in utility grid is depends on size of power generation .small sized distributed generations are integrated into distribution line while large sized integrated in the transmission system.[2] Green energy concept is outcome of renewable energy sources and this is really great transition in energy sector[3] and hence, since 2013-14, investment in this sector has significantly increased. According to Bloomberg New Energy finance is latest energy investment report, world's largest investor in renewable sector is China, while US is on second rank. Globally, solar & Wind power capacity is grew from 74GW in 2013 to 100 GW in 2014. While record break renewable energy generation is undertaken by countries like Denmark, U.K., Germany, Scotland, Ireland.[12] Many researchers have developed different models to improve renewable generation systems, step by step. According to P. Madhu Prabhuraj, R.M. Sasiraja,power can be generated by locally available sources such as sunlight, wind and biogas. Controller designed is used to switch sources of generation as per preference,surplus energy generated is absorbed in dump load.[4] Hang-Seok Choi, Y.J. Cho, J.D. Kim and B.H.Cho uses advanced ZCS technique to reduce switching losses, so as to extract maximum power from solar array.[5] Minjie Chen,Xutao Lee,Yoshihara Tsutomu developed model of Novel soft switching PV inverter with ZVT PWM boost converter. This system gives 97% efficiency due to fast switching frequency upto 100 kHZ.[2] V. Rama Rao, B. Kali Prasanna,Y.T.R. Palleshwari described simulation and modelling of solar and hydro hybrid system.[6] Fuel cell energy is new concept now-a-days. B. Haritha, P. Dhanamajaya discussed about grid connected fuel cell using boost converter,which gives low cost and compact model.[7] Chintlapally Saidulu and K.Rajani developed boost inverter based fuel cell model for standalone applications.[8]By taking in the reference above all literature, many researchers have investigated hybrid models of PV and Wind for reducing their power quality issues so as to improve system's efficiency. But in this paper , focus is on developing hybrid system using PV,Wind and along with this fuel cell plant . Results thus obtained will give information about parameters at output point of each individual system, at load and also at the point of common coupling.

2 Proposed System

The model of proposed system is shown in fig (1), which describes Hybrid Electric Power System. This hybrid system consist of three types of renewable systems involved i.e.solar PV generation system, Wind power generation system and fuel cell power plant. These three systems are interconnected to utility grid featured as 50Hz,400/11kV.

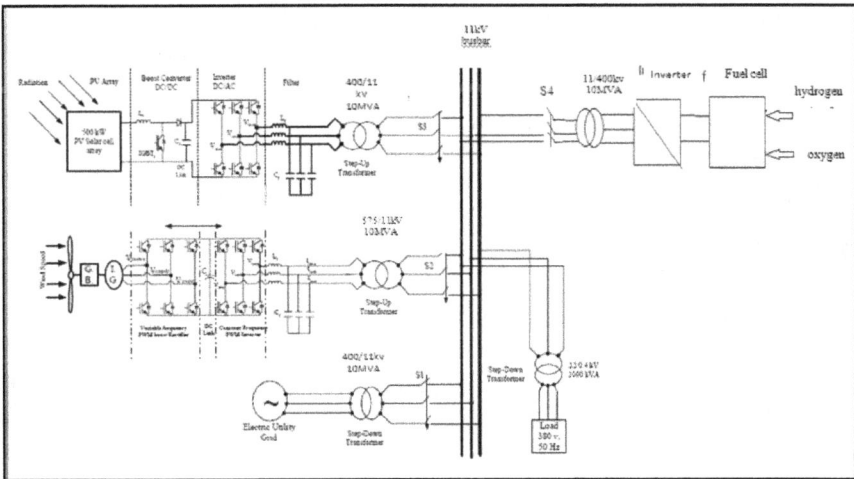

Figure 1. Power and control circuit of PV/wind/ fuel cell hybrid power generation system interconnected with electrical utility

2.1 Solar Power Generation

PV i.e. PhotoVoltaic technology is invented to convert this solar energy in the form of solar radiation into electricity. For this,solar cells are used which are semiconductor devices with p-n junction or layered structure. Light energy is converted to electricity by Photoelectric phenomenon. If we consider, one solar cell has potential upto 0.4V, then 30 cells of 0.4 V need to be connected in series. Voltage generated is in DC. So ,it is converted to AC by using inverter. To obtain power number of solar cells are connected in series and parallel combination as per requirement. System described in this paper is supported by MPPT and P&O technique to extract maximum power from solar panel.[11]

2.2 Wind Power Generation

This system uses naturally available wind energy to rotate turbine of wind mill and mechanical energy such produced can be used for electricity generation. Wind generators are used to convert this mechanical energy to electrical energy. Generators are mainly of two types, (a) Asynchronous generators (b) Synchronous generators. Asynchronous doubly fed induction generator is used here.These types of generators have advantage that they can be run slightly above or below their natural synchronous speed. So, this feature is useful for wind system as speed of air changes suddenly, hence one need to adjust speed of generator so as to get constant output power. But to avoid such output voltage and power fluctuations, AC electricity generated by wind generator is converted to DC and it again converted to AC by using boost rectifier and constant frequency PWM inverter respectively.

2.3 Fuel Cell Plant

Fuel cell is a device which converts chemical energy into electricity with respective amount of heat liberation. In this paper,model used is Hydrogen Fuel cell model which allows simulation of Proton Exchange Membrane Fuel Cell (PEMFC). The required voltage level is achieved by connecting number of fuel cells together to form fuel cell stack. The generated electricity is in DC and it is converted to AC by using inverter.

2.4 Other circuitry

The model essentially consist of other power electronics circuitry such as boost converter, to extract maximum power from renewable and fed it to dc-link. DC link acts as common energy storage element for both rectifier and inverter and also protects the system from high switching transients. Inverter used for all three subsystems is constant frequency inverter which incorporates SPWM technique.

2.5 Sinusoidal pulse width modulation

SPWM technique is used to generate required sequence of voltage pulses by on and off of power switches. Sinusoidal Pulse Width Modulation   in-volves,constant amplitude pulses with different duty cycle for each peri-od.There are two types of SPWM- (a) SPWM with bipolar switching (b) SPWM with unipolar switching.[9].

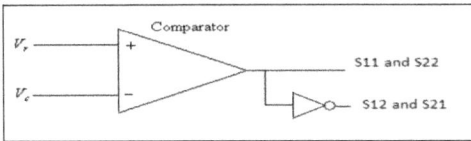

Figure 2. Bipolar SPWM generator

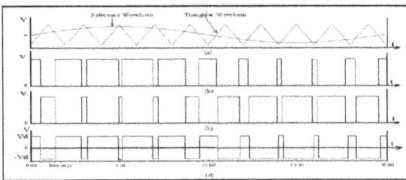

Figure 3. SPWM with Bipolar voltage switching

In bipolar switching, comparator produces one reference waveform to com-pare with triangular carrier signal and produces bipolar switching signal.

While in Unipolar switching, comparator   produces two reference wave-forms to compare with triangular carrier signal. One reference signal is positive while another is negative.[9]

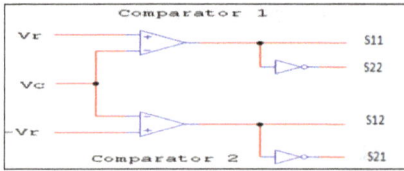

Figure 4. Unipolar PWM generator

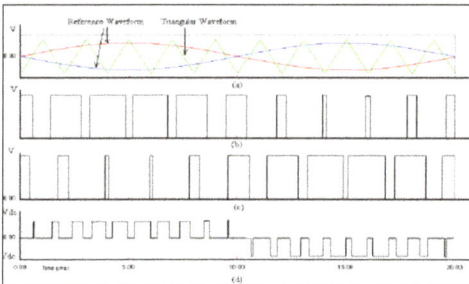

Figure 5. Waveform for SPWM with Unipolar voltage switching

3 Simulation results

Fig (6) shows simulink model for HEPS involving PV, Wind and Fuel Cell inter-connected to utility grid. Fig (7)-(19) shows all the simulation results. Fig (7)(8)(9)(10) shows voltage output waveforms of PV, Wind, Fuel cell subsystems. Transients can be seen at the instant of 1sec because at that instant, fuel cell comes in operation. In fig (6) we can clearly observe that there is constant magnitude and sinusoidal voltage waveform at PCC. Fig (11)-(14) shows current waveforms. little distorted waveforms are observed as PWM inverter is used for DC to AC conversion. Fig (14) shows sinusoidal constant amplitude current injected by HEPS at PCC. Fig (15)-(19) shows power factor at each level. Power factor of WTG is near to unity. Whatever reactive power needed by the grid is supplied by PV system and fuel cell plant. Hence , power factor of PV system is noted as 0.4 (leading) and power factor of fuel cell plant is noted 0.19(leading). Fig (19) shows power factor at PCC and is 0.9(leading). Hence , we can see that quality power is delivered to the load through HEPS and proposed system is efficient one.

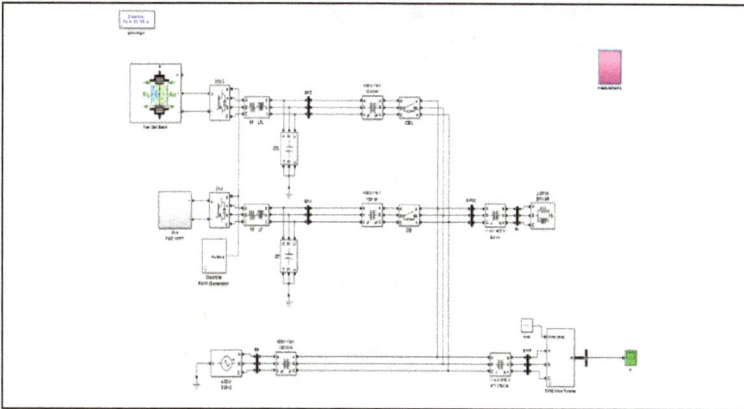

Figure 6. Simulink model of Hybrid PV/WIND/FUEL CELL Interconnected to Utility Grid

Figure 7. Simulation result of voltage output waveform of PV system

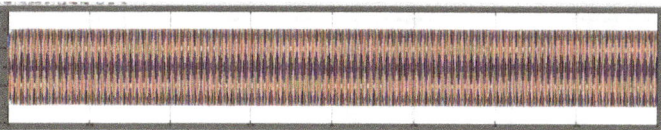

Figure 8 : Simulation result of voltage output waveform of WF system

Figure 9. Simulation result for voltage output waveform of Fuel Cell

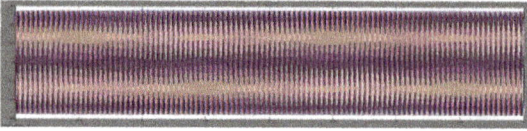

Figure 10: Simulation result of voltage profile at PCC

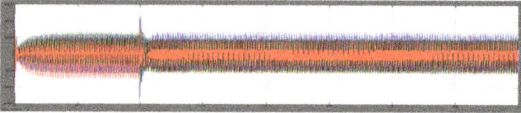

Figure 11. Simulation result of inverter line current from PV system in the grid

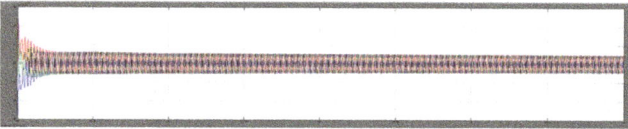

Figure 12. Simulation result of inverter line current from WF in the grid

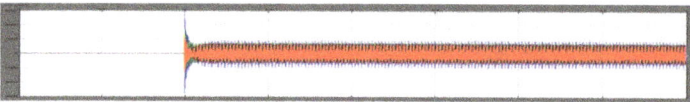

Figure 13. Simulation result of inverter line current from Fuel Cell in the grid

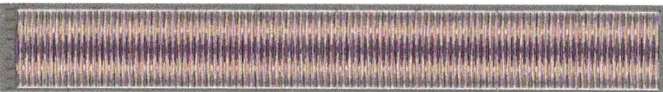

Figure 14. Simulation result of grid line current at PCC

Figure 15. Simulation result for pf of PV system

Figure 16. Simulation result for pf of WF

Figure 17. simulation result for pf of Fuel Cell

Figure 18.: Simulation result for pf at load

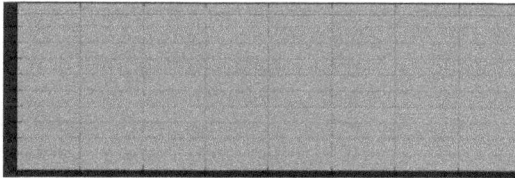

Figure 19. Simulation result of pf at PCC

Conclusion

As per studies conducted and results obtained, conclusions can be made as: Model for hybrid PV/Wind/fuel cell systems interconnected with electrical utility grid is designed and analysed in Matlab simulink environment. System is designed by considering all radiation, temperature for PV system , wind speed and variation for wind farm and fuel supplied for fuel cell. Sinusoidal Pulse width modulation technique is used to control VSI inverter. The system is designed to interconnect utility featured as 50Hz, 11kV busbar. All the three systems working well to assist utility grid by load sharing. Power factor at the point of common coupling observed near about 0.988(leading), which proves that model is efficient one. Future scope includes power quality monitoring in detail and techniques to manage surplus power if generated. Electricity generation using fuel cell is one of the upcoming technologies, which has lot of future scope to improve techniques.

References

1 Conference And Exhibition Indonesia – New, Renewable Energy And Energy Conservation" Techno- Economic Simulation Of Grid Connected PV System Design As Specifically Apllied To Residential",2014.
2 Minje Chen, Xutao Lee And Yoshihara ,"A Novel Soft Switching Grid Connected PV Inverter And Its Implementation", IEEE PEDS 2011, Singapore ,Dec 2011.
3 Er. Mamatha Sandhu, Dr. Tilak Thakur, "Issues, Challenges, Causes, Impacts And Utilization Of Renewable Energy Sources- Grid Integration" , Int. Journal Of Engineering Research And Applications.(IJERA) ,March 2014.
4 P.Madhu Prabhuraj, R.M. Sasiraja.,"Controller For Standalone Hybrid Renewable Power Generation" Int Journal Of Engineering Trends And Technology(IJETI), June 2013.
5 "Grid-Connected Photovoltaic Inverter With Zero –Current – Switching" Article January 2001 Publication at http://www.researchgate.net/publication/228997147 ,January 2001.
6 P.V.V. Rama Rao, B. Kali Prasnna, " Modeling And Simulation Of Utility Interfaced PV/ Hydro Hybrid Electric Power System" Engineering And Technology International Journal Of Electrical, Computer, Energetic , Electronic And Communication Engineering, Vol 8, 2014.
7 B.Haritha, P.Dhanamajaya, "A Grid Connected Fuel Cell Based On Boost Inverter System "International Journal Of Innovative Research In Electrical, Electronics, Instrumentation And Control Engineering ,Vol 2 ,Aug 2014.
8 Chinthapally Sadulu And K.Rajani, "Grid Connected Fuel Cell Based Boost Inverter For Standalone Applications" International Journal For Modern Trends In Science And Technology, Vol 2 ,January 2016.
9 Pankaj H Zope, Pravin G.Bhangale, Prashant Sonare ,S. R.Suralkar, Design And Implementation Of Carrier Based Sinusoidal PWM Inverter, International Journal Of Advanced Research In Electrical, Electronics And Instrumentation Engineering, Vol 1,Oct 2012.
10 Dr. Abu Tariq, mohammed Asim, Mohd.Tariq, "Simulink Based Modeling ,Simulation And Performance Evaluation Of An MPPT For Maximum Power Generation On Resistive Load", 2nd International Conference On Environmental Science And Technology IPCBEE Vol. 6. ,2011.
11 http://ecowatch.com

Lipika Nanda[1] and Pratap Bhanu Mishra[2]

Analysis of Wind Speed Prediction Technique by hybrid Weibull-ANN Model

Abstract: Due to the intermittency and non-stationary characteristics, Wind Speed in general is quite difficult to predict. In order to predict this uncertainty of the wind, a number of forecasting techniques are now available. In the recent years there has been a lot of research going on to predict wind speed with several statistical and biologically inspired computing techniques to reduce the prediction error. This is an attempt to analyze a technique that combines both the above mentioned techniques to create a hybrid model for predicting wind speed with more precision.

Keywords: Weibull distribution, artificial neural networks, Wind speed, Wind power density, Hybrid Model.

1 Introduction

The benefit we get from the clean wind energy comes with the challenge of accurate prediction of wind speed for managing the electricity grids effectively. Wind speed prediction has also been an essential part of weather forecasting, air traffic control, satellite launch, ship navigation, Missile testing, etc.[1]. The power output of a wind turbine generator varies directly with the cube of the wind speed, so a minute change in wind speed can result in a huge power output change of the generator.

This paper reports a Weibull distribution model to analyze wind data and then uses the result of that analysis to train an Artificial Neural Network for daily average wind speed prediction, which uses a back propagation algorithm. For the advancement of this project, wind speed data of Bhubaneswar, Odisha was collected from Air Force Datsav3 station number(USAF) - 429710 (VEBS) provided by tutiempo on a daily basis from 2010 to 2012[2].

---

1 KIIT University/School of Electrical Engineering, Bhubaneswar, India
E-mail: lipika2k6@gmail.com
2 Advanced Micro Devices, Hyderabad, India
E-mail: er.pratapbhanu@gmail.com

2 Background of Weibull and Ann Method

Detailed description of Weibull Distribution, ANN, back propagation algorithm and other wind speed prediction methodologies are available in literature [3], [4], [5], [6]. An outline of the used Weibull and ANN model for wind speed prediction is given below.

2.1 Weibull Model

In the present study, daily average wind speed data of Bhubaneswar, Odisha for six consecutive years have been statistically analyzed using the Weibull distribution model. The data collected was matched with the Weibull distribution in the context of both wind power density and wind speed distribution for accuracy purpose. The probability density function of a Weibull distribution is given by the following equation:

$$f(v_1) = \begin{cases} k/\lambda(x/\lambda)^{k-1}e^{-(x/\lambda)^k}, & x \geq 0 \\ 0, & x < 0 \end{cases} \tag{1}$$

Where, k is the shape parameter, which affects the shape of the distribution rather than simply shifting it. λ= scale parameter, which stretches/shrinks the distribution curve. v_1= wind speed (m/s).

In the present paper, first the k is taken to be 2 as per Rayleigh distribution and then λ and k are calculated using the average wind speed of the whole year. Also k is changed by equaling the average wind power density with the wind power density found out by Weibull method.

The wind power density can be calculated by the following equation:
$$P(v_1) = \frac{1}{2}\rho v_1^3 \tag{2}$$
Where, $P(v_1)$ is the power of the wind per unit area (W/m²).
where, ρ is the air density (kg/ m³), is the mass per unit volume of Earth's atmosphere.

If $f(v_1)$ is the Weibull density function, then the mean power density for the Weibull function becomes:
$$P_w = \frac{1}{2}\rho v^3 f(v_1) \tag{3}$$
Where P_w is the mean power density obtained by the Weibull function (W/m²).

2.2 Artificial Neural Network

The neural network based approach yields some valuable features over traditional methods in wind speed prediction, such as adaptive learning, distributed association, nonlinear mapping, as well as the ability to handle imprecise data. For wind speed prediction, a neural network model can be trained by taking a set of past measurement data. If there is a change in conditions, it can learn the change overtime, and adjust itself for a more accurate prediction [7].

A multi-layer back propagating neural network is used to predict the daily average wind speed using the output data of the Weibull model for each year. The "seasonal component" in the data is removed before training the neural network.

The back-propagation algorithm uses supervised learning. The idea of the back propagation algorithm is mainly to reduce this error, until the ANN learns the training data. The training begins with random weights, and the goal is to adjust them so that the error will be minimal. The activation function of the artificial neurons in ANNs implementing the back-propagation algorithm is a weighted sum (the sum of the inputs x_i multiplied by their respective weights w_{ji}):

$$v_j(\overline{x}, \overline{w}) = \sum_{i=0}^{n} x_i w_{ji} \qquad (4)$$

3 Database

First, the daily average wind speed of Bhubaneswar region was collected. Daily variation curve of average wind speed of Bhubaneswar for three years are given below:

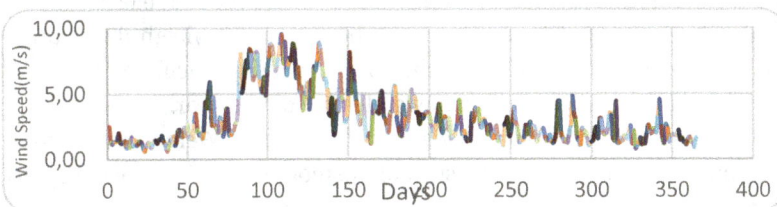

Figure 1. Average Daily wind speed variation, Bhubaneswar for 2010

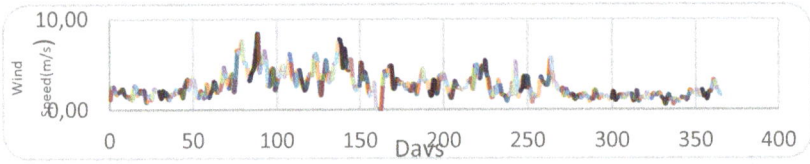

Figure 2. Average Daily wind speed variation, Bhubaneswar for 2011

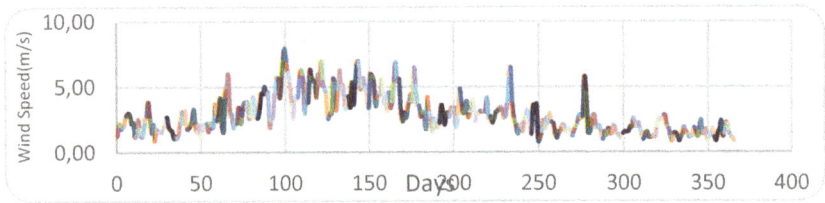

Figure 3. Average Daily wind speed variation, Bhubaneswar for 2012

4 Proposed Hybrid Method

Preparing for modeling, the data was normalized to make their distribution approximately linear, this is because the Time Series methods that are going to be applied for data analysis later on are based on a basic assumption that the random noises follow Gaussian distributions, refer to [8].

The wind speed data was divided into linear bins and the frequency at which they occurred on an average in that year was calculated. The histograms of the data of the year 2010, 2011 and 2012 are plotted in Fig 4 to Fig 6 in series 1, alongside the Weibull prediction curve in series 2, where the shape factor (k) for Weibull distribution was at first assumed to be 2 as per Rayleigh distribution and then was changed to match the wind power density calculated by using Weibull distribution with actual wind power density of that year.

The result obtained from Wind Power Density-matched Weibull distribution was put through the time series analysis tool (ntstool) for predicting wind speed of the three years under study more accurately. The ntstool consists of three (input, hidden and output) layer back propagating neural network. Levenberg–Marquardt (LM) algorithm used in static fitting problems is used for training. The input parameters to ANN model for improving the predictions are: Frequency of sample wind speeds, percentage of occurrence of the wind speeds in the whole year, Weibull Function of the corresponding wind speeds, WPD by Weibull for the corresponding wind speeds, shape factor and scale factor of the particular year. The data are randomly divided into 55% training, 20% testing and 25% validation. The training data, adjust network weight according to error. The validation data, measures network generalization and stop training when generalization stops improving. The testing data have no effect on training and provide an independent measure of network performance during and after training.

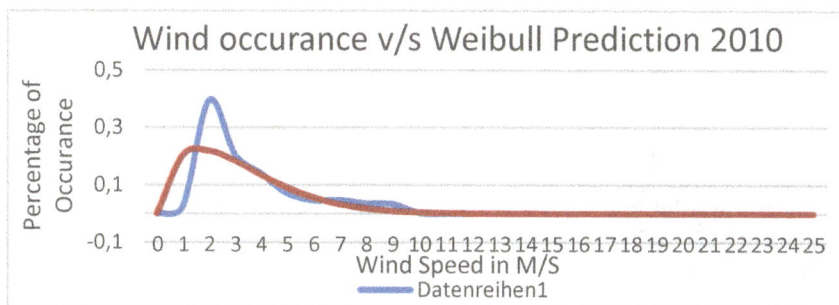

Figure 4. WPD matched wind speed v/s Weibull distribution curve of 2010

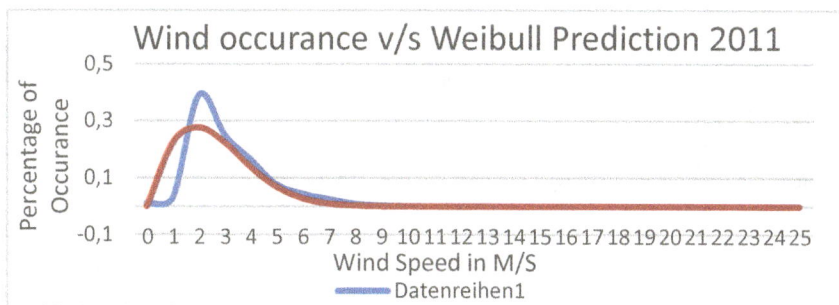

Figure 5. WPD matched wind speed v/s Weibull distribution curve of 2011

ANN extracts information from data to develop complex relationship between input and output. The inputs variables are multiplied by connection weights and its products, biases are added and passed through transfer functions for generating output. Calculation of Hidden Layers was done using the following equation:

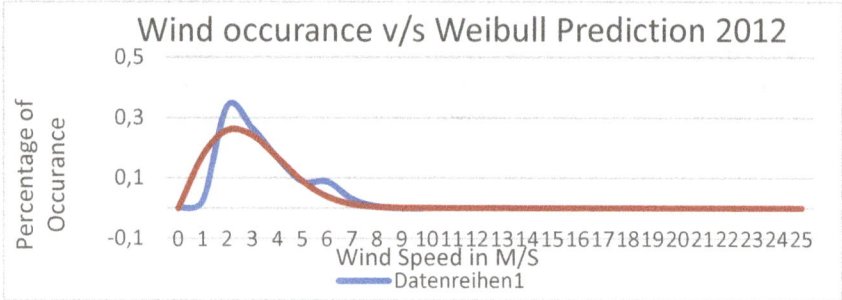

Figure 6. WPD matched wind speed v/s Weibull distribution curve of 2012

$$H_n = \frac{I_n + O_n}{2} + \sqrt{S_n} \qquad (2)$$

where, Hn and Sn are number of hidden layer neurons and number of data samples used in ANN model, In and On denotes number of input and output parameters respectively [9].

The training automatically stops when generalization stops improving as indicated by an increase in the mean square error of the validation data samples. The multilayer perceptron (MLP) neural network architecture (6-13-1) with best validation performance is used for prediction of wind speed for Bhubaneswar location for the year 2010, 2011 and 2012.

The sensitivity test was performed to validate the number of hidden layer neurons by changing the no. of hidden layers by ±1 from hidden layer neurons calculated by equation (5). For the year 2010, 2011 and 2012, for training of ANN model 114, 100 and 101 set of data points of average daily wind speed for Bhubaneswar location were used and for testing, data points from 60 to 114 were utilized from Fig.1, 2 and 3 respectively.

5 Accuracy Evaluation of Hybrid Weibull-Ann Model

The performance plots of ANN model for the year 2010, 2011 and 2012 demonstrate that mean square error becomes minimum as number of epochs increases in the Fig. 7, 8, 9 respectively. The epoch is one complete sweep of training, testing and validation.

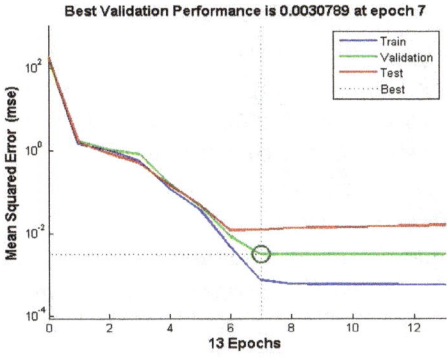

Figure 7. Performance plot of 2010

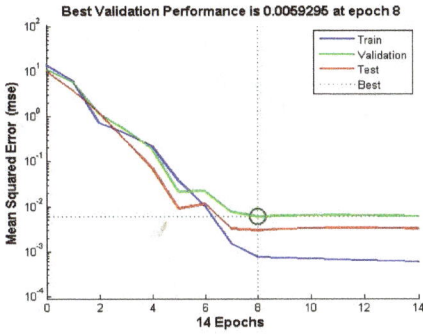

Figure 8. Performance plot of 2011

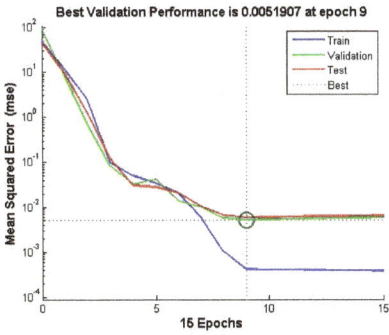

Figure 9. Performance plot of 2012

The test set error and validation set error have comparable characteristics and no major over fitting happens near epoch 7 for data set of 2010 (where best validation performance has taken place).

The test set error and validation set error have comparable characteristics and no major over fitting happens near epoch 8 for 2011 data set.

The test set error and validation set error have comparable characteristics and no major over fitting happens near epoch 9 for 2012.

The correlation coefficient (R-value) shows the association among outputs and target value of ANN model. R value of 1and 0 measures a strong, random association respectively.

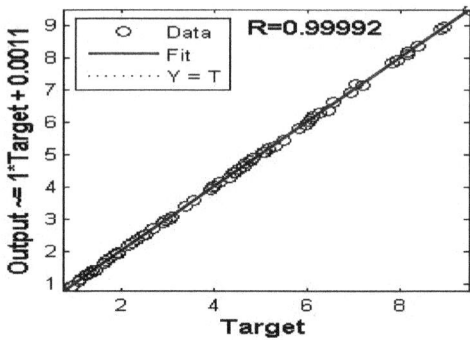

Figure 10. Regression Plot for 2010

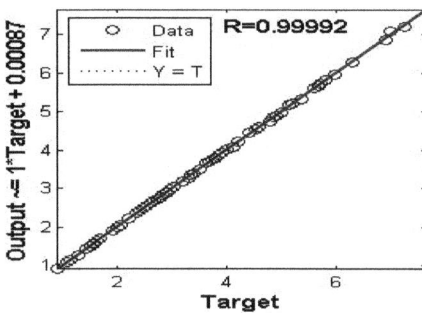

Figure 11. Regression Plot for 2011

The R-value of 0.99992 and slope 1 was achieved for 2010during whole dataset as shown above [Fig. 10].

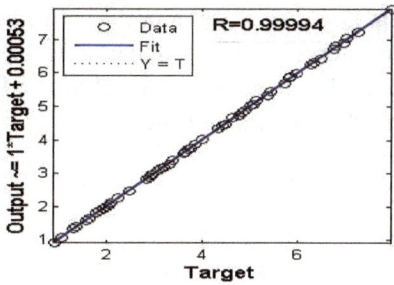

Figure 12. Regression Plot for 2012

The R-value of 0.99992 and slope 1 was achieved for 2011 during whole dataset as shown above [Fig. 11].

The R-value of 0.99994 and slope 1 was achieved for 2012 during whole dataset as shown above [Fig. 12].

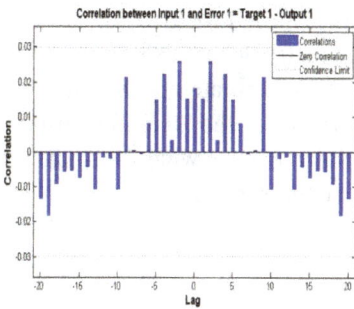

Figure 13. Input-Error C-C Plot for 2010

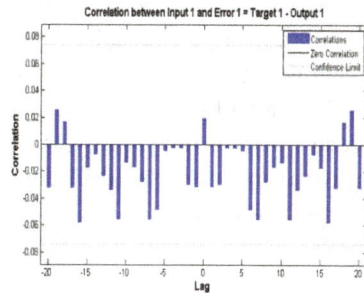

Figure 14. Input-Error C-C Plot for 2011

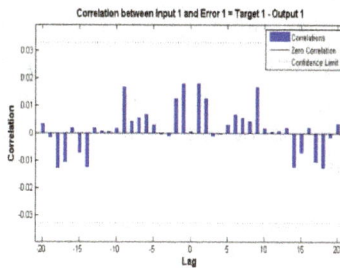

Figure 15. Input-Error Cross-Correlation Plot for 2012

The Input-Error cross-correlation plot in [Fig. 13, 14, 15] shows how the networks error at any given time is correlated with networks input at different time lags for the years 2010, 2011 and 2012 respectively. The bars in this plot should fall within the confidence limit shown by two red dotted lines for better result.

Conclusion

In the present study, daily average wind speed data of Bhubaneswar, Odisha for three consecutive years have been statistically analyzed using the Weibull distribution model. The data collected was matched with the Weibull distribution in the context of both wind power density and wind speed distribution for accuracy purpose. Then an ANN model is developed with Time-series tool (ntstool) for improving wind speed curves created by Weibull distribution. Model was validated by the results obtained in performance plot and also the Input-Error cross correlation was found to be within the confidence limits. The correlation coefficient (R-value) of 0.99992 and slope 1 was achieved for both 2010 and 2011 while an R-value of 0.99994 and slope 1 was achieved for 2012 for the whole data set, showing high prediction accuracy of the developed ANN Model.

The result obtained from the ANN was more accurate and had less error than the Weibull analysis. Data analysis confirms that the Weibull distribution is an adequate technique for describing the daily average wind speed distribution. But using a Hybrid ANN method gives superior result and helps in more accurate forecasting.

Future Work

In future the equations of Weibull distribution can be merged with other non-linear equations for handling nonlinearity and stochastic uncertainty problems associated with wind speed data and then hybridizing it with ANN will be more effective. Addition of any GUI which helps in predicting future data with MATLAB will be helpful for validating the results as well as finding the MAPE (Mean Average Prediction Error).

References

1   K. Sreeklakshmi, and P. Ramakanth kumar, "Performance evaluation of short term wind speed prediction techniques," *International Journal of Computer Science and Network Security*, Vol.8, issue 8, pp.162-169, Aug.2008.
2   www.tutiempo.net
3   Yuehua, Liu, Jiang Yingni, and Gong Qingge. "Analysis of wind energy potential using the Weibull model at Zhurihe", *International Conference on Consumer Electronics Communications and Networks (CECNet)*, 2011.
4   Wang, Ruigang, Wenyi Li, and B. Bagen. "Development of Wind Speed Forecasting Model Based on the Weibull Probability Distribution", *International Conference on Computer Distributed Control and Intelligent Environmental Monitoring*, 2011.
5   Liang Wu. "A study on wind speed prediction using artificial neural network at Jeju Island in Korea", *Transmission & Distribution Conference & Exposition Asia and Pacific*, 10/2009.
6   Bhaskar, Melam, Amit Jain, and N. Venkata Srinath. "Wind speed forecasting: Present status", *International Conference on Power System Technology*, 2010.
7   Xiao-Hua Yu. "Applications of Neural Networks to Dynamical System Identification and Adaptive Control", *Studies in Computational Intelligence*, 2008.
8   B. G. Brown, R. W. Katz, and A. H. Murphy, "Time series models to simulate and forecast wind speed and wind power," *Journal of Climate and Applied Meteorology*, vol. 23, no. 8, pp. 1184–1195, 1984.
9   P. Ramasamy, S.S. Chandel, Amit Kumar Yadav. "Wind speed prediction in the mountainous region of India using an artificial neural network model", *Renewable Energy*, Volume 80, August 2015, Pages 338–347.

K.Navatha[1], Dr. J.Tarun Kumar[2] and Pratik Ganguly[3]

An efficient FPGA Implementation of DES and Triple-DES Encryption Systems

Abstract: Efficient FPGA implementation of DES and Triple Data Encryption Standard (TDES) algorithm, used mainly in cryptographic applications has been presented in this paper. Design of digital cryptographic circuit based system and implementation on a Vertex 6 series target device with the use of optimized VHDL have been taken into consideration. Thorough simulation and synthesis process for different devices was carried out for confirmation of the expected outcome of Triple-DES based system.

The main contributions of this research work are (i) Thorough simulation and synthesis process of the proposed design targeting different FPGA devices, (ii) Complete hardware implementation of Triple-DES algorithm in Virtex 6 series device based FPGA (iii) Comparison of the experimental and implementation results reported so far are the main part to be focused on.

Keywords: FPGA Implementation, Triple DES algorithm, Data Encryption Standard, VHDL, Implementation

1 Introduction

Necessity of information transfer and storage with proper security mechanism is increasing day by day,

Generally huge amount of personal data transfer needs high end security. Therefore, each transmitted bit of information should be converted in an unrecognized form as security measure. Data enciphering should take place in real time and cryptography is the main procedure for this purpose. Different encryp-

---

1 Principal Investigator, Sumathi Reddy Institute of Technology for Women, Warangal,Andhra Pradesh,India

2 Associate Professor, Dept.of. Electronics and Communications Engineering, Sumathi Reddy Institute of Technology for Women, Warangal, T.S

3 Associate Professor, Dept.of. Electronics and Communications Engineering, SR College of Engineering and Technology
Email: Pratik.ganguly1@gmail.com

tion algorithms have been already developed among which one of the most popular is the Triple Data Encryption Standard (DES) algorithm.

The Triple DES algorithm is very popular and widely used because it is reasonably secure and pretty fast.

It's almost impossible to break DES system however because DES is only a 64-bit (eight characters) block cipher, a vast search of 255 steps average can retrieve the key which is used in encryption process. A much more secured version of DES is known as Triple-DES (TDES), that is essentially equivalent to use DES on plaintext for 3 times with the help of 3 different keys. Generally, Triple DES is 3 times slower than the DES but it is much more secured compared to DES.

Implementation of DES and Triple DES algorithm by use of a high-level hardware description language i.e; VHDL combined with advanced FPGA technology has been taken into consideration in this research. The design was synthesized efficiently for different FPGA devices of Spartan and Virtex series, viz Spartan 3, Spartan 3AN, Virtex E, Virtex 6 etc. The design is implemented and verified on a Virtex 6 FPGA development board from Xilinx. The rest of the paper is categorized as follows: Section II lists previous works and implementation of Crypto algorithm i.e. DES and Triple DES and also gives a brief introduction to the Data Encryption Standard and Triple Data Encryption Standard algorithms. Section III gives experimental framework and results. Section IV gives the conclusions.

2 Background

2.1 Previous Work

A vast research & development are ongoing on DES and Triple DES. Triple-DES is already implemented on Spartan devices. The design and implementation of TDES was reported in [7][12]. Handel-C has also been used for development of DES.

2.2 Data Encryption Standard

DES is mainly a block cipher which operates on blocks that are of 64-bits in size. A 64-bit block input of plaintext is encrypted into a 64-bit block output of cipher text format. For encryption and decryption mechanism the same algorithm and key are being used. The security of DES mechanism is in 56-bit key. The DES algorithm functions as follows [1-7] [11] [20].

The plaintext block is considered and an initial permutation is carried out on this. The key is also taken at the same time and presented in a 64-bit block with every 8th bit as parity check. The extracted 56-bit key is ready for use at that moment. The 64-bit plaintext is generally divided into 32- bit two parts. The two parts of the plaintext are combined with data from the key which is called Function F. There are totally 16 rounds of Function f, after which the two parts are recombined into 64-bit block, which then goes through a final permutation to complete the algorithm operation and a 64-bit cipher text block is generated.

2.3 Triple Data Encryption Standard

A brief representation of Triple Data Encryption Algorithm is described here.

TDES is a block cipher which operates on 64-bit data. There are different forms, each of which uses DES cipher for three times. TDES has choice to work with one, two or three 56-bit keys. In fact, plain text is encrypted three times. Different TDES modes have already been proposed:
–   DES-EEE3 is Three DES encryptions with three different keys.
–   DES-EDE3 is Three DES operations in the sequence encrypt-decrypt-encrypt with three different keys.
–   DES-EEE2 and DES-EDE2 are the Same as the previous formats except that the first and third operations use the same key.

3 Experimental Framework and Results

3.1 Synthesis and Implementation

The design was synthesized & implemented with the use of VHDL using Xilinx ISE. Simulation was performed by Xilinx ISE simulator and Modelsim XE simulator. Figure 2 represents the flow which was followed for the digital implementation. The RTL architectures of DES and TDES are shown in Figure 3 and 4 respectively below:

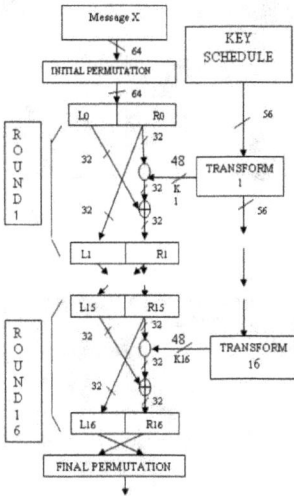

Figure 1. Triple DES diagram

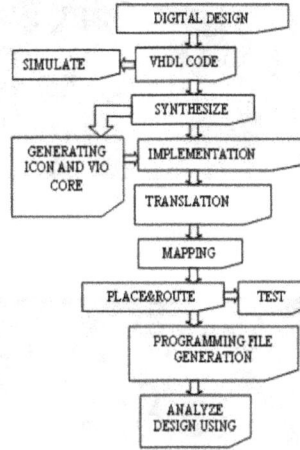

Figure 2. Implementation flowchat

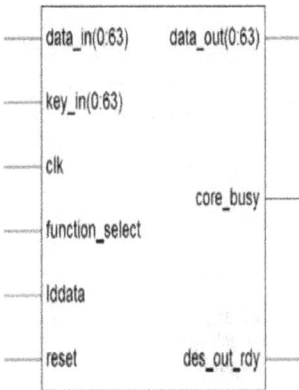

Figure 3. DES RTL Schematic

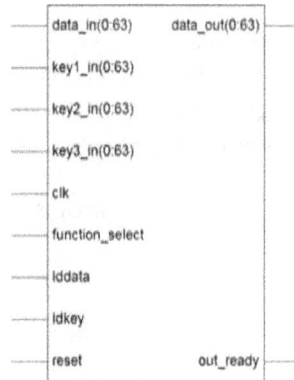

Figure 4. Triple DES RTL Schematic

Code optimization and resource map techniques for DES and Triple-DES have been used efficiently which results in delay trade-offs and minimal area utilization.

Figure 5 and Figure 6 presents the implemented components in the chip. The interconnections of the components are also shown here.

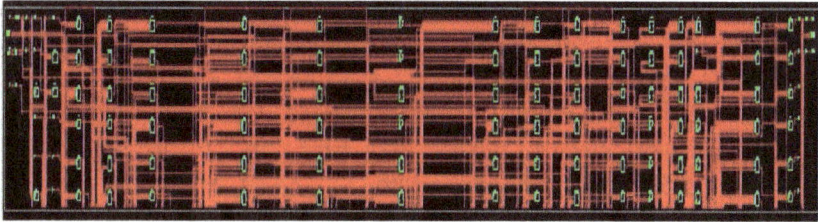

Figure 5. DES Schematic

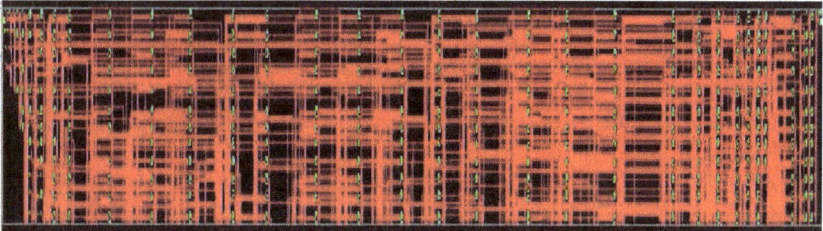

Figure 6. TDES Schematic

FPGA implementation of DES and TDES were accomplished on a Virtex6 device using Xilinx ISE Foundation 10.1i. Table 1 and Table 2 show the performance figures for DES hardware implementations. Table 3 and Table 4 show synthesis results of Triple DES implementations.

Table 1: Synthesis result for DES on Spartan Device

	Spartan 3 (Target device xc3s400,Package fg320,Speed -5)		Spartan 3AN (Target device xc3s700AN,Package fgg484,Speed -5)	
Logic Utilization	Used	Utilization	Used	Utilization
No. of Slices	442 out of 28800	5%	461 out of 11264	4%
No. of Slice Flip Flops	281 out of 28800	1%	273 out of 22528	1%
No. of 4 input LUTs	789 out of 15681	5%	827 out of 22528	3%
No. of	190 out	48	190 out of 502	37%

bonded IOBs	of 391	%		
No. of GCLKs	1 out of 8	12%	1 out of 24	4%

Table 2: Synthesis result for DES on Vertex device

	Vertex 5 (Target device XC5VLX50,Package ff1676,Speed -1)		Vertex 6 (device xc6vlx240t, package ff1156)	
Logic Utilization	Used	Utilization	Used	Utilization
No. of Slice Registers	266 Out of 28800	0%	286 Out of 301440	1%
No. of Slice LUTs	527 Out of 28800	1%	433 Out of 150720	1%
No. of fully used LUT-FF pairs	112 Out of 681	42%	286 Out of 681	42%
No. of bonded IOBs	190 Out of 440	43%	189 Out of 600	32%
No. of BUFG/BUFGC TRLs	1 Out of 32	3%	1 Out of 32	3%

Table 3: Synthesis result for TDES on Spartan device

	Spartan 3 (Target device xc3s1000,Package fg676,Speed -5)		Spartan 3AN (Target device xc3s1400AN,Package fgg676, Speed -5)	
Logic Utilization	Used	Utilization	Used	Utilization
No. of Slices	1585 out of 7680	20%	1622 out of 11264	14%
No. of Slice Flip	1254 out of	8%	1230 out of 22528	5%

Flops	1536 0			
No. of 4 input LUTs	2494 out of 1536 0	16%	2593 out of 22528	11%
No. of bonded IOBs	302 out of 391	77%	302 out of 502	37%
No. of GCLKs	1 out of 8	12%	1 out of 24	4%

Table 4: Synthesis result for TDES on Virtex device

	Vertex 5 (Target device XC5VLX50,Package ff1676,Speed -1)		Vertex 6 (xc6vlx240t, package ff1156)	
Logic Utilization	Used	Utilization	Used	Utilization
No. of Slice Registers	1206 out of 28800	4%	920 Out of 301440	1%
No. of Slice LUTs	1690 out of 28800	5%	1384 Out OF 150720	%
No. of fully used LUT-FF pairs	447 out of 2449	18%	632 Out of 2,266	7%
No. of bonded IOBs	302 out of 440	68%	302 Out of 600	0%
No. of BUFG/BUFGCTRLs	1 out of 32	3%	1 Out of 32	%

From the comparison result we can infer that the proposed implementation scheme is pretty compact and efficient in all aspects. The synthesis is carried out with the Virtex 6 series device that results in minimal consumption of hardware components.

Conclusion

Very high speed performance and efficient hardware implementation have been proposed here for DES and Triple DES based systems. This is the most efficient and flexible solution for providing advanced security in crypto systems and wireless protocols secured layers. The proposed and previous hardware implementations have been compared here by measuring results between those which shows quite enough performance and efficiency improvement for the proposed system.

Future Work

Further research can be carried out for optimization techniques for HDL code of algorithms which can be integrated for efficient and more compact design.

Acknowledgement

One of the authors K.Navatha is thankful to DST, Ministry of Science and Technology, New Delhi, India for the financial assistance (SR/WOS - A/ET-22/2012) and to the Management and Principal of SRIT, Warangal for the encouragement.

References

1   William C. Barker, "Recommendation for the Triple Data Encryption Algorithm (TDEA) Block Cipher", Revised 19 May 2008, NIST Special Publication 800-67, Version 1.1.
2   Fiolitakis Antonios, Petrakis Nikolaos, Margaronis Panagiotis, Antonidakis Emmanouel, "Hardware Implementation of Triple-DES Encryption/ Decryption Algorithm", International Conference on Telecommunications and Multimedia, 2006
3   Fábio Dacêncio Pereira, Edward David Moreno Ordonez, Rodolfo Barros Chiaramonte , "VLIW Cryptoprocessor: Architecture and Performance in FPGAs", IJCSNS International Journal of Computer Science and Network Security, VOL.6 No.8A, August 2006.

4   Dr. V. Kamakoti, G. Ananth and U.S. Karthikeyan, "Cryptographic Algorithm Using a Multi-Board FPGA Architecture", Nios II Embedded Processor Design Contest—Outstanding Designs 2005.

5   Toby Schaffer, Member, Alan Glaser, Member, and Paul D. Franzon, "Chip-Package Co-Implementation of a Triple DES Processor", IEEE Transactions on Advanced Packaging, Vol. 27, No. 1, February 2004.

6   Andrew S. Tanenbaum, "Computer Networks", 2003

7   D. Stinson. "Cryptography: Theory and Practice", 2nd Edition, Chapman and Hall/CRC, 2002

8   Vikram Pasham and Steve Trimberger, "High-Speed DES and Triple DES Encryptor/Decryptor", Xilinx Application Note: Virtex-E Family and Virtex-II Series, XAPP270 (v1.0) August 03, 2001

9   F. Hoornaert, J. Goubert, and Y. Desmedt, "Efficient hardware implementation of the DES," in Proc. Adv. Cryptol. (CRYPTO'84), 1984, pp. 147–173.

10 "Data Encryption Standard (DES) ", Federal Information Processing Standard Publication, FIPS PUB 46-3, National Bureau of Standards, 1977.

Sunil Kumar Jilledi[1] and Shalini J[2]

A Novelty Comparison of Power with Assorted Parameters of a Horizontal Wind Axis Turbine for NACA 5512

Abstract: Utilization of energy is increasing day by day, in order to save the fossil fuels. Renewable energy sources became a one of the most prominent sources of energy. Solar energy, wind energy, tidal energy and geothermal energy etc are playing a vital role. Wind energy is one of the most prominent sources of energy. A lot of research is going to improve the power from the wind energy and to improve the quality of the power. A small case study has been performed on horizontal wind axis plant. By using XFOIL the blade has been developed by using Circular Foil and NACA 5512.The main scope of the paper is running on the power, Coefficient of Power, Axial induction factor, Tip Speed Ratio. etc., .Here the power profiles have been compared with various parameters of the wind plant. And all the comparative simulation results have been presented clearly by using software tool.

Keywords: Coefficient of power, axial induction factor, wind power, XFOIL, NACA 5512.

1 Introduction

Wind energy occupies outstanding room within the global energy production. Wind energy has been extracted from the wind over hundreds of years ago; it is constructed by wood to pull up the water for irrigation purpose. Then it has been extended for generation of electric power in the year, 1850 by the researchers Daniel Halladay and John Burnham, the technology is improving day by day now the wind energy is playing vital role in the generation of electrical energy. we can have a fleeting look of information of wind energy, during the

---

1 PhD scholar,Omprakash jogender sharma university, India
Email: sunilkumarjelledi@gmail.com, sunil.kumar@astu.edu.et
2 Assistant Professor, Dairy Technology, S.V. V. University, India
Email: aparanjishalu@gmail.com

first quarter of 2014, the U.S. wind industry installed 133 turbines, totaling 214 megawatts (MW).

Most of the wind turbines are operated by Horizontal wind axis turbine and vertical wind axis turbines. HWAT are playing a major role in power generation as compared with VWAT. HWAT are most efficient models for bulk power generation. But most of the Major research is taking place in designing the blades for HWAT, to reduce the sizes of the plants and to get more and efficient power from the plant. Tip Speed Ratio (TSR) one of the important parameter which affects the efficiency of the wind turbine. The important parameter is aerodynamic shape, assortment and huge research work is going which is straightforwardly and circuitously relates to the improvement of power generated by the wind turbine [13].

2 Aerodynamics and Blade

BEM theory plays a major role in evaluating the forces on the wind turbine [12]. Ozge Polat and Ismail H.Tuncer has done research how to optimize blade shape at prescribed wind speed, rotor speed, rotor diameter and number of blades of four digit NACA profile of a wind turbine by using parallel genetic algorithm [6]. National Advisory Committee for Aeronautics is playing a key role in airfoil, NACA it indicates how many number of airfoil has to use. In generally 20 foils have been used, 3 are circular and 17 are NACA foils [3]. Among then NACA 5512 is getting more prominent, because it is having good lift, it increases the power generation. The power generated by the wind turbine is dependent on the wind speed, Area of the wind turbine rotor and air density and it is given as [8]

$$\text{wind power} = \frac{1}{2} A V^3 \rho \dots (1)$$

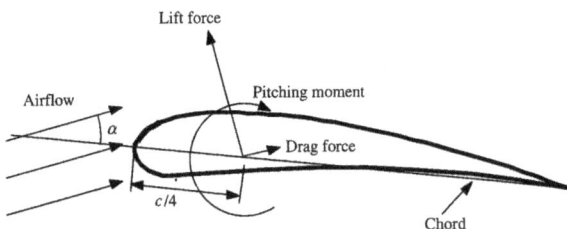

Figure 1. overview of Lift, Drag force, Angle of Attack chord of a blade

Coefficient of power is defined as the ratio between Electricity produced by wind turbine to the Total energy available in the wind. It has been explained by the Betz a German physicist, as per the theoretical aspect good turbines will 35-45% only. The term tip speed (λ) is given as

$$\lambda = \Omega R/U \tag{2}$$

The induction factor is defined is given as

$$a = 1/(\frac{4\sin\phi}{\sigma C_n} + 1) \tag{3}$$

The Axial induction factor is

$$a' = \frac{1}{[[\frac{4\cos\phi}{\sigma C_L}]-1]} \tag{4}$$

The drag coefficient of the horizontal wind axis turbine is given by

$$C_d = \frac{D}{\frac{\rho}{2}A_r V^2} \tag{5}$$

The maximum value of Cp according to Betz limit is 0.593[2]. Now days the modern turbines are operating up to 0.5 and this value are more optimized value for the wind turbine. The expression for Cp is given

$$Cp = \frac{P}{\frac{1}{2}\rho A V^3} \tag{6}$$

In addition, optimization algorithms to have an optimal distribution of blade twist angles and chord lengths to get a chosen speed ratio (TSR) are usually applied. For Horizontal wind axis turbines the blade twist is optimized such that every sections faces the relative wind at an angle which offers the best glide ratio on the chosen design. And in designing the blade the Reynolds number is taken as 1000000[7, 9],

3 Step by Step Analysis

The air craft industry started the design of blades to get more efficient aero planes, the same technology has been implementing for design of the blades. Many evaluations tools are there in the research environments like CFD, RANS and Vortex models but Blade Element Momentum (BEM) [1] are most resourceful methods to forecast the Horizontal Axis Wind Turbines (HWAT) in the wind industry [4].Based on the BEM the XFOIL is running. The software XFOIL, developed by Drela and Gilesat MIT, this program is used to investigate and calculate the flow around subsonic isolated airfoils. C. J. Bai1, F. B. Hsiao, etal, has designed the 10KW wind turbine by using the BEM model but they didn't done any concentration the power of the turbine but the major concentration is on design of the blade. In this paper an extension of the design of blade has been done and the performance curves of the power compared with the important terms like Cp, Ct, Kp etc.

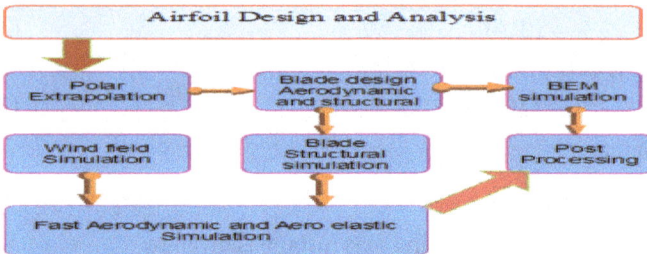

Figure 2. Flowchart of the simulation software and working flow style

By using the XfOIL, developed the NACA 5512. Many of the researchers has been presented the analysis of the NACA 4412[3] The researcher Bhaskar Upadhyay Aryalin his paper he designed the wind blade by using NACA 5512 but the concentration is on structural analysis and design.[5].Many papers has focused on the design of wind blades and they presented the structural analysis. [2]. But here by using the NACA 5512 in XFOIL, the power profiles has been compared with various parameters up to now there is no paper has been developed on this idea. The author Miguel Toledo Velázquez et.al has presented clear modeling of the wind turbine by using the BEM models but authors concern parameters are very less like tip speep, torque, etc, but here input parameters are more and the analysis concerned more regarding the comparison of power with respective to different mechanical terms.

3.1 Implementation of Software tool

The wind turbine will be modeled in the software, Airfoil design then the same will be used for the XFOIL analysis. In the simulation software the main module is 360 degree here it is extrapolated in the form of AoA. Here two methods will be available but the most will be modeled using Montgomery and viterna. By using the software we can design the blade, rotor and turbine. Here the main focus is comparison of the power profiles with the parameters like wind speed, Coefficient of Power, Thrust, Rotor Torque, Blade Bending Moment, Wind Speed, Tip Speed Ratio, Rotational Speed, Ct, Cm, Dimension less power coefficient(Kp), Pitch angle.

4 Data for simulation

To simulate the XFOIL and the data required for the software were taken and calculate by using the mathematical modeling. The data considered for the analysis is listed in table.01 and table.02.

Table 1: XFOIL data analysis for NACA 5512

S.No	PoS(m)	Chord (m)	Twist (deg)	S.No	PoS(m)	Chord (m)	Twist (deg)
1.	0	0.40	24.74	4.	2.50	1.40	24.74
2.	0.62	0.40	24.74	5.	5.00	0.70	5.00
3.	1.25	1.40	24.74	6.	7.50	0.53	2.00

Table 2: Wind turbine ratings

Rotor Diameter = 80m	Cut out Speed = 25m/sec	Tip speed Ratio start = 1
Number of blades= 3	Blade length = 38 m	Tip Speed Ratio end =15
Hub height = 61.5 m	Chord length =1.58 m	Over hang = 3.7 m
Tower height = 60m	Density = 1830 kg/m3	Iterations =100
Cut in speed =4m/sec	Tit angle = 4 degrees	

5 Simulation Results

The results have been presented clearly for the Horizontal wind axis turbine for the different parameters. In designing the HWAT turbine by considering the standard values as mentioned above for the NACA 5512 model the graphical results has been presented. The Graphs01 relates to the power coefficient (Cp) of a wind blade to the Thrust coefficient (Ct), Moment Coefficient (Cm), Dimension less power coefficient (Kp), Tip Speed ratio. According to the Betz law the theoretical calculated Power Coefficient is in the range of 0.593 so the Cp is in the range. The optimum valve of the power coefficient for the NACA 5512 is 0.32.

For the NACA 5512 the XFOIL analysis is considered for 5°to 25° and an increment of 1° The Graphical results Graph-02 Shows the graphical results of Axial induction factor to Angle of Attack Alpha, Axial blade force coefficient, Blade twist angle Theta, Inflow Angle Phi

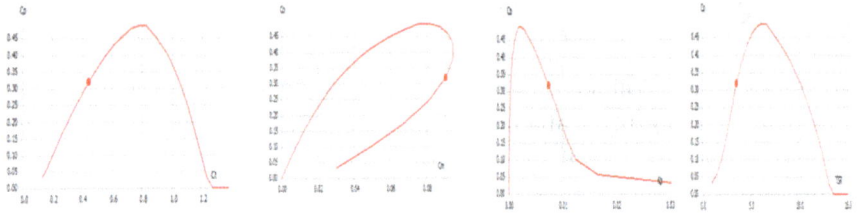

Graph 1 Graphical results for Power Coefficient Vs Thrust Coefficient (Cp), Moment Coefficient (Cm), Dimension less power coefficient (Kp), Tip speed ratio

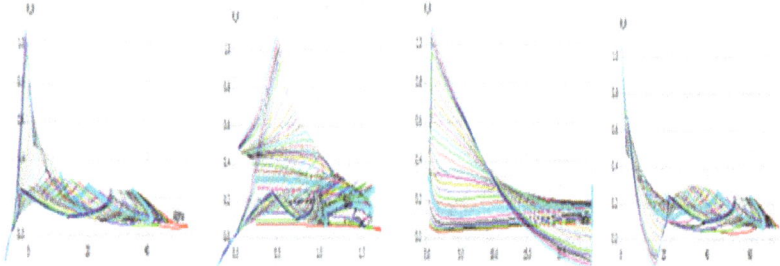

Graph 2. Graphical results for Axial Induction Factor Vs Angle of Attack Alpha, Axial blade force coefficient, Blade twist angle Theta, Inflow Angle Phi

The graphical results Graphs.03 these can estimate the power of the wind turbine, shows clearly the comparative output for power Vs Thrust, Rotor Torque, Blade Bending Moment, Wind speed, Tip Speed Ratio, Rotational speed, Pitch angle, Cp, Ct.

From the above graphical result.01, 02, 03 of NACA 5512 by XFOIL analysis using BEM theory has done, the wind blade has designed for the length of 38m, chord length of 1.58m as think about in the table.02 the maximum power produced by this NACA 5512 blades is $5.0*10^5$ W is produced by this model. So by comparing the NACA 4412 it is having better performance analysis [3].

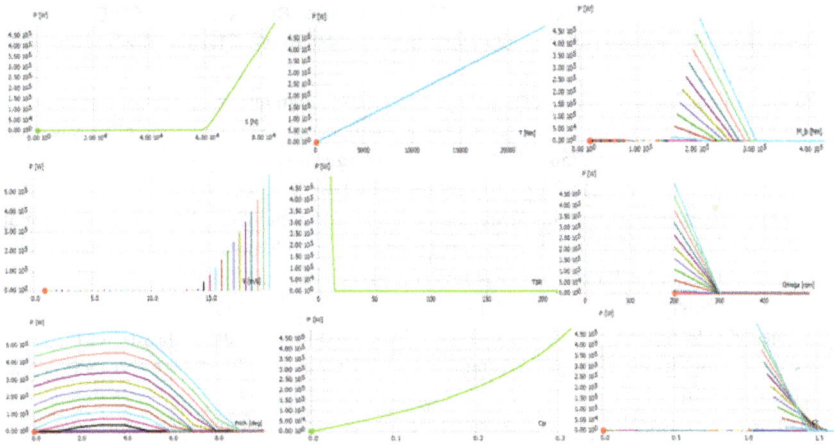

Graph 3. Graphical results for Power Vs Thrust, Thrust, Rotor Torque, Blade Bending Moment, Wind speed, Tip Speed Ratio, Rotational speed, Pitch angle, Cp,Ct

Conclusion

The XFOIL analysis for NACA 5512 by using BEM theory has been developed in Q-blade software. The NACA 5512 is very reliable and more efficient as compared with other NACA models. Here NACA 5512 is used for developing a Horizontal Wind Axis Turbine. The power output of the turbine through multiple iterations has been presented clearly. Various analysis for the HWAT has been presented clearly like Power Vs Cp, Ct, TSR. The maximum power produced by the NACA 5512 model for the above design is $5\times 10^5\ W$. This set of analysis is very useful for the industry and research development those who are working on NACA 5512.Even it can simulate for Vertical Wind axis turbines. It is very feasible for the simulation analysis to compare the practical system.

References

1   C. J. Bai1, F. B. Hsiao2,*, M. H. Li3, G. Y. Huang4, Y. J. Chen5,Design of 10 kW Horizontal-Axis Wind Turbine (HAWT) Blade and Aerodynamic Investigation Using Numerical Simulation, 7th Asian-Pacific Conference on Aerospace Technology and Science, 7th APCATS 2013, www.sciencedirect.com.
2   G. H. Farooq Ahmad Najar, "Blade Design and Performance Analysisof Wind Turbine," in International Conference on Global Scenario in Environment and Energy, India, 2013.

3 Sandip. A. Kale et al, INTERNATIONAL JOURNAL of RENEWABLE ENERGY RESEARCH . ,Vol. 4, No. 1, 2014 Aerodynamic Design of a Horizontal Axis Micro Wind Turbine Blade Using NACA 4412 Profile

4 Miguel Toledo Velázquez, Marcelino Vega Del Carmen, Juan Abugaber Francis, Luis A. Moreno Pacheco, Guilibaldo Tolentino Eslava, Design and Experimentation of a 1 MW Horizontal Axis Wind Turbine, Journal of Power and Energy Engineering, 2014, 2, 9-16 Published Online January 2014 (http://dx.doi.org/10.4236/jpee.2014.21002)

5 Bhaskar Upadhyay Aryal et al.: Design and Analysis of a Small Scale Wind Turbine Rotor at Arbitrary Conditions Rentech Symposium Compendium, Volume 4, September 2014

6 Ozge Polat, Ismail H.Tuncer,"Aerodynamic shape optimization of Wind Turbine Blades using a parallel Genetic Algorithm", Procedia Engineering, 61, 2013, pp. 28-31

7 Ahmed MR, Narayan S, Zullah MA, Lee YH. , "Experimental and numerical studies on a low Reynolds number airfoil for wind turbine blades", Journal of Fluid Science and Technology 2011;6:357-71.

8 Badr MA, Maalawi KY. , "A practical approach for selecting optimum wind rotors", Renewable Energy 2003;28:803-22

9 Ronit K. Singh, M. Rafiuddin Ahmed, , Mohammad Asid Zullah, Young-Ho Lee, "Design of a low Reynolds number airfoil for small horizontal axis wind turbines", Renewable Energy 42 (2012) 66-76

10 P. D. Clausen and D. H. Wood, "Research and development issues for small wind turbines", Elsevier Academic Press, Renewable Energy Journal 16, 1999, pp.922-927.

11 David Wood, Small Wind Turbine – Analysis, Design and Application , Springer ,2011.

12 Grant Ingram, Wind Turbine Blade Analysis using the Blade Element Momentum Method, 2011

13 J. F. Manwell, J.G. McGowan, A.L. Rogers, Wind Energy Explained: Theory, Design and Application, John Wiley and Sons, Ltd, 2002, pp 247-317

14 Sunil Kumar J1, Shalini J2, Birtukan Teshome 3, Milkias Berhanu Tuka4, Fikadu Wakijira5 ,Improvement of Active and Reactive Power at the Wind Based Renewable Energy Sources: A case study on ADAMA wind power plant, International Journal of Scientific & Engineering Research, Volume 4, Issue 9, September-2013

Naghma Khatoon[1] and Amritanjali[2]

Retaliation based Enhanced Weighted Clustering Algorithm for Mobile Ad-hoc Network (R-EWCA)

Abstract: Mobile ad-hoc networks (MANETs) are wireless networks comprised of mobile computing devices connected via wireless media which works without the aid of any fixed infrastructure like access point or base station of cellular networks. MANET attracts attention of researchers because of its inevitable characteristics. However in order to improve efficiency of MANET it is important to cope up with the challenges imposed due to dynamic and wireless nature of such networks. Clustering is an important approach to improve scalability of MANET. It divides the network into smaller subgroups called clusters having a cluster head in each cluster which acts as a local coordinator for its cluster. Energy constrained is still one of the challenging issue which make the nodes to adopt a natural tendency to conserve its energy by not forwarding the packets of another nodes in the network. In this paper we have proposed a method to force selfish nodes to behave normal. For this we have used score and punishment factor with the concept of clustering to force cluster heads and gateways not to behave selfishly and cooperate with other cluster heads in the network so as to improve the durability of stable clusters in the network.

Keywords: clustering, cluster head, retaliation, stability factor, punishment factor, score

1 Introduction

MANET is a cooperative infra-structureless network of mobile nodes. Clustering is an important approach for achieving scalability in MANET with large network having a large number of mobile nodes. In clustering nodes are given different status based on the responsibilities assigned to them i.e. cluster heads (CHs),

---

1 BIT Mesra/Department of Computer Science & Engineering, Ranchi, India
E-mail: naghma.bit@gmail.com
2 BIT Mesra/Department of Computer Science & Engineering, Ranchi, India
E-mail: amritanjali@bitmesra.ac.in

gateways and normal nodes. CHs and gateway nodes are responsibe for inter-cluster routing and only CHs are responsible for intra-cluster routing [1,2,3].

In MANET, traditionally all clustering algorithms work assuming that all the nodes in the network are cooperative in nature. However this is actually not the case. A node may become selfish in order to save its energy and bandwidth and starts dropping the packets of other nodes in the network. If all nodes starts behaving selfishly, then the basic concept on which MAENT lies i.e. cooperation among all nodes in the network will not be achieved. Packet drops may also occur due to congestion in the path which is sorted out by prudently selecting alternate path to the desired destination. Our aim is to find out the selfish nodes in the network and force it to behave normal by giving some stricter punishment to them. For this we have proposed a Retaliation based Enhanced Weighted Clustering Algorithm (R-EWCA) which combines the enhanced version of weighted clustering algorithm with the concept of retaliation to give punishment to selfish nodes to force it to behave as normal node and cooperate with all other nodes in the network.

2 Proposed work

In this section we have presented the Retaliation based Enhanced Weighted Clustering Algorithm (R-EWCA) to enforce cooperation by the selfish nodes so as to provide a reliable approach of clustering in MANET.

2.1 Some key features of R-EWCA

Given are the particularization of contrubutions of the proposed R-EWCA in mobile ad-hoc network:
(i)   The score is used for gateways and CHs to monitor its behavior.
(ii) The punishment factor enforces a selfish node to behave normal and be helpful in packet forwarding of other nodes.
(iii) Non-eligible nodes to become a CH are eliminated at the initial stage of clustering.
(iv) The stability factor of a node depends upon distance and mobility of neighboring nodes make clusture structure stable for longer time period.

The Retaliation based Enhanced Weighted Clustering Algorithm (R-EWCA) is based on the ideas proposed by Akhtar [4] with transfigurations made for our applications. R-EWCA executes in two phases: the setup phase and the monitoring phase. The setup phase is invoked only once during initial clustering setup. We have modified the stability factor parameter and used it with the enhanced WCA algorithm [7] to create the clusters. On the other hand the monitoring phase include retaliation phase (periodic in nature) and re-clustering phase(non-periodic in nature). It takes care of any changes in clusters as well as surveillance of gateways and CHs to monitor its behaviour in a protected mode (promiscuous mode ON). The retaliation phase not simply eliminate the selfish nodes but force it to behave properly. Thus no node will maliciously drop out the packets of genuine node to conserve its energy because if it do so, it will be given punishment by dropping its packets too by other nodes. The re-clustering is invoked only when the number of clusters is deviated from the ideal number of clusters for the given network which depends on the network size. During re-clustering one additional parameter i.e. score is also included to find out the combined weight and the node with maximum combined weight is selected as CH.

2.1.1 Description of retaliation model in clustering

As we know that cooperation among nodes is a prerequisite for working of MANET to fulfill its goal. However, nodes working in such wireless platform have the natural tendency of sefeguarding itself by conserving its battery power and bandwidth. Nodes start behaving selfishly and drop packets of another nodes. We are implementing the concept of retaliation with clustering approach to enforce cooperation among nodes by punishing the misbehaved (selfish) nodes. We have used Score (S) and Punishment Factor (PF) for each gateway/CH in the network. Score indicates the degree of selfishness of a gateway/CH and Punishment Factor indicates the number of packets to be dropped by an honest gateway/CH against a selfish node to give the latter a stricter punishment to force it to behave normal. In this way the selfish node tend to minimize its selfishness nature of non-cooperation because now it knows that by dropping packets of other node, its packets will also be dropped and it has to exhaust more energy in rebroadcasting its same packet again and again.

In the retaliation model, each CH and gateway monitors the behavior of its neighboring CHs/gateways by listening the neighbor traffic in the protected mode (promiscuous mode ON) for a threshold period of time. During this phase

CHs and gateways are in surveillance mode. Each CH/gateway monitors the number of packets received for forwarding (NPRF) and the number of packets actually forwarded (NPF) by its neighboring CHs or gateways for a specified time period in a Behavior Information Table. By taking these two criteria the Packet Forwarding Indicator(PFI) is calculated using equation 1. The mean of PFI gives the Score (S) value of the CH/gateway under surveillance as calculated in equation 2. Punishment Factore (PF) is calculated by subtracting the Score from one and multiplying it with ten to give the integer value which shows the number of packets of the misbehaved node to be dropped by the normal nodes. At first when a node is initiated, its Score is assigned to one means it is an honest node and its PF is initialized to zero. Subsequently, by overhearing the traffic of the neighboring CHs and gateways in the protected mode, the values of NPRF and NPF is updated and Score and PF is calculated. An illustrative example is shown in figure 1. A cluster with CH-1 starts monitoring the behavior of its neighboring gateways (2,3,4 and 5) to eliminate the selfish node if any, from the routing path and retaliate it to force it to behave normal. It captures the PFI of all its gateway nodes and calcute their Score and PF as shown in table 1. A normal gateway will drop the packets to or from the selfish node exponential to its PF. Each time a packet is dropped by a normal node or a node overhears this dropping, it decrements the PF by one. It is continued untill the PF of the selfish node becomes zero. In this way this model gives stricter punishment to the misbehaved nodes because when a node become selfish to conserve its energy and drop packets of other node it has to destroy more energy in rebroadcasting its packets again and again untill its PF becomes zero.

$$PFI = 1 - [(NPRF - NPF)/NPRF] \qquad (1)$$

$$Score = (\textstyle\sum_{i=1}^{k} PFI_i)/k \qquad (2)$$

$$PF = (1 - Score) \times 10 \qquad (3)$$

Where

PFI = Packet Forwarding Indicator
$NPRF$ = No. of Packets Received for Forwarding
NPF = No. of Packets Forwarded
k = No. of clusters it connects for gateway, 1 for CH
PF = Punishment Factor

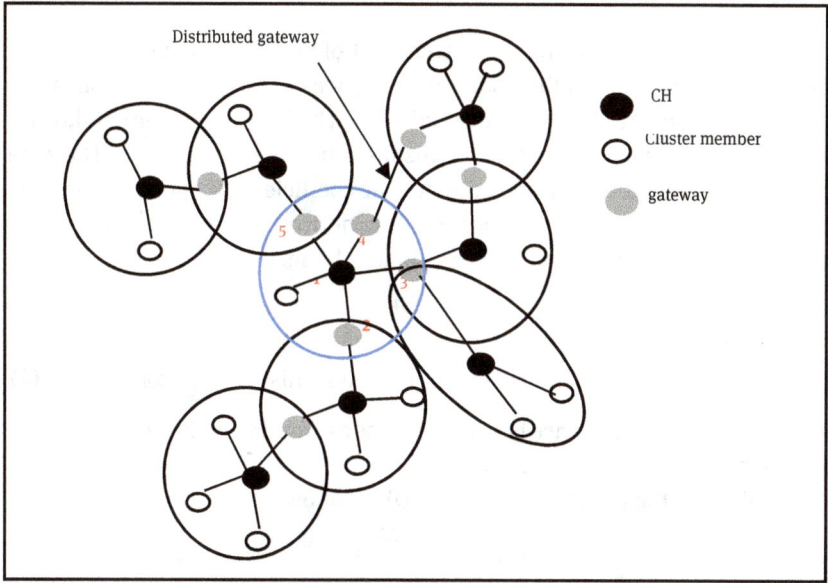

Figure 1. Cluster illustration for calculation of score and selfishness factor for CH-1

Table 1: Behavior Information Table for CH-1

Node ID	PFI	Score	PF
2	0.8, 0.7	0.8	2
3	0.7, 0.5, 0.9	0.7	3
4	0.6, 0.8	0.7	3
5	0.6, 0.5	0.6	4

2.2 CH election parameters

This section describes different metrics used for the selection of most appropriate node to act as CH in the proposed R-EWCA.

2.2.1 Stability Factor (STF)

This factor is used for reducing the detachment of nodes and increasing cluster stability [6,7]. According to the node's mobility we can divide the transmission zone of a node v_i as trusted zone or risked zone. The inner circle with radius $\alpha_1 r$ forms the trusted zone of node v_i with transmission range "r" (figure 2). The zone having width $r(\alpha_2-\alpha_1)$ forms the risked zone. The coefficients α_1 and α_2 are suitably selected based on the mobility of nodes in the network. Zone Factor (ZF) determines where in the transmission range a node is lying.

$$ZF\,(v_i, v_j) = \begin{cases} 1, & \text{if dist } (v_i, v_j) \leq \alpha_1 r \\ 1 - [(\text{dist } (v_i, v_j) - \alpha_1 r) / (\alpha_2 - \alpha_1) r], & \text{if } \alpha_1 r < \text{dist } (v_i, v_j) < \alpha_2 r \end{cases} \tag{4}$$

Where dist (v_i,v_j) is the distance between the nodes v_i and v_j, $\alpha_1 r$ is the inner radius and $\alpha_2 r$ is the outer radius of the circles.
Now we calculate the Effective Distance (ED) as follows:

$$ED\,(v_i, v_j) = ZF\,(v_i, v_j) \times \text{dist}\,(v_i, v_j) \tag{5}$$

Then we calculate the Cumulative Effective Distance (CED) from the node v_i to all its neighboring nodes:

$$CED\,(v_i, v_j) = \sum_{j=1}^{n} ED\,(v_i, v_j) \tag{6}$$

On the basis of equations 4,5 and 6, Stability Factor (STF) is calculated as:

$$STF\,(v_i) = CED(v_i)/N(v_i) \tag{7}$$

Where $N(v_i)$ is the degree of connectivity of node vi.

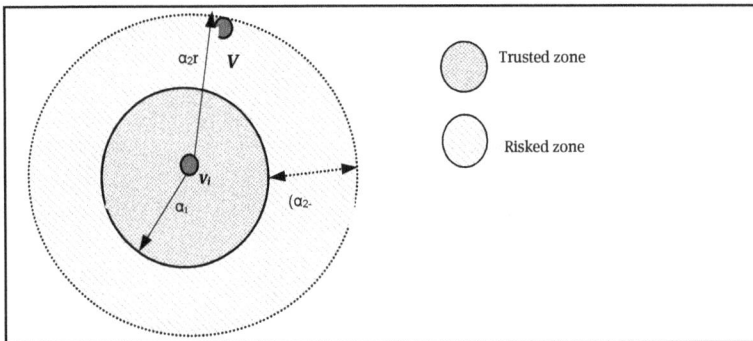

Figure 2. Transmission range zones

2.2.2 Degree of Connectivity (C)

It is the total number of directly connected one-hop neighbors of a node.

$$C(v_i) = |N(i)| \qquad\qquad\qquad (8)$$

2.2.3 Sum of distances (D)

It is the sum of distances from a node to all its neighbors, used to estimate the energy consumption of the node.

$$D(v_i) = \sum_{j=1}^{n} dist(v_i, v_j) \qquad\qquad\qquad (9)$$

2.2.4 Remaining Battery Energy (RBE)

A node with higher remaining battery lifetime is chosen to act as a CH [9].

Algorithm The Retaliation phase of R-EWCA

Begin:

(1) CH_i starts monitoring the behavior of its neighboring gateways/CHs in protected mode for threshold period of time and calculate their score (S) by using equation 2.

(2) IF $Score_i$=1 THEN node$_i$ is normal

(3) ELSE IF $Score_i$ < 1 THEN node$_i$ is a selfish; punish it

(4) REPEAT:

(5) Drop the packets of selfish node exponential to its PF (equation 3) and decrement the PF after each of its packet drop by normal node.

(6) If another node overhears this dropping, decrements the PF of the corresponding selfish node by one after each packet drop.

(7) UNTIL PF=0

(8) ENDIF

End

Conclusion

In this paper, we have presented a new algorithm called Retaliation based Enhanced Weighted Clustering Algorithm (R-EWCA) for mobile ad-hoc network. This algorithm combines the enhanced version of weighted clustering along with the concept of retaliation for misbehaved nodes so as to enhance cooperation for inter-cluster routing. Score is used to determine the degree of selfishness of gateways/CHs to isolate the selfish nodes from the routing path and the punishment factor is used to give stricter punishment to the selfish nodes to force it to behave normal and cooperate with other nodes, thus enhancing the overall performance of clustering in MANET. In future we work to include other parameters to determine the trustworthiness of nodes to make the algorithm more secure and dynamic.

References

1   M. Chatterjee, S. K. Das, and D. Turgut, "An on- demand weighted clustering algorithm (WCA) for Ad hoc networks," Global Telecommunications Conference, GLOBECOM '00. IEEE, vol. 3, pp. 1697-1701, 2000.
2   W. Bednarczyk, and P. Gajewski1, "An Enhanced Algorithm for MANET Clustering Based on Weighted Parameters," Universal Journal of Communications and Network, 1(3), 88-94, 2013.
3   S. Karunakaran, and P. Thangaraj, "A Cluster-Based Service Discovery Protocol for Mobile Ad-hoc Networks," American Journal of Scientific Research, Issue 11, pp. 179-190, 2011.
4   A. K. Akhtar and G. Sahoo, "A Novel Methodology To Overcome Routing Misbehavior In Manet Using Retaliation Model," International Journal of Wireless & Mobile Networks (IJWMN) Vol. 5, No. 4, pp. 187-202, August 2013.
5   A. Dahane, A. Loukil, B. Kechar and N. Berrached, "Energy Efficient and Safe Weighted Clustering Algorithm for Mobile Wireless Sensor Networks,"Hindawi Publishing Corporation, vol.2015, pp. 1-18, 2015.
6   M. Aissa, and A. Belghith, "An efficient scalable weighted clustering algorithm for mobile Ad Hoc networks," International Conference on Information Technology and e-Services (ICITeS), pp. 1-6, 2013.
7   M. Aissa, and A. Belghith, "A node quality based clustering algorithm in wireless mobile AdHoc networks," Procedia Computer Science 32, pp. 174 – 181, 2014.
8   Hosseini-Seno, T. C. Wan, and R. Budiarto, "Energy Efficient Cluster Based Routing Protocol for MANETs," International Conference on Computer Engineering and Applications IP-CSIT vol. 2, pp. 380-384, ACSIT Press, Singapore, 2011.

Dr.K.Meenakshi Sundaram[1] and
Sufola Das Chagas Silva Araujo[2]
Chest CT Scans Screening of COPD based Fuzzy Rule Classifier Approach

Abstract: Chronic Obstructive Pulmonary Disease (COPD) is a name that refers to two lung diseases - chronic bronchitis and emphysema. These diseases are characterized by an impediment to airflow that interferes with normal breathing. Researchers have developed different techniques to improve the performance of automatic screening process. This paper improves the accuracy over the existing techniques using the adaptive region growing property and Fuzzy Rule based (FRB) classifier. The input image is pre-processed using median filtering technique to remove the noise. The contours of the image will be obtained using region growing technique. The FRB classifier is then used to confirm the suspected COPD cavities. The proposed technique is implemented in MATLAB and the performance is compared with the existing techniques. Results show that the proposed method achieves more accuracy as compared with existing techniques.

Keywords: Chronic obstructive pulmonary disease, FRB classifier, Median filtering, Local Gabor XOR Pattern (LGXP), Region Growing Technique

1 Introduction

The pathological changes in patients with chronic obstructive pulmonary disease are composite and are in four dissimilar compartments of the lungs: the central large airways, the small peripheral airways; the lung parenchyma and the pulmonary vasculature. Generally, in all such structural functional studies, pathological changes associate feebly with both clinical and functional patterns of the disease [8]. COPD occurrence is generally higher than recognized by health authorities [4]. For example, in the USA National Health and Nutrition Examination Survey III, 70% of those with airflow obstruction had never re-

1 Department of Computer Engineering, Agnel Institute of Technology and Design, Goa, India
Email: meenaksji@gmail.com
2 Department of Computer Engineering, Padre Conceicao College of Engineering, Goa, India
Email: sufolachagas100@rediffmail.com

ceived the diagnosis of COPD [10]. Recently, the Nippon COPD Epidemiology (NICE) study in Japan, presented the current series, had a similar finding [3].

From the patient's viewpoint, it is also a disease that has a reflective effect on quality of life [9]. The burden of COPD can be assessed in a number of ways such as mortality, morbidity, prevalence, disability-adjusted life years, cost and quality of life. A number of authors have reviewed this topic in detail elsewhere [11, 1]. More recently, lung Computed Tomography (CT) scanning has been used to quantify emphysema in life and has been related to both lung morphology and function [4].

In this paper, input image is pre-processed; the lung region is segmented from that image, the cavity region is segmented from lung region, extract some features for training the classifier and used the FRB classifier to identify the COPD affected lung.Pre-processing is done using the median Filter [6] to avoid the noise in the input image and to increase the image quality. The lung segmentation is done by comparing the region growing technique and the Local Gabor XOR pattern (LGXP) based region growing technique. The cavity segmentation is done by evaluating the pixel range in the segmented lung region and setting a threshold value and comparing every pixel with that threshold value. Parameters are chosen to train the classifier.The classifier used in proposed technique is FRB classifier. The advantage of psycho visual redundancy and the dependency of a pixel on its surrounding neighbours is the correlation between a pixel and its neighbours decides whether it is located in smooth area or in complicated area [7].

2 Proposed Technique for Identification of Cavity

The block diagram of the proposed approach is shown in figure 1. In this chest CT scan images are taken with COPD and without COPD [4]. The sample images are pre-processed and then send for segmentation. After segmentation, some parameters are chosen to train the classifier. After that the chosen parameters are given to the classifier, where the Fuzzy Rule based classifier is used. The FRB classifier then identify whether the input chest CT scan image is affected by COPD .The input image is loaded to the MATLAB environment where the median filter is used for pre-processing to reduces the noise. The RGB image is converted into grey scale image.

Lung segmentation is a process of segmenting the lungs from the chest CT scan image. Initially choose a default pixel and set a threshold value for comparison. The default pixel value is compared with the adjacent pixel values.

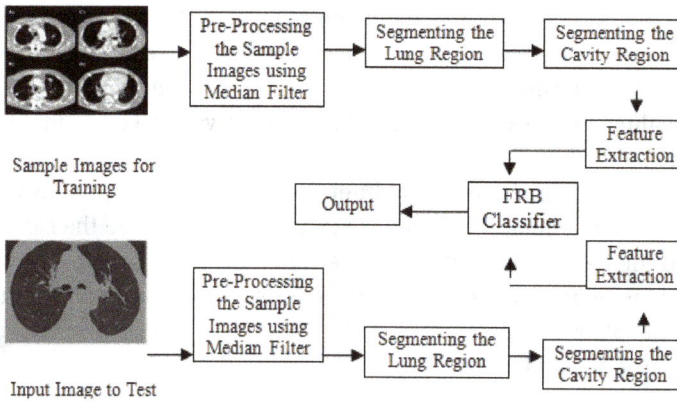

Figure 1. Block Diagram for Proposed Technique

If the difference is greater than the threshold value, then the adjacent pixel is excluded and if not the pixel is included for region growing. The approach uses supervised learning where the class labels are based on measured lung function instead of manually annotated regions of interest (ROIs)[12].

The process of normal region growing technique is shown in the Figure 2.

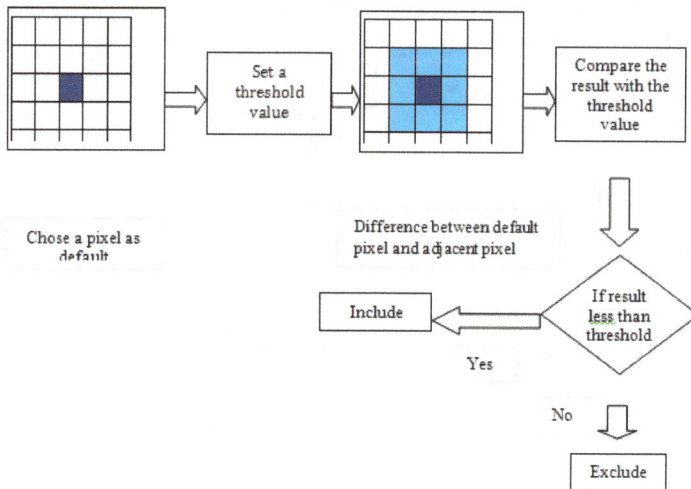

Figure 2. Block Diagram of normal Region Growing Technique

Local Gabor XOR Pattern (LGXP) based region growing technique is used to segment the lungs from the chest CT scan image. The Gabor Phase Technique will exchange all the pixel values to phase values (0 to 360). Later than converting all the pixel values to phase values, find these phase values comes under which quadrant[15].

The first, second, third and fourth quadrant have value as, zero, one ,two and three respectively. Put adjacent pixel's value as zero which have the same quadrant value of the default pixel. If the above condition is not satisfied, put adjacent pixel's value to one. Convert binary pixel format to decimal value and apply decimal value to the default pixel.

The LGXP technique is shown in the Figure 3.

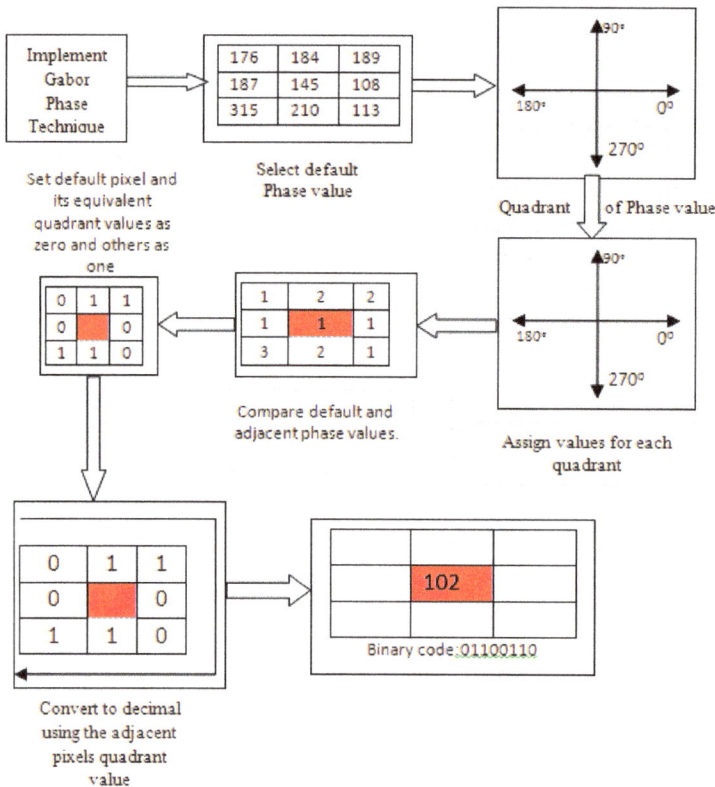

Figure 3. Block diagram of LGXP Technique

Phase value of the pixels from the LGXP process are compared to the normal region growing technique. If the difference between the adjacent pixel and the default pixel got the value less than the threshold value, include that adjacent pixel for region growing or else exclude the pixel.

2.1 Local Gabor XOR Pattern (LGXP)

A properly tuned Gabor filter can be used to effectively preserve the ridge structures while reducing noise [2]. In LGXP encoding method, the phase is quantized into 4 ranges. From the obtained matrix, the binary value obtained is 01011101 and its equivalent decimal value is 93. The LGXP in binary and decimal is as follows:

$$LGXP_{\mu v}(P_c) = \left[LGXP_{\mu v}^{N}, LGXP_{\mu v}^{N-1}, \ldots \ldots LGXP_{\mu v}^{1} \right]_{binary}$$

$$= \left[\sum_{i=1}^{N} 2^{i-1} LGXP_{\mu v}^{i} \right]_{decimal}$$

$$LGXP_{\mu v}^{i} \, (i = 1,2,3 \ldots N)$$

$LGXP_{\mu v}^{i} \, (i = 1,2,3 \ldots N)$ denotes the pattern calculated between P_c and its neighbour P_i, which is computed as follows:

$$LGXP_{\mu v}^{i} = q(\phi_{\mu v}(P_c)) \otimes q(\phi_{\mu v}(P_i)), i = 1,2,\ldots N$$

Where $\phi \mu v$ denotes the phase, $\otimes$ denotes the LXP operator, which is based on XOR operator, q denotes the quantization operator which calculates the quantized code of the phase according to the number of phase ranges.

$$a \otimes b = \begin{cases} 0, if a = b \\ 1, else \end{cases}$$

$$q(\phi_{\mu v}(.)) = i, if \frac{360 * i}{e} \le \phi_{\mu v}(.), < \frac{360 * (i+1)}{e},$$

$$i = 0,1,\ldots \ldots \ldots, b-1$$

Where, e is the number of phase ranges .Each pattern map is split into m non overlapping sub blocks and the histograms of the sub blocks of scales and the orientations are concatenated to form LGXP descriptor of input image

$$H = \left[H_{\mu 0 v 0 1} \ldots H_{\mu 0 v 0 m}; \ldots \ldots; H_{\mu 0 - 1 v s - 1} 1; H_{\mu 0 - 1 v s - 1 m} \right]$$

Where $H_{\mu v}(i = 1,2,....., m)$ denotes the histogram of the i[th]sub block of the LGXP map with scale v and orientation.

2.2 Cavity Segmentation

The cavities present in the lung region are an essential thing to identify the COPD affected lung [16]. To identify the cavity in the lung, set an adaptive thresholding that separates the foreground from the background with non-uniform illumination. After that, compare the threshold with all the pixels. While comparing the pixels to the threshold value, if the pixel value is greater than the threshold value then it would be the cavity region and if the pixel value is less than the threshold value then it would be the lung region. The diagnosis has a value by the capnogram shape, and assessed presently by qualitatively, also by visual inspection [13].

2.3 Feature Extraction

To discover the disease in the lung, have to feed the extracted feature into the classifier, because these features give vital information about the region which is used to train the classifier. The COPD quantification of lung images as a multiple instance learning (MIL) problem is more suitable for weakly labelled data[14].In this paper an FRB classifier is used for feature extraction. The features need to extract are number of cavities in the lung region, minimum area of cavity region, maximum area of cavity region, total number of pixels in each cavity, maximum repeated pixel intensity in the cavity region and maximum repeated pixel in the lung region to find the total number of cavities in the lung region. Because the normal lung would also have some cavities present in its region [17].This classifier shows more accurate value and it took minimum time for an execution.

2.4 Fuzzy Rule-Based Classifier (FRB)

FRBCSs give an interpretable replica by means of linguistic labels in their rules.
Consider m labelled patterns
p = 1, 2, . . . ,m where $x_p = (x_{p1},....., x_{p1})$, x_{pi} is the ith attribute

value (i = 1, 2, . . . , n).A set of linguistic values and their membership functions are there to describing each and every attribute. Use fuzzy rules of the following form:

Rule Rj: If x1 is Aj1 and . . . and xnis Ajn then Class =Cj with RWj ----- (1)
where Rjis the label of the jth rule, x=(x1....xn) is an n-dimensional pattern vector, Aj1is an antecedent fuzzy set on behalf of a linguistic term,Cj is a class label, and RWjis the rule weight. Specially, in this paper the rule weight is computed using the Penalized Certainty Factor defined in [4] as:

$$ PCF = \frac{\sum\limits_{X_p \in ClassicC} \mu A_j(x_p) - \sum\limits_{X_p \notin ClassicC} \mu A_j(x_p)}{\sum\limits_{p=1}^{m} \mu A_j(x_p)} (2)$$

Let $x_p=(x_{p1}... ... ... x_{pn})$ be a new pattern, L denotes the number of rules in the rule base and M the number of classes of the problem; then, the steps of the FRM [10] are as follows:

Matching degree, is the strength of activation of the if-part for all rules in the rule base with the patternx_p.A conjunction operator (t-norm) is functional in order to carry out this computation.

$$\mu_{Aj}(x_p) = T(\mu_{Aj1}(x_{p1}),......\mu A_{jn}(x_{pn})) \qquad j=1,.....L \quad (3)$$

Association degree computes the degree of the pattern x_p with the M classes according to each rule in the rule base.

$$b_j^k = h(\mu_{Aj}(x_p), RW_j^k) \qquad k=1,......,M, j=1,......l \quad (4)$$

Classification. This function will determine the class label l corresponding to the maximum value.

$$F(Y_{1,.....} Y_M) = \arg\max(Y_k), k = 1,...., M$$

3 Training And Testing Using FRB Classifier

Some of the data features are to be taken to identify the normal lung region and COPD affected lung by this the classifier is trained [18].These data features values, are given to the classifier. For instance, choosing three normal CT scan images and three abnormal CT scan images, need to calculate all the six data features separately for all the CT scan images had chosen. The classifier is trained to identify the normal and abnormal lung. New CT scan image is given to the classifier to find whether it has COPD or not. Afterwards, the six data features are calculated for the new image. The computed values are then give to the FRB classifier. The FRB classifier compares these values with the stored values to identify whether the given CT scan image comes under normal or abnormal category.

4 Results and Discussion

The experiment is conducted in MATLAB. The figure 5 shows the normal and abnormal lung images.

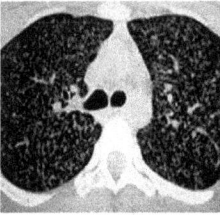

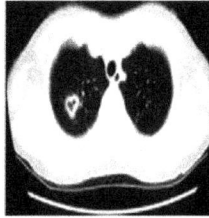

Figure 5.1: Normal Image "Lung" Figure 5.2: Abnormal Image "Lung"

Figure 5. Example Images

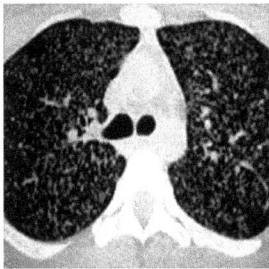

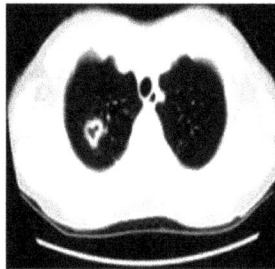

Figure 6.1: Normal Image "Lung"   Figure 6.2: Abnormal Image "Lung"

Figure 6. Results of using median filtering technique

The sample images are filtered using median filter to improve the quality of the images as shown in the figure 6.

The filtered lung image is segmented and is shown in Figure 7.

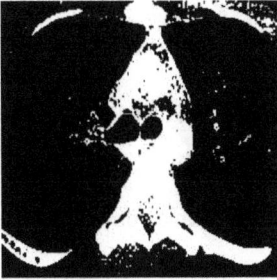

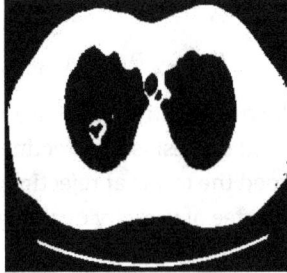

Figure 7.1: Normal Image "Lung"   Figure   7.2:   Abnormal   Image "Lung"

Figure 7. Example Images acquired after segmentation

The figure 8 shows a sample image of segmented cavities and segmented cavities with CT scan image for the COPD affected lung.

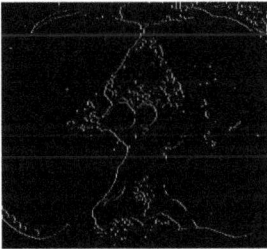

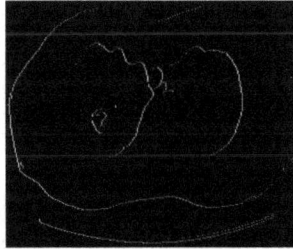

Figure 8.1: Normal Image "Lung"   Figure 8.2: Abnormal Image "Lung"

Figure 8. Example Images after Cavity Segmentation

5 Performance Analysis Using Evaluation Metrics

The evaluation of the COPD identification of the images is carried out using the following metrics,

$$Sensitivity = TP/(TP + FN)$$

$$Accuracy = (TN + TP)/(TN + TP + FN + FP)$$

Where, True Positive TP, True Negative TN, False Negative FN, False Positive FP
Sensitivity shows how good the test is at detecting a disease.
Specificity shows how good the test is at rejecting a disease.
Accuracy measures the degree of veracity of a diagnostic test on a condition.

Table 1: Comparative analysis of existing technique with proposed technique

Techniques	P	N	P	N	Sensitivity (%)	Specificity (%)	Accuracy (%)
SVM Technique	19	6	7	18	76	72	74
ELM Technique	20	5	5	20	80	80	80
Proposed FRB Technique	22	3	2	23	88	92	90

Table 1 shows that the proposed technique gives better performance than the existing techniques.

Conclusion

This paper proposes an efficient technique for the detection of COPD. The proposed technique contains pre-processing, lung segmentation, cavity segmentation, feature extraction, training and testing using FRB classifier. The FRB classifier is efficient and simple in nature and is also more accurate and faster than other techniques. The performance of the proposed technique and the existing technique is analysed using evaluation metrics. To evaluate these metrics, should need some terms like True Positive, True Negative, False Positive and False Negative. After evaluating these metrics it shows that the performance of proposed technique is better when compared to the existing technique in terms of accuracy. The result shows that the accuracy of proposed technique higher than existing techniques.

References

1   Anto M., Vermeire P., Vestbo J and Sunyer J., "Epidemiology of Chronic Obstructive Pulmonary Disease," *The European Respiratory Journal,* vol. 17, no.5, pp. 982–994, 2001.
2   Ashraf El-Sisi, "Design and Implementation Biometric Access Control System Using Fingerprint for Restricted Area Based on Gabor Filter", *The International Arab Journal of Information Technology*, Vol. 8, No. 4, October 2011.
3   Fukuchi Y1, Nishimura M, Ichinose M, Adachi M, Nagai A, Kuriyama T, Takahashi K, Nishimura K, Ishioka S, Aizawa H, Zaher C, "COPD in Japan: the Nippon COPD Epidemiology Study", *Journal of Respirology*, 9: pp. 458–465, 2004.
4   GOLD Guidelines 2003. *www.goldcopd.com*. Date last accessed: December 2 2005.
5   MacNee W., Gould G., and Lamb D., "Quantifying Emphysema by CT Scanning: Clinical Pathological Correlates," *Annals of New Year Academy of Sciences*; vol. 624: pp. 179–194, 1991.
6   Magesh B., Vijayalakshmi P., and AbiramiM.,"Computer Aided Diagnosis System for the Identification andClassification of Lesions in Lungs",*International Journal of Computer Trends and Technology*, pp. 110-114,2011.
7   Moazzam Hossain, Sadia Al Haque, and FarhanaSharmin, "Variable Rate Steganography in Gray Scale Digital Images using Neighborhood Pixel Information",*International Arab Journal of Information Technology*, vol.7, no.1, pp. 34-38, 2010.
8   Pare PD., and Hogg JC, "Lung Structure Function Relationships," *In: Calverley P, Pride N, eds. Chronic Obstructive Pulmonary Disease, Chapman & Hall, London*, pp. 35–45,1996.
9   Rennard .S, M. Decramer, P.M.A. Calverley, N.B. Pride,J.B. Soriano, P.A. Vermeire and J. Vestbo,"Impact of COPD in North America and Europe in 2000: Subjects," Perspective of Confronting COPD International Survey, *TheEuropean Respiratory Journal*, vol. 20, pp.799–805, 2002.
10 Stang P., Lydick E., Silberman C., Kempel A., and Keating ET., "The Prevalence of COPD: using Smoking Rates to Estimate Disease Frequency in the General Population",*Chest Journal*, vol.117: Suppl. 2, pp. 354S–359S, 2000.
11 Viegi G., Scognamiglio A., Baldacci S., Pistelli F., and Carrozzi L., "Epidemiology of Chronic Obstructive Pulmonary Disease (COPD)",*Respiration*, vol. 68: pp. 4–19,2001.
12 Lauge Sorensen., Mads Nielsen., Pechin Lo., Haseem Ashraf., Jesper H. Pedersen and Marleen de Bruijne., "Texture-Based Analysis of COPD: A Data-Driven Approach", *IEEE Transactions on Medical Imaging*, Vol31(1): pp 70 – 78, 2012.
13 Rebecca J. Mieloszyk, George C. Verghese, Kenneth Deitch, Brendan Cooney, Abdullah Khalid, Milciades A. Mirre-González, Thomas Heldt and Baruch S. Krauss, "Automated Quantitative Analysis of Capnogram Shape for COPD–Normal and COPD–CHF Classification",*IEEE Transactions on Biomedical Engineering*, Vol 61 (12), pp 2882 – 2890, 2014.
14 VeronikaCheplygina, LaugeSørensen, David M. J. Tax, Jesper Holst Pedersen, Marco Loog and Marleen de Bruijne,"Classification of COPD with Multiple Instance Learning", *International Conference on Pattern Recognition (ICPR)*, pp1508 – 1513, 2014.

15 K.MeenakshiSundaram and Ravichandran CS, "An Optimized ANFIS Classifier Approach for Screening of COPD from Chest CT Scans with Adaptive Median Filtering",*International Journal of Computer Science and Information Technologies*, Vol. 5 (2), pp 1949-1957, 2014.

16 K.MeenakshiSundaram and Ravichandran CS, "An Efficient ANFIS Based Approach for Screening of Chronic Obstructive Pulmonary Disease from Chest CT Scans with Adaptive Median Filtering", *Decision Making and Knowledge Decision Support SystemsLecture Notes in Economics and Mathematical Systems*, Vol675 pp 125-141, 2014.

17 K.MeenakshiSundaram and Ravichandran CS, "Classification of Lung Regions using Morphometrics for Chest CT Scans",*International Journal of Computer Applications*,Vol 87(5) pp0975 –8887, 2014.

18 K.MeenakshiSundaram and Ravichandran CS, "An Adaptive Fuzzy Rule based Approach with Laplacian Gaussian Filtering for Screening of Chest CT Scans", Vol1 (2) pp 08-16, 2013.

Author Index

www.ingramcontent.com/pod-product-compliance
Lightning Source LLC
Chambersburg PA
CBHW052117230326
41598CB00079B/3778